AF615109

EMS Tracts in Mathematics 21

EMS Tracts in Mathematics

This series includes advanced texts and monographs covering all fields in pure and applied mathematics. *Tracts* will give a reliable introduction and reference to special fields of current research. The books in the series will in most cases be authored monographs, although edited volumes may be published if appropriate. They are addressed to graduate students seeking access to research topics as well as to the experts in the field working at the frontier of research.

For a complete listing see our homepage at www.ems-ph.org.

6 Erich Novak and Henryk Woźniakowski, *Tractability of Multivariate Problems. Volume I: Linear Information*
7 Hans Triebel, *Function Spaces and Wavelets on Domains*
8 Sergio Albeverio et al., *The Statistical Mechanics of Quantum Lattice Systems*
9 Gebhard Böckle and Richard Pink, *Cohomological Theory of Crystals over Function Fields*
10 Vladimir Turaev, *Homotopy Quantum Field Theory*
11 Hans Triebel, *Bases in Function Spaces, Sampling, Discrepancy, Numerical Integration*
12 Erich Novak and Henryk Woźniakowski, *Tractability of Multivariate Problems. Volume II: Standard Information for Functionals*
13 Laurent Bessières et al., *Geometrisation of 3-Manifolds*
14 Steffen Börm, *Efficient Numerical Methods for Non-local Operators. $\mathcal{H}^2$-Matrix Compression, Algorithms and Analysis*
15 Ronald Brown, Philip J. Higgins and Rafael Sivera, *Nonabelian Algebraic Topology. Filtered Spaces, Crossed Complexes, Cubical Homotopy Groupoids*
16 Marek Janicki and Peter Pflug, *Separately Analytical Functions*
17 Anders Björn and Jana Björn, *Nonlinear Potential Theory on Metric Spaces*
18 Erich Novak and Henryk Woźniakowski, *Tractability of Multivariate Problems. Volume III: Standard Information for Operators*
19 Bogdan Bojarski, Vladimir Gutlyanskii, Olli Martio and Vladimir Ryazanov, *Infinitesimal Geometry of Quasiconformal and Bi-Lipschitz Mappings in the Plane*
20 Hans Triebel, *Local Function Spaces, Heat and Navier–Stokes Equations*

Kaspar Nipp
Daniel Stoffer

Invariant Manifolds in Discrete and Continuous Dynamical Systems

Authors:

Kaspar Nipp
ETH Zürich
Seminar for Applied Mathematics (SAM)
CH-8092 Zürich
Switzerland

E-mail: nipp@math.ethz.ch

Daniel Stoffer
ETH Zürich
Department of Mathematics
CH-8092 Zürich
Switzerland

E-mail: stoffer@math.ethz.ch

2010 Mathematical Subject Classification: 37-02; 37Cxx, 37Dxx, 34Cxx, 34Dxx, 65Lxx, 65P10

Key words: Discrete and continuous dynamical systems, invariant manifolds, foliation, singular perturbations, geometric numerical integration, differential algebraic equations

ISBN 978-3-03719-124-8

The Swiss National Library lists this publication in The Swiss Book, the Swiss national bibliography, and the detailed bibliographic data are available on the Internet at http://www.helveticat.ch.

Contact address:

European Mathematical Society Publishing House
Seminar for Applied Mathematics
ETH-Zentrum SEW A27
CH-8092 Zürich
Switzerland

Phone: +41 (0)44 632 34 36
Email: info@ems-ph.org
Homepage: www.ems-ph.org

Typeset using the authors' TEX files: I. Zimmermann, Freiburg
Printed in Germany

9 8 7 6 5 4 3 2 1

Preface

In this book dynamical systems are investigated from a geometric viewpoint. A strong geometric property of a dynamical system is admitting an invariant manifold. In this case essential dynamics of the system takes place on some lower dimensional surface. Invariant manifolds are an important tool in a broad range of applications such as mechanical systems, chemical reaction dynamics, fluid mechanics, electronic circuit theory and singular perturbation theory. The theory of invariant manifolds for dynamical systems is well established. It goes back to the work of Hadamard [47] and Perron [105] and it was further developed by many authors. We mention Fenichel [38], [39], [40], Hirsch, Pugh, Shub [55], Kelley [63], Carr [23], Wiggins [129], Chaperon [24], [25]. The aim of this book is to present rigorous results on invariant manifolds in dynamical systems and to give examples of possible applications. The book is targeted at researchers in the field of dynamical systems interested in precise theorems easy to apply. Part III might also serve as an underlying text for a student seminar in mathematics.

Our approach to invariant manifolds for discrete dynamical systems is based on the so-called graph transform, already used by Hadamard [47]. Assume that the dynamical system is given by a map $P: X \times Y \to X \times Y$ where X, Y are nonempty open sets in some Banach spaces. We consider manifolds described as the graph M of a function $\sigma: X \to Y$. Under certain conditions the image of M under the map P is again a graph of a function $\bar{\sigma}: X \to Y$. This induces an operator $\mathcal{F}$ in the space of functions considered. An invariant manifold is obtained as a fixed point of the operator $\mathcal{F}$. The existence of a fixed point is established by the contraction principle.

We also consider continuous dynamical systems given by an ordinary differential equation (ODE). The time-T map of an ODE is a discrete dynamical system. The graph transform approach for maps carries over to ODEs, if applied to the time-T map. We give conditions on the vector field implying the existence of an invariant manifold for the ODE. Invariant manifold results for ODEs can also be derived by different approaches without using the graph transform. We mention Perron [105], Kelley [63], Knobloch, Kappel [72], Knobloch [69], Yi [130].

In the existing literature on invariant manifolds there is a strong tendency to formulate the results in a rather general setting. We aim at invariant manifold results in a simple setting, easy to apply and providing quantitative estimates. As in Kirchgraber, Lasagni, Nipp, Stoffer [66] we formulate conditions easy to verify and leading to sharp results if the coordinates are chosen in an appropriate way. In the discrete case we give conditions on the Lipschitz constants of the map and in the continuous case conditions on the derivatives of the vector field.

The book is organized as follows. In Part I discrete dynamical systems in Banach spaces are considered. We derive results on the existence of attractive and repulsive invariant manifolds that may be described as the graph of some function. We also treat manifolds described in several charts. In addition, we state results on the smoothness of the invariant manifold, on perturbations of the manifold and on the foliation of the adjacent space. In Part II we establish analogous results for continuous dynamical systems in finite dimensions. In Part III we apply the results of the first two parts. The emphasis is on applications to numerical analysis and to singularly perturbed systems of ODEs. In an appendix the hypotheses and conditions used in the theorems are arranged to help to navigate in the bulk of assumptions made.

We want to thank several people who supported us in this book project. Special thanks go to Urs Kirchgraber who many years back introduced us to the topic of invariant manifolds for maps and to the graph transform approach. We also thank Peter Szmolyan who initiated our research on geometric singular perturbation theory for maps as presented in Chapter 13. The finish of the material of this book as well as part of the final editing was done on a sabbatical leave at the TU Berlin in the summer of 2011. We thank the ETH for this support and Volker Mehrmann for hosting us at his institute and for providing a good working atmosphere. We also thank Marianne Pfister for transforming a large part of our handwriting into LATEX code and Olivier Barbey for drawing part of the figures. The EMS publishing house and in particular Manfred Karbe and Irene Zimmermann have efficiently handled our manuscript, many thanks.

Zürich, June 2012

Kaspar Nipp
Daniel Stoffer

Contents

Part I

Discrete Dynamical Systems – Maps

In Part I we investigate discrete dynamical systems in Banach spaces given by some map. We make assumptions on the map such that the dynamical system admits an invariant manifold. Invariant manifold results for maps are considered, e.g., in Hadamard [47], Hirsch, Pugh, Shub [55], Chaperon [25], De la Llave [33], Broer, Osinga, Vegter [21], Ma, Kuepper [86], Nipp, Stoffer [99], Osinga [103], Stuart, Humphries [125].

The maps we consider are of the form

$$P\colon \begin{pmatrix} x \\ y \end{pmatrix} \longmapsto \begin{pmatrix} \bar{x} \\ \bar{y} \end{pmatrix} = \begin{pmatrix} F(x,y) \\ G(x,y) \end{pmatrix}, \quad (x,y) \in X \times Y,$$

where X and Y are open sets in Banach spaces. We assume that P is inflowing with respect to Y, i.e., $G(x,Y) \subset Y$ for all $x \in X$, and outflowing with respect to X, i.e., $F(X,y) \supset X$ for all $y \in Y$. A typical situation is that P is contracting in y-direction and expanding in x-direction. For illustrating the graph transform approach we take the well-known Hénon map

$$\begin{aligned} \bar{x} &= 1 - ax^2 + y, \\ \bar{y} &= bx \end{aligned}$$

for $a = -3/4$ and $b = -1$. Note that for $b = -1$ the Hénon map is orientation and area preserving. It has a hyperbolic fixed point at $(2,-2)$ with eigenvalues $\lambda_1 = 1/\delta$, $\lambda_2 = \delta$ and eigenvectors $v_1 = (1,-\delta)^T$, $v_2 = (-\delta,1)^T$ where $\delta = (3-\sqrt{5})/2$. Its local stable and local unstable manifold, respectively, contains (locally) all points approaching the fixed point under the map and all points approaching it under the inverse map, respectively. The graph transform is illustrated in Figure 1. The graph of

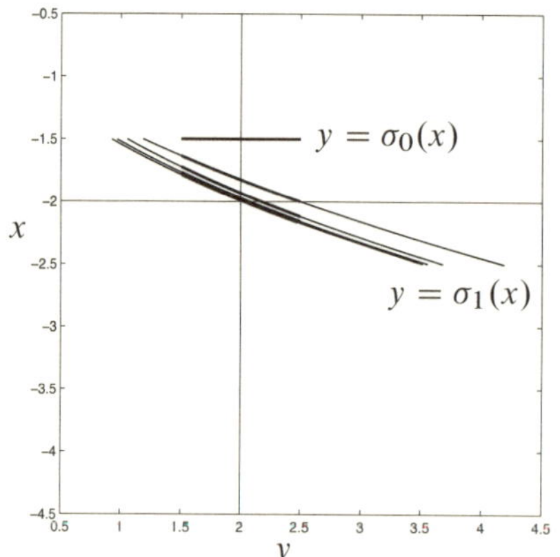

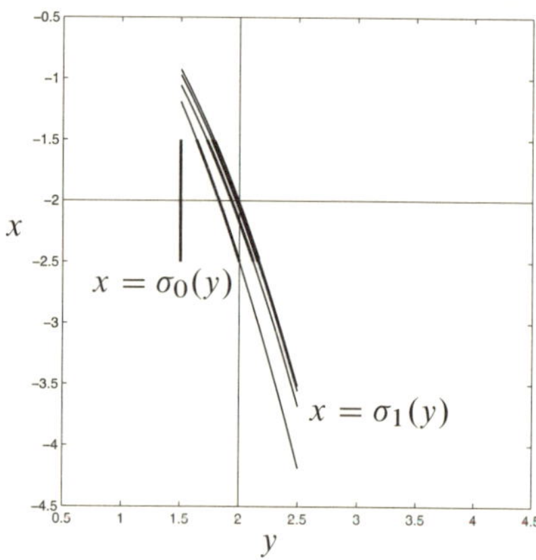

Figure 1. The graph transform illustrated on the left for the local unstable and on the right for the local stable manifold of the fixed point $(2,-2)$ of the Hénon map.

the function $y = \sigma_0(x) \equiv -1.5, 1.5 \le x \le 2.5$, is mapped to the graph $y = \sigma_1(x)$. Since the map is outflowing with respect to x the graph of $\sigma_1(x)$ covers a larger interval. Restricting $\sigma_1(x)$ to $1.5 \le x \le 2.5$ and applying again the map yields a graph $y = \sigma_2(x)$. If this graph transform process is iterated the graphs of $\sigma_j(x)$, $j \to \infty$, approach the local unstable manifold of the fixed point as seen in Figure 1 on the left. Under iterations of the inverse Hénon map

$$\bar{x} = y/b,$$
$$\bar{y} = -1 + x + ay^2/b^2$$

the iterates of the graph $x = \sigma_0(y) \equiv 1.5, -1.5 \ge y \ge -2.5$ approach the local stable manifold of the fixed point $(2, -2)$, cf. Figure 1 on the right. We come back to this example in Section 8.1 where we rigorously prove the existence of the local stable and local unstable manifold of $(2, -2)$ by applying the general manifold theorem of Chapter 1.

For the general manifold results for maps of Part I we formulate conditions on the Lipschitz constants of the functions F and G of the map P. We investigate attractive, repulsive and hyperbolic manifolds. An attractive manifold M is described as the graph of some function s_A, i.e., $M = \{(x, y) \mid x \in X, y = s_A(x)\}$ and a repulsive manifold N is given as $N = \{(x, y) \mid y \in Y, x = s_R(y)\}$. A hyperbolic manifold K is the intersection of an attractive manifold M and a repulsive manifold N. In addition to the existence results we provide assertions on smoothness, perturbation, approximation and foliation.

Our approach requires to have well adjusted coordinates. Most of our results are concerned with maps described in a single coordinate system. However, we also treat the case where the map P is given in several charts. Here the assumption that P is outflowing with respect to X is replaced by the assumptions that P is "flowing" from one chart to the next one.

Part I is organized as follows. In Chapter 1 we prove existence results for attractive, repulsive and hyperbolic manifolds. In Chapter 2 we derive some perturbation and approximation results. In Chapter 3 we show that under certain conditions the manifolds are of class C^k if the map is of class C^k. An attractive or a repulsive invariant manifold gives rise to a foliation of the adjacent space. We prove the existence of such invariant families of fibers in Chapter 4. Finally, in Chapter 5 we prove that under additional conditions the fibers depend smoothly on their base point on the manifold.

Chapter 1
Existence

Our applications in Part III mainly deal with attractive negatively invariant manifolds. Since the repulsive case is slightly simpler to derive we first give a detailed derivation of our existence result on repulsive positively invariant manifolds in Section 1.1. In Section 1.2 we prove the corresponding result for attractive negatively invariant manifolds. In Section 1.3 we combine the results of Sections 1.1 and 1.2 to treat hyperbolic invariant manifolds. In Section 1.4 we derive an existence result for attractive outflowing invariant manifolds of a map defined in several charts and we provide some tools used in Chapter 13.

1.1 Repulsive positively invariant manifolds

We consider a map which is expanding in some directions and contracting or less expanding in the complementary directions. Using the graph transform approach we show that under certain conditions the map admits a repulsive positively invariant manifold. More precisely, let $\mathcal{B}_x$, $\mathcal{B}_y$, $\mathcal{B}_\vartheta$ be Banach spaces and let $X \subset \mathcal{B}_x$, $Y \subset \mathcal{B}_y$, $E \subset \mathcal{B}_\vartheta$ be open subsets of the given spaces. We consider a family of maps P_ϑ of the form

$$P_\vartheta : X \times Y \ni \begin{pmatrix} x \\ y \end{pmatrix} \longmapsto \begin{pmatrix} \bar{x} \\ \bar{y} \end{pmatrix} = \begin{pmatrix} F(x, y, \vartheta) \\ G(x, y, \vartheta) \end{pmatrix} \in \mathcal{B}_x \times \mathcal{B}_y, \tag{1.1}$$

$\vartheta \in E$ being the family parameter. In what follows we denote norms by $|\,.\,|$ independently of the spaces considered. We make the following assumptions for the map P_ϑ.

Hypothesis HM

The functions $F \in C^0(X \times Y \times E, \mathcal{B}_x)$, $G \in C^0(X \times Y \times E, \mathcal{B}_y)$ have the following properties.

a) *P_ϑ is inflowing with respect to Y, i.e., $G(x, y, \vartheta) \in Y$ holds for all $(x, y, \vartheta) \in X \times Y \times E$.*

b) *P_ϑ is outflowing with respect to X, i.e., for every $\bar{x} \in X$, $y \in Y$, $\vartheta \in E$ there is $x \in X$ such that $F(x, y, \vartheta) = \bar{x}$.*

c) *There are nonnegative constants Γ_{11}, L_{12}, L_{13}, L_{21}, L_{22} and L_{23} such that for*

$x, x_1, x_2 \in X$, $y, y_1, y_2 \in Y$, $\vartheta, \vartheta_1, \vartheta_2 \in E$ the functions F and G satisfy

$$\begin{aligned}|F(x_1, y, \vartheta) - F(x_2, y, \vartheta)| &\geq \Gamma_{11}|x_1 - x_2|,\\ |F(x, y_1, \vartheta_1) - F(x, y_2, \vartheta_2)| &\leq L_{12}|y_1 - y_2| + L_{13}|\vartheta_1 - \vartheta_2|,\\ |G(x_1, y_1, \vartheta_1) - G(x_2, y_2, \vartheta_2)| &\leq L_{21}|x_1 - x_2| + L_{22}|y_1 - y_2|\\ &\quad + L_{23}|\vartheta_1 - \vartheta_2|.\end{aligned}$$

Hypothesis HMR

There is $x^* \in X$ such that the function $F(x^*, \cdot, \cdot): Y \times E \to \mathcal{B}_x$ is bounded.

Remark 1.1. (1) *Notation:* We denote hypotheses and conditions by a sequence of letters as follows:

H for **H**ypothesis (first letter),
C for **C**ondition (first letter),
M for **M**ap (second letter),
R for **R**epulsive manifold (third letter),
A for **A**ttractive manifold (third letter).

(2) In applications it is important to allow the numbers Γ_{11}, L_{12}, L_{13}, L_{21}, L_{22}, L_{23} to depend on ϑ. In particular, in this case L_{13}, L_{23} satisfy

$$L_{13}(\vartheta) \geq \sup_{x\in X,\, y\in Y} \limsup_{\vartheta^* \longrightarrow \vartheta} \frac{|F(x, y, \vartheta) - F(x, y, \vartheta^*)|}{|\vartheta - \vartheta^*|},$$

$$L_{23}(\vartheta) \geq \sup_{x\in X,\, y\in Y} \limsup_{\vartheta^* \longrightarrow \vartheta} \frac{|G(x, y, \vartheta) - G(x, y, \vartheta^*)|}{|\vartheta - \vartheta^*|}$$

for $\vartheta, \vartheta^* \in E$.

We want to find a repulsive positively invariant manifold for the map P_ϑ of the form

$$N_\vartheta := \{(x, y) \mid y \in Y,\ x = s_R(y, \vartheta)\},$$

where $s_R: Y \times E \to X$ is uniformly Lipschitz continuous, cf. Figure 1.1. N_ϑ being positively invariant means that $P_\vartheta(N_\vartheta) \subset N_\vartheta$ holds. Consider the space of bounded functions σ in $C^0(Y \times E, \mathcal{B}_x)$ having a global Lipschitz constant. Equipped with the supremum norm ($|\sigma| = \sup_{y\in Y,\vartheta\in E} |\sigma(y, \vartheta)|$) this space is a Banach space. Let $C_{\alpha,\beta}$ be the following closed subset of this space:

$$C_{\alpha,\beta} := \{\sigma \in C^0(Y \times E, X) \mid \sigma \text{ is bounded, uniformly } \alpha\text{-Lipschitz continuous with respect to } y \text{ and uniformly } \beta\text{-Lipschitz continuous with respect to } \vartheta\}.$$

Note that $C_{\alpha,\beta}$ is a complete metric space. For $\sigma \in C_{\alpha,\beta}$ consider the manifold $N_{\vartheta,\sigma} = \{(x, y) \mid y \in Y,\ x = \sigma(y, \vartheta)\}$. We look for conditions on the map P_ϑ and

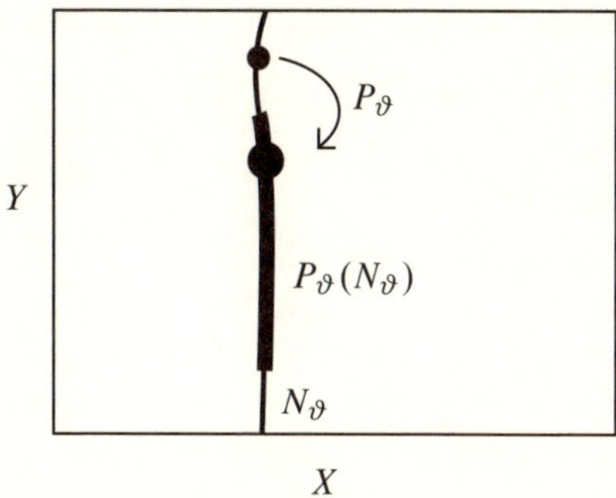

Figure 1.1. The positively invariant manifold N_ϑ.

on the constants α, β such that there exists a function $\bar{\sigma} \in C_{\alpha,\beta}$ with the property that the graph $N_{\vartheta,\bar{\sigma}}$ of $\bar{\sigma}$ is mapped to $N_{\vartheta,\sigma}$. This situation is sketched in Figure 1.2. The function $\bar{\sigma}$ has to satisfy

$$F(\bar{\sigma}(y,\vartheta), y, \vartheta) = \sigma(G(\bar{\sigma}(y,\vartheta), y, \vartheta), \vartheta), \quad (y,\vartheta) \in Y \times E. \tag{1.2}$$

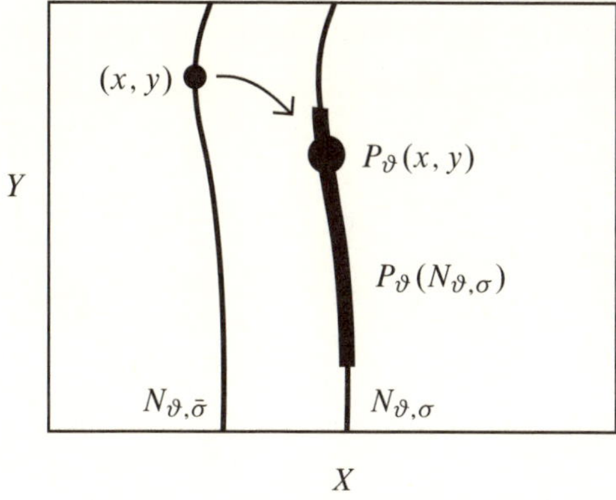

Figure 1.2. The graph $N_{\vartheta,\bar{\sigma}}$ of $\bar{\sigma}$ is mapped to the graph $N_{\vartheta,\sigma}$ of σ.

We consider the operator $\mathcal{K}^0$ mapping the function σ to $\bar{\sigma}$ according to equation (1.2). A fixed point s_R of the operator $\mathcal{K}^0$ describes a manifold $N_{\vartheta,s_R} = \{(x,y) \mid y \in Y,\ x = s_R(y,\vartheta)\}$ which is positively invariant under the map P_ϑ. (Note that if $Y = \mathcal{B}_y$ then typically $P_\vartheta(N_{\vartheta,s_R}) = N_{\vartheta,s_R}$ meaning that N_{ϑ,s_R} is invariant under the map P_ϑ). Instead of the operator $\mathcal{K}^0$ we introduce an operator $\mathcal{K}$ having the same fixed points as $\mathcal{K}^0$. This modification simplifies the following investigations and leads to the same results. Let $\mathcal{K}$ be the operator taking a function $\sigma \in C_{\alpha,\beta}$ to a function $\bar{\sigma}$ according to the equation

$$F(\bar{\sigma}(y,\vartheta), y, \vartheta) = \sigma(G(\sigma(y,\vartheta), y, \vartheta), \vartheta), \quad (y,\vartheta) \in Y \times E. \tag{1.3}$$

The operator $\mathcal{K}$ is well defined due to Hypothesis HM: Let $(y,\vartheta) \in Y \times E$. Hypothesis HM a) implies $G(\sigma(y,\vartheta), y, \vartheta) \in Y$. It follows that $\sigma(G(\sigma(y,\vartheta), y, \vartheta), \vartheta) \in X$. By Hypothesis HM b) there is $\bar{\sigma}(y,\vartheta)$ satisfying equation (1.3) which is unique by Hypothesis HM c). We want to show that under appropriate conditions the operator $\mathcal{K}$ maps $C_{\alpha,\beta}$ into itself and is a contraction in $C_{\alpha,\beta}$ and hence has a unique fixed point.

Lemma 1.2. *Let the map P_ϑ satisfy Hypotheses HM and HMR. Let the conditions*

$$L_{21}\alpha^2 - (\Gamma_{11} - L_{22})\alpha + L_{12} \le 0, \tag{1.4}$$
$$\Gamma_{11} - L_{21}\alpha - 1 > 0, \tag{1.5}$$
$$(L_{13} + L_{23}\alpha)/(\Gamma_{11} - L_{21}\alpha - 1) \le \beta \tag{1.6}$$

be satisfied. Then

i) $\mathcal{K}\colon C_{\alpha,\beta} \to C_{\alpha,\beta}$.

ii) *$\mathcal{K}$ is a contraction in $C_{\alpha,\beta}$ with contractivity constant* $(1 + L_{21}\alpha)/\Gamma_{11} < 1$.

Proof. i) By Hypothesis HM c) and by the definition of $\mathcal{K}$ we find for $(y_i, \vartheta_i) \in Y \times E$, $i = 1, 2$,

$$\begin{aligned}
&\Gamma_{11}|\bar{\sigma}(y_1,\vartheta_1) - \bar{\sigma}(y_2,\vartheta_2)| - L_{12}|y_1 - y_2| - L_{13}|\vartheta_1 - \vartheta_2| \\
&\quad \le |F(\bar{\sigma}(y_1,\vartheta_1), y_1, \vartheta_1) - F(\bar{\sigma}(y_2,\vartheta_2), y_2, \vartheta_2)| \\
&\quad = |\sigma(G(\sigma(y_1,\vartheta_1), y_1, \vartheta_1), \vartheta_1) - \sigma(G(\sigma(y_2,\vartheta_2), y_2, \vartheta_2), \vartheta_2)| \\
&\quad \le \alpha[L_{21}(\alpha|y_1 - y_2| + \beta|\vartheta_1 - \vartheta_2|) + L_{22}|y_1 - y_2| + L_{23}|\vartheta_1 - \vartheta_2|] \\
&\qquad + \beta|\vartheta_1 - \vartheta_2|.
\end{aligned}$$

We conclude that

$$\begin{aligned}
|\bar{\sigma}(y_1,\vartheta_2) - \bar{\sigma}(y_2,\vartheta_2)| \le \frac{1}{\Gamma_{11}}&(L_{21}\alpha^2 + L_{22}\alpha + L_{12})\,|y_1 - y_2| \\
&+ \frac{1}{\Gamma_{11}}\big(\beta(1 + L_{21}\alpha) + L_{13} + L_{23}\alpha\big)|\vartheta_1 - \vartheta_2|.
\end{aligned}$$

It follows that $\bar{\sigma}$ is α-Lipschitz continuous with respect to y if $(L_{21}\alpha^2 + L_{22}\alpha + L_{12})/\Gamma_{11} \le \alpha$ holds which is equivalent to Condition (1.4). The function $\bar{\sigma}$ is β-Lipschitz continuous with respect to ϑ if $\big(\beta(1 + L_{21}\alpha) + L_{13} + L_{23}\alpha\big)/\Gamma_{11} \le \beta$ holds which follows from Conditions (1.5) and (1.6). It remains to show that $\bar{\sigma}$ is bounded.

By Hypothesis HMR there is $x^* \in X$ such that the function $h^*: (y,\vartheta) \mapsto F(x^*, y, \vartheta)$ is bounded. By Hypotheses HM c) and by the definition of $\mathcal{K}$ we find

$$\begin{aligned}
\Gamma_{11}(|\bar{\sigma}(y,\vartheta)| - |x^*|) &\le \Gamma_{11}|\bar{\sigma}(y,\vartheta) - x^*| \\
&\le |F(\bar{\sigma}(y,\vartheta), y, \vartheta) - F(x^*, y, \vartheta)| \\
&\le |\sigma(G(\sigma(y,\vartheta), y, \vartheta), \vartheta)| + |F(x^*, y, \vartheta)| \\
&\le |\sigma| + |h^*|.
\end{aligned}$$

We thus have

$$|\bar{\sigma}(y,\vartheta)| \le |x^*| + (|h^*| + |\sigma|)/\Gamma_{11},$$

implying that $\bar{\sigma}$ is bounded.

ii) Let $\sigma_1, \sigma_2 \in C_{\alpha,\beta}$ and define $\bar{\sigma}_1 := \mathcal{K}\sigma_1$, $\bar{\sigma}_2 := \mathcal{K}\sigma_2$. By Hypothesis HM c) and by the definition of $\mathcal{K}$ we have for $(y,\vartheta) \in Y \times E$,

$$\begin{aligned}
\Gamma_{11}|\bar{\sigma}_1(y,\vartheta) - \bar{\sigma}_2(y,\vartheta)| &\le |F(\bar{\sigma}_1(y,\vartheta), y, \vartheta) - F(\bar{\sigma}_2(y,\vartheta), y, \vartheta)| \\
&= |\sigma_1(G(\sigma_1(y,\vartheta), y, \vartheta), \vartheta) - \sigma_2(G(\sigma_2(y,\vartheta), y, \vartheta), \vartheta)| \\
&\le |\sigma_1(G(\sigma_1(y,\vartheta), y, \vartheta), \vartheta) - \sigma_2(G(\sigma_1(y,\vartheta), y, \vartheta), \vartheta)| \\
&\quad + |\sigma_2(G(\sigma_1(y,\vartheta), y, \vartheta), \vartheta) \\
&\qquad - \sigma_2(G(\sigma_2(y,\vartheta), y, \vartheta), \vartheta)| \\
&\le (1 + \alpha L_{21})|\sigma_1 - \sigma_2|.
\end{aligned}$$

We conclude that $|\bar{\sigma}_1 - \bar{\sigma}_2| \le |\sigma_1 - \sigma_2|\,(1 + \alpha L_{21})/\Gamma_{11}$. Due to Condition (1.5) the constant $(1 + \alpha L_{21})/\Gamma_{11}$ is less than 1. □

By means of the contraction principal it follows that if there is $\alpha, \beta > 0$ such that Conditions (1.4), (1.5) and (1.6) hold then the operator $\mathcal{K}$ has a unique fixed point $s_R \in C_{\alpha,\beta}$. By construction it then follows that the map P_ϑ admits a positively invariant manifold $N_\vartheta = \{(x,y) \mid y \in Y,\ x = s_R(y,\vartheta)\}$.

We now look for conditions on the constants Γ_{11}, L_{12}, L_{21} and L_{22} under which there is $\alpha \ge 0$ satisfying Conditions (1.4), (1.5).

We first consider the case $L_{21} > 0$. Condition (1.5) is satisfied if

$$\alpha \in \Big(-\infty, \frac{\Gamma_{11} - 1}{L_{21}} \Big).$$

There is an $\alpha \in \mathbb{R}$ satisfying Condition (1.4) if and only if

$$4\,L_{12}L_{21} \le (\Gamma_{11} - L_{22})^2.$$

Under this condition, equation (1.4) holds for all $\alpha \in [\lambda_R, \alpha_{\max}]$ with

$$\lambda_R, \alpha_{\max} = \frac{\Gamma_{11} - L_{22} \mp \sqrt{(\Gamma_{11} - L_{22})^2 - 4L_{12}L_{21}}}{2L_{21}}.$$

Since we want $\alpha \geq 0$ we require

$$2\sqrt{L_{12}L_{21}} \leq \Gamma_{11} - L_{22}, \tag{1.7}$$

implying $\Gamma_{11} - L_{22} \geq 0$. The intersection of the two intervals $(-\infty, (\Gamma_{11} - 1)/L_{21})$ and $[\lambda_R, \alpha_{\max}]$ is nonempty if

$$\lambda_R < \frac{\Gamma_{11} - 1}{L_{21}}. \tag{1.8}$$

Thus, there is $\alpha \geq 0$ satisfying Conditions (1.4) and (1.5) if Conditions (1.7) and (1.8) hold. If these conditions are satisfied, $\alpha = \lambda_R$ is the smallest possible choice. As a result Conditions (1.4), (1.5) of Lemma 1.2 can be deduced from the following two conditions.

Condition CM

$$2\sqrt{L_{12}L_{21}} < \Gamma_{11} - L_{22}.$$

Condition CMR

$$1 < \Gamma_{11} - \Delta,$$

where

$$\Delta = \frac{2L_{12}L_{21}}{\Gamma_{11} - L_{22} + \sqrt{(\Gamma_{11} - L_{22})^2 - 4L_{12}L_{21}}}.$$

For Condition CM we have replaced the $\leq$-sign by the $<$-sign in Condition (1.7). This slightly sharper condition is needed in Chapter 4. Condition CM is a "weak coupling condition" for the map P_ϑ, Condition CMR reflects that P_ϑ has to be expanding with respect to x.

It is easy to verify that in the case $L_{21} = 0$, Conditions CM and CMR imply Conditions (1.4), (1.5) with $\alpha = \lambda_R = L_{12}/(\Gamma_{11} - L_{22})$.

We are now able to formulate the existence result for the positively invariant manifold N_ϑ together with additional properties of N_ϑ.

Theorem 1.3. *Let the map P_ϑ given in equation* (1.1) *satisfy Hypotheses HM and HMR. Moreover, assume that the constants Γ_{11}, L_{12}, L_{21}, L_{22} satisfy Conditions CM and CMR.*

Then there exists a bounded function $s_R\colon Y \times E \to X$ such that the following assertions hold for $\vartheta \in E$.

i) *The set $N_\vartheta = \{(x, y) \mid y \in Y,\ x = s_R(y, \vartheta)\}$ is a positively invariant manifold of the map P_ϑ, i.e., $P_\vartheta(N_\vartheta) \subset N_\vartheta$. The function s_R satisfies the invariance equation*

$$F(s_R(y, \vartheta), y, \vartheta) = s_R(G(s_R(y, \vartheta), y, \vartheta), \vartheta) \tag{1.9}$$

for all $y \in Y$.

ii) *The function s_R is in C_{λ_R,ν_R} with*

$$\lambda_R = \frac{2L_{12}}{\Gamma_{11} - L_{22} + \sqrt{(\Gamma_{11} - L_{22})^2 - 4L_{12}L_{21}}}, \quad \nu_R = \frac{L_{13} + L_{23}\lambda_R}{\Gamma_{11} - L_{21}\lambda_R - 1},$$

i.e., s_R is bounded, uniformly λ_R-Lipschitz continuous with respect to y and uniformly ν_R-Lipschitz continuous with respect to ϑ.

iii) *The positively invariant manifold N_ϑ is uniformly repulsive with repulsivity constant*

$$\chi_R = \Gamma_{11} - \Delta > 1,$$

$$\Delta = \frac{2L_{12}L_{21}}{\Gamma_{11} - L_{22} + \sqrt{(\Gamma_{11} - L_{22})^2 - 4L_{12}L_{21}}} = L_{21}\lambda_R,$$

i.e., if $(x_0, y_0) \in X \times Y$ and $(x_1, y_1) := P_\vartheta(x_0, y_0)$ then the inequality

$$|x_1 - s_R(y_1, \vartheta)| \geq \chi_R\, |x_0 - s_R(y_0, \vartheta)|$$

holds.

iv) *If the set $\Lambda \subset X \times Y$ is bounded with respect to x and positively invariant under the map P_ϑ then Λ is contained in N_ϑ, i.e., if $\Lambda \subset X_0 \times Y$ with bounded $X_0 \subset X$ and if $P_\vartheta(\Lambda) \subset \Lambda$ then $\Lambda \subset N_\vartheta$.*

v) *If there is a map $\kappa\colon Y \to Y$ such that F is κ-invariant and G is κ-equivariant then s_R is κ-invariant, i.e., if for all $(x, y) \in X \times Y$,*

$$F(x, \kappa y, \vartheta) = F(x, y, \vartheta),$$
$$G(x, \kappa y, \vartheta) = \kappa G(x, y, \vartheta),$$

then $s_R(\kappa y, \vartheta) = s_R(y, \vartheta)$ holds.

vi) *If the function F has the form $F(x, y, \vartheta) = B(x, y, \vartheta)x + \hat{F}(x, y, \vartheta)$ with B invertible and if for all $y \in Y$ the estimate $|B(s_R(y, \vartheta), y, \vartheta)^{-1}| \leq b < 1$ holds then*

$$|s_R(y, \vartheta)| \leq \frac{b}{1-b} \sup_{y \in Y} |\hat{F}(s_R(y, \vartheta), y, \vartheta)|.$$

Remark 1.4. (1) In the case $L_{22} \geq 1$, which may occur in applications, the two conditions CM, CMR may be replaced by the single condition CM. This follows from

$$\Delta = \frac{1}{2}\left(\Gamma_{11} - L_{22} - \sqrt{(\Gamma_{11} - L_{22})^2 - 4L_{12}L_{21}}\right)$$

$$< \frac{1}{2}(\Gamma_{11} - L_{22}) < \Gamma_{11} - L_{22}.$$

(2) If, e.g., the map κ in assertion v) is taken to be $\kappa\colon y \mapsto y + z$ for some fixed $z \in Y$ one obtains that if the map P_ϑ is z-periodic with respect to y then s_R is also z-periodic with respect to y.

If, e.g., $\kappa\colon y \mapsto -y$, one gets that if F is even in y and G is odd in y then s_R is even in y.

Proof of Theorem 1.3. We have already proved assertions i) and ii).

iii) By the invariance equation (1.9) we get

$$\begin{aligned}|x_1 - s_R(y_1,\vartheta)| &= |F(x_0,y_0,\vartheta) - s_R(G(x_0,y_0,\vartheta),\vartheta)|\\ &\geq |F(x_0,y_0,\vartheta) - F(s_R(y_0,\vartheta),y_0,\vartheta)|\\ &\quad - |s_R(G(s_R(y_0,\vartheta),y_0,\vartheta),\vartheta) - s_R(G(x_0,y_0,\vartheta),\vartheta)|\\ &\geq (\Gamma_{11} - \lambda_R L_{21})\,|x_0 - s_R(y_0,\vartheta)|.\end{aligned}$$

iv) Define $D := \sup_{(x,y)\in\Lambda} |x - s_R(y,\vartheta)|$. Note that $D < \infty$. The proof is by contradiction. Assume that $D > 0$. Then there is $(x,y) \in \Lambda$ with

$$|x - s_R(y,\vartheta)| > \frac{D}{\chi_R},$$

since $\chi_R > 1$. The set Λ being positively invariant implies that $(\bar{x},\bar{y}) = P_\vartheta(x,y) \in P_\vartheta(\Lambda) \subset \Lambda$. By assertion iii) we have $|\bar{x} - s_R(\bar{y},\vartheta)| \geq \chi_R\,|x - s_R(y,\vartheta)| > D$ contradicting the definition of D.

v) We restrict the operator $\mathcal{K}$ to the space

$$C_{\lambda_R,\nu_R}\big|_\kappa := \{\sigma \in C_{\lambda_R,\nu_R} \mid \sigma(\kappa y,\vartheta) = \sigma(y,\vartheta)\}$$

of κ-invariant functions which is a closed subset of C_{λ_R,ν_R}. It suffices to show that the operator $\mathcal{K}$ maps $C_{\lambda_R,\nu_R}\big|_\kappa$ into itself. By definition of the operator $\mathcal{K}$ (cf. equation (1.3)) we have

$$F((\mathcal{K}\sigma)(\kappa y,\vartheta),\kappa y,\vartheta) = \sigma(G(\sigma(\kappa y,\vartheta),\kappa y,\vartheta),\vartheta).$$

Since σ and F are κ-invariant with respect to y and since G is κ-equivariant with respect to y it follows that

$$F((\mathcal{K}\sigma)(\kappa y,\vartheta),y,\vartheta) = \sigma(G(\sigma(y,\vartheta),y,\vartheta),\vartheta).$$

By means of equation (1.3) defining the operator $\mathcal{K}$ we conclude $(\mathcal{K}\sigma)(\kappa y,\vartheta) = (\mathcal{K}\sigma)(y,\vartheta)$.

vi) Due to equation (1.9) we have for $y \in Y$, $\vartheta \in E$,

$$s_R(G(s_R(y,\vartheta),y,\vartheta),\vartheta) = B(s_R(y,\vartheta),y,\vartheta)s_R(y,\vartheta) + \hat{F}(s_R(y,\vartheta),y,\vartheta).$$

Multiplying this equation with B^{-1} from the left and taking norms we get

$$|s_R(y,\vartheta)| \leq b\,|s_R(G(s_R(y,\vartheta),y,\vartheta),\vartheta)| + b\,|\hat{F}(s_R(y,\vartheta),y,\vartheta)|.$$

Taking the supremum first on the right-hand side then on the left-hand side we obtain

$$\sup_{y\in Y} |s_R(y,\vartheta)| \le b \sup_{y\in Y} |s_R(y,\vartheta)| + b \sup_{y\in Y} |\widehat{F}(s_R(y,\vartheta), y, \vartheta)|.$$

This immediately yields the given estimate for $|s_R|$. □

1.2 Attractive negatively invariant manifolds

As in Section 1.1 we consider a family of maps P_ϑ of the form

$$P_\vartheta : X \times Y \ni \begin{pmatrix} x \\ y \end{pmatrix} \longmapsto \begin{pmatrix} \bar{x} \\ \bar{y} \end{pmatrix} = \begin{pmatrix} F(x,y,\vartheta) \\ G(x,y,\vartheta) \end{pmatrix} \in \mathcal{B}_x \times \mathcal{B}_y, \quad \vartheta \in E \subset \mathcal{B}_\vartheta, \tag{1.10}$$

where $\mathcal{B}_x$, $\mathcal{B}_y$, $\mathcal{B}_\vartheta$ are Banach spaces and $X \subset \mathcal{B}_x$, $Y \subset \mathcal{B}_y$, $E \subset \mathcal{B}_\vartheta$ are open subspaces. In this section we suppose that P_ϑ is contracting in y-direction and less contracting or expanding in x-direction.

As in Section 1.1 we assume that Hypothesis HM holds. Hypothesis HMR is replaced by Hypothesis HMA.

Hypothesis HM

The functions $F \in C^0(X \times Y \times E, \mathcal{B}_x)$, $G \in C^0(X \times Y \times E, \mathcal{B}_y)$ have the following properties.

a) *P_ϑ is inflowing with respect to Y, i.e., $G(x,y,\vartheta) \in Y$ holds for all $(x,y,\vartheta) \in X \times Y \times E$.*

b) *P_ϑ is outflowing with respect to X, i.e., for every $\bar{x} \in X$, $y \in Y$, $\vartheta \in E$ there is $x \in X$ such that $F(x,y,\vartheta) = \bar{x}$.*

c) *There are nonnegative constants Γ_{11}, L_{12}, L_{13}, L_{21}, L_{22} and L_{23} such that for $x, x_1, x_2 \in X$, $y, y_1, y_2 \in Y$, $\vartheta, \vartheta_1, \vartheta_2 \in E$ the functions F and G satisfy*

$$\begin{aligned} |F(x_1,y,\vartheta) - F(x_2,y,\vartheta)| &\ge \Gamma_{11}|x_1 - x_2|, \\ |F(x,y_1,\vartheta_1) - F(x,y_2,\vartheta_2)| &\le L_{12}|y_1 - y_2| + L_{13}|\vartheta_1 - \vartheta_2|, \\ |G(x_1,y_1,\vartheta_1) - G(x_2,y_2,\vartheta_2)| &\le L_{21}|x_1 - x_2| + L_{22}|y_1 - y_2| \\ &\quad + L_{23}|\vartheta_1 - \vartheta_2|. \end{aligned}$$

Hypothesis HMA

There is $y^* \in Y$ such that the function $G(\cdot, y^*, \cdot) : X \times E \to Y$ is bounded.

We show that under these hypotheses there exists an attractive negatively invariant manifold for the map P_ϑ provided the following two conditions are satisfied. The "weak coupling condition" CM is as in Section 1.1. The "repulsivity condition" CMR is replaced by the "attractivity Condition" CMA.

Condition CM

$$2\sqrt{L_{12}L_{21}} < \Gamma_{11} - L_{22}.$$

Condition CMA

$$L_{22} + \Delta < 1,$$

where

$$\Delta = \frac{2L_{12}L_{21}}{\Gamma_{11} - L_{22} + \sqrt{(\Gamma_{11} - L_{22})^2 - 4L_{12}L_{21}}}.$$

Theorem 1.5. *Let the map P_ϑ given in equation* (1.10) *satisfy Hypotheses HM and HMA. Moreover, assume that the constants Γ_{11}, L_{12}, L_{21}, L_{22} satisfy Conditions CM and CMA.*

Then there exists a bounded function $s_A\colon X \times E \to Y$ such that the following assertions hold for $\vartheta \in E$.

i) *The set $M_\vartheta = \{(x, y) \mid x \in X,\ y = s_A(x, \vartheta)\}$ is a negatively invariant manifold of the map P_ϑ, i.e., $P_\vartheta(M_\vartheta) \supset M_\vartheta$ and $M_\vartheta = P_\vartheta(M_\vartheta) \cap X \times Y$. The function s_A satisfies the invariance equation*

$$G(x, s_A(x, \vartheta), \vartheta) = s_A(F(x, s_A(x, \vartheta), \vartheta), \vartheta) \tag{1.11}$$

for all x with $F(x, s_A(x, \vartheta), \vartheta) \in X$.

ii) *The function s_A is in C_{λ_A, ν_A} with*

$$\lambda_A = \frac{2L_{21}}{\Gamma_{11} - L_{22} + \sqrt{(\Gamma_{11} - L_{22})^2 - 4L_{12}L_{21}}}, \qquad \nu_A := \frac{L_{23} + L_{13}\,\lambda_A}{1 - L_{22} - L_{12}\,\lambda_A},$$

i.e., s_A is bounded, uniformly λ_A-Lipschitz continuous with respect to x and uniformly ν_A-Lipschitz continuous with respect to ϑ.

iii) *The negatively invariant manifold M_ϑ is uniformly attractive with attractivity constant*

$$\chi_A := L_{22} + \Delta < 1,$$

$$\Delta = \frac{2L_{12}L_{21}}{\Gamma_{11} - L_{22} + \sqrt{(\Gamma_{11} - L_{22})^2 - 4L_{12}L_{21}}} = L_{12}\lambda_A,$$

i.e., if $(x_0, y_0) \in X \times Y$ and $(x_1, y_1) := P_\vartheta(x_0, y_0) \in X \times Y$ then the inequality

$$|y_1 - s_A(x_1, \vartheta)| \le \chi_A\,|y_0 - s_A(x_0, \vartheta)|$$

holds.

iv) *If the set* $\Lambda \subset X \times Y$ *is bounded with respect to* y *and negatively invariant under the map* P_ϑ *then* Λ *is contained in* M_ϑ, *i.e., if* $\Lambda \subset X \times Y_0$ *with bounded* $Y_0 \subset Y$ *and if* $\Lambda \subset P_\vartheta(\Lambda)$ *then* $\Lambda \subset M_\vartheta$.

v) *If there is a map* $\kappa\colon X \to X$ *such that* F *is* κ*-equivariant and* G *is* κ*-invariant then* s_A *is* κ*-invariant, i.e., if for all* $(x, y) \in X \times Y$,

$$F(\kappa x, y, \vartheta) = \kappa F(x, y, \vartheta),$$
$$G(\kappa x, y, \vartheta) = G(x, y, \vartheta),$$

then $s_A(\kappa x, \vartheta) = s_A(x, \vartheta)$ *holds.*

vi) *If the function* G *has the form* $G(x, y, \vartheta) = B(x, y, \vartheta)y + \widehat{G}(x, y, \vartheta)$ *and if for all* $x \in X$ *the estimate* $|B(x, s_A(x, \vartheta), \vartheta)| \le b < 1$ *holds then*

$$|s_A(x, \vartheta)| \le \frac{1}{1-b} \sup_{x \in X} |\widehat{G}(x, s_A(x, \vartheta), \vartheta)|.$$

vii) *The map* $P_\vartheta\big|_{M_\vartheta}$ *is invertible, i.e., for every* $(\bar{x}, \bar{y}) \in M_\vartheta$ *there is a unique* $(x, y) \in M_\vartheta$ *such that* $P_\vartheta(x, y) = (\bar{x}, \bar{y})$.

Remark 1.6. (1) Analogously to Remark 1.4.1 it holds that in the case $\Gamma_{11} \le 1$, which may occur in applications, the two conditions CM, CMA may be replaced by the single condition CM.

(2) The invariance equation (1.11) and assertion iii) may be extended to all $x \in X$ in the following way. Let $\bar{M}_\vartheta = P_\vartheta(M_\vartheta)$, $\bar{X}_\vartheta = \{x \in \mathcal{B}_x \mid \text{there is } y \in Y \text{ such that } (x, y) \in \bar{M}_\vartheta\}$ be the projection of $\bar{M}_\vartheta$ into $\mathcal{B}_x$. For $\vartheta \in E$ the set $\bar{M}_\vartheta$ is the graph of some function $\bar{s}_A(\cdot, \vartheta)\colon \bar{X}_\vartheta \to Y$. Let $\bar{S} := \bigcup_{\vartheta \in E} (\bar{X}_\vartheta, \vartheta) \subset \mathcal{B}_x \times \mathcal{B}_\vartheta$. For $\bar{s}_A\colon \bar{S} \to Y$ one has $\bar{s}_A|_{X \times E} = s_A$ and $\bar{s}_A$ satisfies the invariance equation

$$G(x, \bar{s}_A(x, \vartheta), \vartheta) = \bar{s}_A(F(x, \bar{s}_A(x, \vartheta), \vartheta), \vartheta) \quad \text{for } x \in X, \vartheta \in E. \tag{1.12}$$

Assertion iii) may be extended to

$$|y_1 - \bar{s}_A(x_1, \vartheta)| \le \chi_A \, |y_0 - \bar{s}_A(x_0, \vartheta)|$$

for all $(x_0, y_0) \in X \times Y$.

(3) In the applications in Part III we often have $X = \mathcal{B}_x$. In this case the set M_ϑ is an invariant manifold, i.e., $P(M_\vartheta) = M_\vartheta$. The invariance equation (1.11) holds for all $x \in X$.

(4) If, e.g., the map κ in assertion v) is taken to be $\kappa\colon x \mapsto x + z$ for some fixed $z \in X$ one obtains that if the map P_ϑ is z-periodic with respect to x then s_A is also z-periodic with respect to x.

If, e.g., $\kappa\colon x \mapsto -x$ one gets that if F is odd in x and G is even in x then s_A is even in x.

(5) Assertion vi) is often useful to get good bounds for $|s_A|$. In addition the following statement holds for the Lipschitz constant of s_A with respect to x. Let L_1^B, L_2^B be the

Lipschitz constants of $B(x, y, \vartheta)$ with respect to x and y, respectively, and let $\widehat{L}_{21}$, $\widehat{L}_{22}$ be the Lipschitz constants of $\widehat{G}(x, y, \vartheta)$. If $\gamma := \Gamma_{11} - b - |s_A| L_2^B - \widehat{L}_{22} > 0$ and $\gamma^2 > 4L_{12}\,(|s_A|\, L_1^B + \widehat{L}_{21})$ then s_A is $\hat{\lambda}_A$-Lipschitz continuous with respect to x with

$$\hat{\lambda}_A := \frac{2\big(|s_A| L_1^B + \widehat{L}_{21}\big)}{\gamma + \sqrt{\gamma^2 - 4L_{12}\,(|s_A|\, L_1^B + \widehat{L}_{21})}}.$$

It is not difficult to derive a similar bound for the Lipschitz constant of s_A with respect to ϑ.

Proof of Theorem 1.5. i), ii) We define the following set of functions:

$$\begin{aligned} C_{\lambda_A,\nu_A} := \big\{\sigma \in C^0(X \times E, Y) \mid \sigma \text{ is bounded, uniformly } \lambda_A\text{-Lipschitz continuous} \\ \text{with respect to } x \text{ and uniformly } \nu_A\text{-Lipschitz} \\ \text{continuous with respect to } \vartheta\big\}. \end{aligned}$$

Equipped with the supremum norm the set C_{λ_A,ν_A} becomes a complete metric space. For $\vartheta \in E$, $\sigma \in C_{\lambda_A,\nu_A}$ we define the manifold $M_{\vartheta,\sigma} := \{(x, y) \mid x \in X,\ y = \sigma(x, \vartheta)\}$. We look for a function $s_A \in C_{\lambda_A,\nu_A}$ such that $M_{\vartheta,s_A} \subset P_\vartheta(M_{\vartheta,s_A})$.

Assertion 1.5.1. *For $\sigma \in C_{\lambda_A,\nu_A}$ there is a unique function $\bar{\sigma} \in C_{\lambda_A,\nu_A}$ such that $M_{\vartheta,\bar{\sigma}} \subset P_\vartheta(M_{\vartheta,\sigma})$.*

We first show that $P_\vartheta(M_{\vartheta,\sigma})$ is the graph of some function. Let $z_i = (x_i, \sigma(x_i, \vartheta)) \in M_{\vartheta,\sigma}$, $i = 1, 2$, $x_1 \neq x_2$, be two distinct points in $M_{\vartheta,\sigma}$. The points z_i, $i = 1, 2$, are mapped to $\bar{z}_i = P_\vartheta(z_i) = (\bar{x}_i, \bar{y}_i) = (F(x_i, \sigma(x_i, \vartheta), \vartheta),\ G(x_i, \sigma(x_i, \vartheta), \vartheta))$. We have

$$\begin{aligned} |\bar{x}_1 - \bar{x}_2| &= |F(x_1, \sigma(x_1, \vartheta), \vartheta) - F(x_2, \sigma(x_2, \vartheta), \vartheta)| \\ &\geq |F(x_1, \sigma(x_1, \vartheta), \vartheta) - F(x_2, \sigma(x_1, \vartheta), \vartheta)| \\ &\quad - |F(x_2, \sigma(x_1, \vartheta), \vartheta) - F(x_2, \sigma(x_2, \vartheta), \vartheta)| \\ &\geq (\Gamma_{11} - L_{12}\lambda_A)\,|x_1 - x_2|. \end{aligned} \tag{1.13}$$

Since λ_A is the smaller root of the quadratic equation

$$L_{12}\lambda_A^2 - (\Gamma_{11} - L_{22})\,\lambda_A + L_{21} = 0, \tag{1.14}$$

we have

$$\frac{L_{21} + L_{22}\lambda_A}{\Gamma_{11} - L_{12}\lambda_A} = \lambda_A. \tag{1.15}$$

From $2\Delta = \Gamma_{11} - L_{22} - \sqrt{(\Gamma_{11} - L_{22})^2 - 4L_{12}L_{21}}$ one obtains

$$\Gamma_{11} - \Delta > L_{22} + \Delta \geq 0. \tag{1.16}$$

We conclude from equations (1.13) and (1.16) that $|\bar{x}_1 - \bar{x}_2| > 0$, implying that for fixed ϑ the set $P_\vartheta(M_{\vartheta,\sigma})$ is the graph of some function $\tilde{\sigma}_\vartheta$. We show that $\tilde{\sigma}_\vartheta$ is defined for all $\bar{x} \in X$. We have to show that for arbitrary $\bar{x} \in X$ the equation

$$\bar{x} = F(x, \sigma(x, \vartheta), \vartheta) \tag{1.17}$$

has a solution $x \in X$. We consider equation (1.17) as a fixed point equation of an operator $H : h \in X \mapsto H(h) \in X$ defined by the equation

$$\bar{x} = F(H(h), \sigma(h, \vartheta), \vartheta).$$

The operator H is well defined since by Hypothesis HM b) this equation has a solution $H(h)$ which is unique by Hypothesis HM c). For $h_1, h_2 \in X$ we have by definition of H that

$$\begin{aligned} 0 &= |F(H(h_1), \sigma(h_1, \vartheta), \vartheta) - F(H(h_2), \sigma(h_2, \vartheta), \vartheta)| \\ &\geq \Gamma_{11}|H(h_1) - H(h_2)| - L_{12}\lambda_A|h_1 - h_2| \end{aligned}$$

and hence

$$|H(h_1) - H(h_2)| \leq \frac{L_{12}\,\lambda_A}{\Gamma_{11}}\,|h_1 - h_2|.$$

By means of equation (1.16) we conclude that the operator $H(h)$ is a contraction in X. We denote the unique fixed point of H by h^*. Obviously, $x = h^*$ is a solution of equation (1.17). This means that $\tilde{\sigma}_\vartheta$ is indeed defined for all $\bar{x} \in X$. We now define the function $\bar{\sigma}$ by

$$\bar{\sigma}(\cdot, \vartheta) := \tilde{\sigma}_\vartheta\big|_X.$$

We have shown that for all $\bar{x} \in X$ there is $x \in X$ such that $\bar{x} = F(x, \sigma(x, \vartheta), \vartheta)$ and

$$\bar{\sigma}(\bar{x}, \vartheta) = G(x, \sigma(x, \vartheta), \vartheta). \tag{1.18}$$

We show that $\bar{\sigma} \in C_{\lambda_A,\nu_A}$. For $\bar{x}_1, \bar{x}_2 \in X$, $\vartheta_1, \vartheta_2 \in E$ we have

$$\begin{aligned} |\bar{x}_1 - \bar{x}_2| &= |F(x_1, \sigma(x_1, \vartheta_1), \vartheta_1) - F(x_2, \sigma(x_2, \vartheta_2), \vartheta_2)| \\ &\geq (\Gamma_{11} - L_{12}\lambda_A)\,|x_1 - x_2| - (L_{12}\nu_A + L_{13})\,|\vartheta_1 - \vartheta_2| \end{aligned} \tag{1.19}$$

and

$$\begin{aligned} |\bar{\sigma}(\bar{x}_1, \vartheta_1) - \bar{\sigma}(\bar{x}_2, \vartheta_2)| &= |G(x_1, \sigma(x_1, \vartheta_1), \vartheta_1) - G(x_2, \sigma(x_2, \vartheta_2), \vartheta_2)| \\ &\leq (L_{21} + L_{22}\lambda_A)\,|x_1 - x_2| + (L_{22}\nu_A + L_{23})\,|\vartheta_1 - \vartheta_2|. \end{aligned}$$

From (1.19) one may estimate $|x_1 - x_2|$. Putting the result into the last estimate we obtain

$$\begin{aligned} &|\bar{\sigma}(\bar{x}_1, \vartheta_1) - \bar{\sigma}(\bar{x}_2, \vartheta_2)| \\ &\quad \leq \frac{L_{21} + L_{22}\lambda_A}{\Gamma_{11} - L_{12}\lambda_A}\,|\bar{x}_1 - \bar{x}_2| \\ &\qquad + \left(\frac{L_{21} + L_{22}\lambda_A}{\Gamma_{11} - L_{12}\lambda_A}(L_{12}\nu_A + L_{13}) + L_{22}\nu_A + L_{23}\right)|\vartheta_1 - \vartheta_2|. \end{aligned}$$

Using equation (1.15) we get that $\bar{\sigma}$ is λ_A-Lipschitz continuous with respect to $\bar{x}$ and ν_A-Lipschitz continuous with respect to ϑ.

In order to show that $\bar{\sigma}$ is bounded we estimate using equation (1.18) and y^* from Hypothesis HMA

$$\begin{aligned}|\bar{\sigma}(\bar{x},\vartheta)| &\le |G(x,\sigma(x,\vartheta),\vartheta) - G(x,y^*,\vartheta)| + |G(x,y^*,\vartheta)| \\ &\le L_{22}\,|\sigma(x,\vartheta) - y^*| + |G(\cdot,y^*,\cdot)|.\end{aligned}$$

Since $\sigma \in C_{\lambda_A,\nu_A}$ is bounded it follows that $\bar{\sigma}$ is bounded. This completes the proof of Assertion 1.5.1.

By means of Assertion 1.5.1 we may define the operator $\mathcal{F} : C_{\lambda_A,\nu_A} \to C_{\lambda_A,\nu_A}$ by

$$\mathcal{F}\sigma = \bar{\sigma}.$$

It holds that

$$P_\vartheta(M_{\vartheta,\sigma}) \cap X \times Y = M_{\vartheta,\mathcal{F}\sigma}.$$

For the existence of a negatively invariant manifold M_{ϑ,s_A} we need a function $s_A(x,\vartheta) \in C_{\lambda_A,\nu_A}$ such that $P_\vartheta(M_{\vartheta,s_A}) \cap X \times Y = M_{\vartheta,s_A}$, cf. Figure 1.3. This is equivalent to the requirement that s_A is a fixed point of the operator $\mathcal{F}$. We show that the operator $\mathcal{F}$ is a contraction and hence has a unique fixed point.

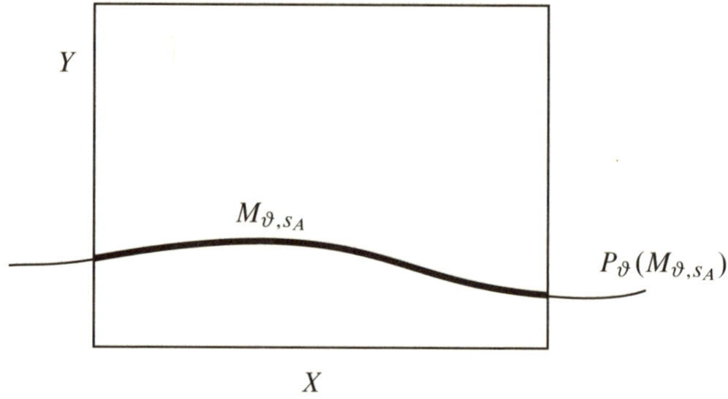

Figure 1.3. The negatively invariant manifold M_{ϑ,s_A}.

Assertion 1.5.2. *$\mathcal{F}$ is a contraction with contractivity constant $\chi_A = L_{22} + \Delta < 1$.*

We show that for $\sigma_i \in C_{\lambda_A,\nu_A}$ and $\bar{\sigma}_i := \mathcal{F}\sigma_i$, $i = 1,2$, the difference $|\bar{\sigma}_1 - \bar{\sigma}_2|$ may be estimated by $|\sigma_1 - \sigma_2|$. For $\bar{x} \in X$ there are $x_1, x_2 \in X$ such that

$$\bar{x} = F(x_i, \sigma_i(x_i,\vartheta), \vartheta), \quad i = 1,2.$$

Taking differences we get

$$\begin{aligned} 0 &= |F(x_1, \sigma_1(x_1, \vartheta), \vartheta) - F(x_2, \sigma_2(x_2, \vartheta), \vartheta)| \\ &\geq \Gamma_{11}\,|x_1 - x_2| - L_{12}\,|\sigma_1(x_1, \vartheta) - \sigma_2(x_2, \vartheta)| \\ &\geq \Gamma_{11}\,|x_1 - x_2| - L_{12}\,(\lambda_A\,|x_1 - x_2| + |\sigma_1 - \sigma_2|) \end{aligned}$$

from which we conclude that

$$|x_1 - x_2| \leq \frac{L_{12}}{\Gamma_{11} - L_{12}\lambda_A}\,|\sigma_1 - \sigma_2|. \tag{1.20}$$

We now compute

$$\begin{aligned} |\bar{\sigma}_1(\bar{x}, \vartheta) - \bar{\sigma}_2(\bar{x}, \vartheta)| &= |G(x_1, \sigma_1(x_1, \vartheta), \vartheta) - G(x_2, \sigma_2(x_2, \vartheta), \vartheta)| \\ &\leq (L_{21} + L_{22}\lambda_A)\,|x_1 - x_2| + L_{22}\,|\sigma_1 - \sigma_2| \end{aligned}$$

and with (1.20) and (1.15) we get

$$|\mathcal{F}\sigma_1 - \mathcal{F}\sigma_2| = |\bar{\sigma}_1 - \bar{\sigma}_2| \leq (L_{22} + L_{12}\lambda_A)\,|\sigma_1 - \sigma_2|. \tag{1.21}$$

Condition CMA and $\Delta = L_{12}\lambda_A$ imply that $\mathcal{F}$ is a contraction. This proves Assertion 1.5.2.

By means of Assertions 1.5.1, 1.5.2 and using the contraction principle we conclude that the operator $\mathcal{F}$ has a unique fixed point $s_A \in C_{\lambda_A, \nu_A}$. This implies that $M_\vartheta = \{(x, y) \mid x \in X,\ y = s_A(x, \vartheta)\}$ is a negatively invariant manifold of the map P_ϑ. The invariance equation (1.11) follows from $\mathcal{F}s_A = s_A$ and (1.18).

iii) Using the invariance equation (1.11) we get

$$\begin{aligned} |y_1 - s_A(x_1, \vartheta)| &= |G(x_0, y_0, \vartheta) - s_A(F(x_0, y_0, \vartheta), \vartheta)| \\ &\leq |G(x_0, y_0, \vartheta) - G(x_0, s_A(x_0, \vartheta), \vartheta)| \\ &\qquad + |s_A(F(x_0, s_A(x_0, \vartheta), \vartheta), \vartheta) - s_A(F(x_0, y_0, \vartheta), \vartheta)| \\ &\leq L_{22}\,|y_0 - s_A(x_0, \vartheta)| + \lambda_A\,L_{12}\,|y_0 - s_A(x_0, \vartheta)|. \end{aligned}$$

iv) Define $D := \sup_{(x,y) \in \Lambda} |y - s_A(x, \vartheta)|$. Note that $D < \infty$. The proof is by contradiction. Assume that $D > 0$. Then there is $(x, y) \in \Lambda$ with

$$|y - s_A(x, \vartheta)| > \chi_A D, \tag{1.22}$$

since $\chi_A < 1$. Since $(x, y) \in \Lambda \subset P_\vartheta(\Lambda)$ it follows that there is $(\tilde{x}, \tilde{y}) \in \Lambda$ such that $P_\vartheta(\tilde{x}, \tilde{y}) = (x, y)$. By assertion iii) we have $|y - s_A(x, \vartheta)| \leq \chi_A\,|\tilde{y} - s_A(\tilde{x}, \vartheta)| \leq \chi_A\,D$ contradicting (1.22).

v) We restrict the operator $\mathcal{F}$ to the space

$$C_{\lambda_A, \nu_A}\big|_\kappa := \{\sigma \in C_{\lambda_A, \nu_A} \mid \sigma(\kappa x, \vartheta) = \sigma(x, \vartheta)\}$$

of κ-invariant functions which is a closed subset of C_{λ_A,ν_A}. It suffices to show that the operator $\mathcal{F}$ maps $C_{\lambda_A,\nu_A}\big|_\kappa$ into itself. We have shown that for $\bar{x} \in X$ there is $x \in X$ with $\bar{x} = F(x, \sigma(x,\vartheta), \vartheta)$, cf. (1.17). Using the equivariance of F and the invariance of G we conclude that $F(\kappa x, \sigma(\kappa x, \vartheta), \vartheta) = \kappa\bar{x}$ and

$$\begin{aligned}(\mathcal{F}\sigma)(\kappa\bar{x}, \vartheta) &= G(\kappa x, \sigma(\kappa x, \vartheta), \vartheta)\\ &= G(x, \sigma(x, \vartheta), \vartheta)\\ &= \big(\mathcal{F}\sigma\big)(\bar{x}, \vartheta),\end{aligned}$$

implying $\mathcal{F}\sigma \in C_{\lambda_A,\nu_A}\big|_\kappa$.

vi) For $\bar{x} \in X$, $\vartheta \in E$ we have using (1.11)

$$\begin{aligned}s_A(\bar{x}, \vartheta) &= s_A(F(x, s_A(x, \vartheta), \vartheta), \vartheta)\\ &= B(x, s_A(x, \vartheta), \vartheta)s_A(x, \vartheta) + \widehat{G}(x, s_A(x, \vartheta), \vartheta).\end{aligned}$$

First taking norms then taking the supremum on the right-hand side and finally on the left-hand side we get

$$\sup_{x\in X} |s_A(x, \vartheta)| \le b \sup_{x\in X} |s_A(x, \vartheta)| + \sup_{x\in X} |\,\widehat{G}(x, s_A(x, \vartheta), \vartheta)|$$

implying

$$\sup_{x\in X} |s_A(x, \vartheta)| \le \frac{1}{1-b} \sup_{x\in X} |\,\widehat{G}(x, s_A(x, \vartheta), \vartheta)|.$$

vii) We have shown that for all $\sigma \in C_{\lambda_A,\nu_A}$ and all $\bar{x} \in X$ equation (1.17) has a unique solution $x \in X$. Hence this holds true for $\sigma = s_A$. □

1.3 Hyperbolic invariant manifolds

In this section we assume that the map considered is contracting in some directions and expanding in some other directions, and that in the remaining directions the map is less contracting and less expanding. We consider the Banach spaces $\mathcal{B}_x$, $\mathcal{B}_z$, $\mathcal{B}_y$ and open subsets $X \subset \mathcal{B}_x$, $Z \subset \mathcal{B}_z$, $Y \subset \mathcal{B}_y$ and the map

$$P : X \times Z \times Y \ni \begin{pmatrix} x\\ z\\ y\end{pmatrix} \longmapsto \begin{pmatrix} \bar{x}\\ \bar{z}\\ \bar{y}\end{pmatrix} = \begin{pmatrix} F(x, z, y)\\ H(x, z, y)\\ G(x, z, y)\end{pmatrix}.$$

For simplicity, we have suppressed a possible dependence on parameters. We suppose that the map P is expanding in x-direction and contracting in y-direction. We want to apply Theorems 1.5 and 1.3 and therefore introduce the following notations:

$$x_A = \begin{pmatrix} x \\ z \end{pmatrix}, \quad y_A = y,$$

$$F_A(x_A, y_A) = \begin{pmatrix} F(x_A, y_A) \\ H(x_A, y_A) \end{pmatrix}, \quad G_A(x_A, y_A) = G(x_A, y_A),$$

$$P_A \colon X_A \times Y_A \ni \begin{pmatrix} x_A \\ y_A \end{pmatrix} \longmapsto \begin{pmatrix} \bar{x}_A \\ \bar{y}_A \end{pmatrix} = \begin{pmatrix} F_A(x_A, y_A) \\ G_A(x_A, y_A) \end{pmatrix} \tag{1.23}$$

and

$$x_R = x, \quad y_R = \begin{pmatrix} z \\ y \end{pmatrix},$$

$$F_R(x_R, y_R) = F(x_R, y_R), \quad G_R(x_R, y_R) = \begin{pmatrix} H(x_R, y_R) \\ G(x_R, y_R) \end{pmatrix},$$

$$P_R \colon X_R \times Y_R \ni \begin{pmatrix} x_R \\ y_R \end{pmatrix} \longmapsto \begin{pmatrix} \bar{x}_R \\ \bar{y}_R \end{pmatrix} = \begin{pmatrix} F_R(x_R, y_R) \\ G_R(x_R, y_R) \end{pmatrix}. \tag{1.24}$$

Note that the maps P_A and P_R, respectively, are different notations for the same map P. We assume that the map P_A satisfies the assumptions of Theorem 1.5 and that the map P_R satisfies the assumptions of Theorem 1.3. It follows that the map P admits both an attractive negatively invariant manifold M and a repulsive positively invariant manifold N. The intersection $K = M \cap N$ of the two manifolds is called a hyperbolic invariant manifold, cf. Figure 1.4.

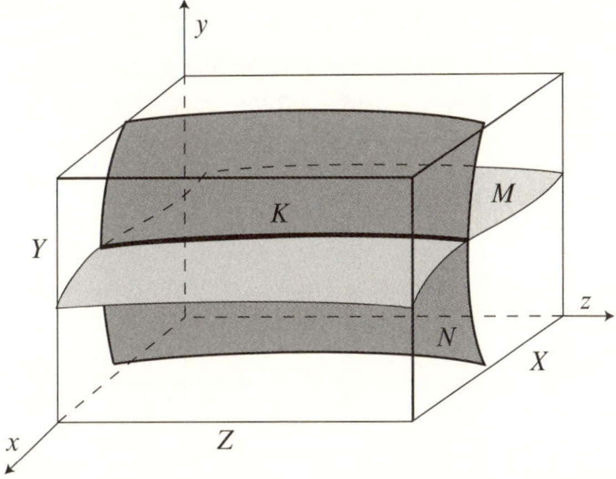

Figure 1.4. The hyperbolic invariant manifold K, its unstable manifold M and its stable manifold N.

We give the precise results in

Theorem 1.7. *Assume that the maps P_A and P_R defined in* (1.23) *and* (1.24)*, respectively, satisfy Hypothesis HM with constants Γ_{11}^A, L_{12}^A, L_{21}^A, L_{22}^A and Γ_{11}^R, L_{12}^R, L_{21}^R, L_{22}^R, respectively. Let P_A satisfy Hypothesis HMA and assume that the constants Γ_{11}^A, L_{12}^A, L_{21}^A, L_{22}^A satisfy Conditions CM and CMA with $\Delta_A := L_{12}^A \lambda_A$ where*

$$\lambda_A := 2L_{21}^A \Big/ \Big(\Gamma_{11}^A - L_{22}^A + \sqrt{(\Gamma_{11}^A - L_{22}^A)^2 - 4L_{12}^A L_{21}^A}\Big).$$

Let P_R satisfy Hypothesis HMR and assume that the constants Γ_{11}^R, L_{12}^R, L_{21}^R, L_{22}^R satisfy Conditions CM and CMR with $\Delta_R := L_{21}^R \lambda_R$ where

$$\lambda_R := 2L_{12}^R \Big/ \Big(\Gamma_{11}^R - L_{22}^R + \sqrt{(\Gamma_{11}^R - L_{22}^R)^2 - 4L_{12}^R L_{21}^R}\Big).$$

Moreover, assume that $\lambda_A \lambda_R < 1$.

Then the following assertions hold.

i) *The map P_A admits an attractive negatively invariant manifold of the form*

$$M = \{(x, z, y) \mid (x, z) \in X \times Z,\ y = s_A(x, z)\}.$$

All assertions of Theorem 1.5 *hold for P_A.*

ii) *The map P_R admits a repulsive positively invariant manifold of the form*

$$N = \{(x, z, y) \mid (z, y) \in Z \times Y,\ x = s_R(z, y)\}.$$

All assertions of Theorem 1.3 *hold for P_R.*

iii) *There are functions $r_x\colon Z \to X$, $r_y\colon Z \to Y$ such that the set $K := M \cap N$ is given as*

$$K = \big\{(x, z, y) \mid z \in Z, (x, y) = \big(r_x(z), r_y(z)\big)\big\} \tag{1.25}$$

and is an invariant manifold of the map P, i.e., $P(K) = K$.

The function r_x is uniformly Lipschitz continuous with Lipschitz constant $\lambda_R(1 + \lambda_A)/(1 - \lambda_A \lambda_R)$; the function r_y has Lipschitz constant $\lambda_A(1 + \lambda_R)/(1 - \lambda_A \lambda_R)$. The functions r_x and r_y satisfy the invariance equations

$$r_x(z) = s_R\big(z, s_A(r_x(z), z)\big), \quad r_y(z) = s_A\big(s_R(z, r_y(z)), z\big). \tag{1.26}$$

iv) *If the set $\Lambda \subset X \times Z \times Y$ is bounded with respect to x and y and is invariant under the map P then Λ is contained in K, i.e., if $\Lambda \subset X_0 \times Z \times Y_0$ with bounded $X_0 \subset X$, $Y_0 \subset Y$ and if $P(\Lambda) = \Lambda$ then $\Lambda \subset K$.*

v) *If there is a map $\kappa\colon Z \to Z$ such that F and G are κ-invariant and H is κ-equivariant then the functions r_x and r_y are κ-invariant, i.e., if for all $(x, z, y) \in X \times Z \times Y$,*

$$F(x, \kappa z, y) = F(x, z, y),$$

$$H(x, \kappa z, y) = \kappa H(x, z, y),$$

$$G(x, \kappa z, y) = G(x, z, y),$$

then $r_x(\kappa z) = r_x(z)$ and $r_y(\kappa z) = r_y(z)$ holds.

Proof. i), ii) These assertions follow immediately from Theorems 1.5 and 1.3, respectively.

iii) If $(x,z,y) \in K = M \cap N$ then $x = s_R(z, s_A(x,z))$ and $y = s_A(s_R(z,y),z)$ holds. We show that for $z \in Z$ these equations have a unique solution $x = r_x(z)$ and $y = r_y(z)$, respectively. Let $C_b^0(Z,X)$ be the space of bounded functions in $C^0(Z,X)$ equipped with the supremum norm. We consider the operators

$$\mathcal{K}_x \colon C_b^0(Z,X) \longrightarrow C_b^0(Z,X), \quad (\mathcal{K}_x r)(z) := s_R(z, s_A(r(z),z)),$$
$$\mathcal{K}_y \colon C_b^0(Z,Y) \longrightarrow C_b^0(Z,Y), \quad (\mathcal{K}_y r)(z) := s_A(s_R(z,r(z)),z).$$

Since

$$|\mathcal{K}_x r_1 - \mathcal{K}_x r_2| \le \lambda_R \lambda_A |r_1 - r_2|, \quad |\mathcal{K}_y r_1 - \mathcal{K}_y r_2| \le \lambda_A \lambda_R |r_1 - r_2|$$

and $\lambda_A \lambda_R < 1$, the operators $\mathcal{K}_x$ and $\mathcal{K}_y$ are contractions and hence have a unique fixed point r_x and r_y, respectively. It follows that K has the form (1.25) and that the invariance equation (1.26) holds.

We show that K is invariant under P. Since $K = M \cap N$ and since N is positively invariant we have

$$P(K) \subset P(N) \subset N \subset X \times Z \times Y$$

and

$$P(K) \subset P(M)$$

implying

$$P(K) \subset P(M) \cap X \times Z \times Y = M$$

and

$$P(K) \subset M \cap N = K, \tag{1.27}$$

i.e., K is positively invariant. To prove that K is negatively invariant we have to show that for every $(\bar{x}, \bar{z}, \bar{y}) \in K$ there is $(x,z,y) \in K$ such that $P(x,z,y) = (\bar{x}, \bar{z}, \bar{y})$. Since M is negatively invariant there is $(x,z,y) \in M$ with $P(x,z,y) = (\bar{x}, \bar{z}, \bar{y})$. For this (x,z,y) we define the set $\Lambda := \{P^n(x,z,y) \mid n \in \mathbb{N}_0\}$. By definition, Λ is positively invariant. Since $P(x,z,y) = (\bar{x}, \bar{z}, \bar{y}) \in K$ it follows with (1.27) that $P(\Lambda) \subset K$ implying that Λ is bounded with respect to x. Theorem 1.3 iv) implies $\Lambda \subset N$ and hence $(x,z,y) \in N$ and therefore also $(x,z,y) \in K$.

We show that r_x and r_y are Lipschitz continuous. These functions satisfy the invariance equations (1.26) and hence the estimates

$$|r_x(z_1) - r_x(z_2)| \le \lambda_R |z_1 - z_2| + \lambda_R \lambda_A \big(|r_x(z_1) - r_x(z_2)| + |z_1 - z_2|\big),$$
$$|r_y(z_1) - r_y(z_2)| \le \lambda_A \lambda_R \big(|z_1 - z_2| + |r_y(z_1) - r_y(z_2)|\big) + \lambda_A |z_1 - z_2|$$

hold, implying

$$|r_x(z_1) - r_x(z_2)| \le \frac{\lambda_R(1+\lambda_A)}{1-\lambda_A\lambda_R}\,|z_1 - z_2|,$$
$$|r_y(z_1) - r_y(z_2)| \le \frac{\lambda_A(1+\lambda_R)}{1-\lambda_A\lambda_R}\,|z_1 - z_2|.$$

iv) The assertion follows immediately from the corresponding properties of the manifolds M and N, cf. Theorems 1.5 and 1.3.

v) We define the maps $\kappa_A\colon \binom{x}{z} \mapsto \binom{x}{\kappa z}$ and $\kappa_R\colon \binom{z}{y} \mapsto \binom{\kappa z}{y}$. From the invariance properties of P_A and P_R, respectively, cf. Theorem 1.5 v) and 1.3 v), we obtain

$$s_A(x, \kappa z) = s_A(x, z) \quad \text{and} \quad s_R(\kappa z, y) = s_R(z, y).$$

The invariance equations (1.26) imply that

$$r_x(\kappa z) = s_R\big(\kappa z, s_A(r_x(\kappa z), \kappa z)\big) = s_R\big(z, s_A(r_x(\kappa z), z)\big),$$
$$r_y(\kappa z) = s_A\big(s_R(\kappa z, r_y(\kappa z)), \kappa z\big) = s_A\big(s_R(z, r_y(\kappa z)), z\big).$$

Since these equations have a unique solution we conclude that $r_x(\kappa z) = r_x(z)$ and $r_y(\kappa z) = r_y(z)$. □

Remark 1.8. Theorem 1.7 includes the case where $\mathcal{B}_z = \{0\}$. In this case the z-coordinate of the map P is omitted. This particular case of a map admitting a hyperbolic fixed point is treated in Section 8.1 in more detail.

1.4 Manifolds defined by several charts

In this section we consider a map admitting an invariant manifold which cannot easily be described as a graph of a single function. Instead, we will describe a manifold $\mathcal{M}$ in several charts.

1.4.1 A general existence result

Let the map

$$P\colon x \longmapsto \bar{x} = F(x), \quad x \in \mathbb{R}^\ell \tag{1.28}$$

be described in charts Φ_i, $i = 1, \dots, \kappa$. A chart Φ_i is a map

$$\begin{aligned} \Phi_i\colon Q_i \subset \mathbb{R}^\ell &\longrightarrow W_i = U_i \times V_i \subset \mathbb{R}^m \times \mathbb{R}^n, \quad m + n = \ell, \\ x &\longmapsto \begin{pmatrix} u_i \\ v_i \end{pmatrix}. \end{aligned} \tag{1.29}$$

In Φ_i the piece $\mathcal{M}_i = \mathcal{M} \cap Q_i$ of a manifold $\mathcal{M}$ is described in W_i as the graph M_i of a function $\sigma_i \in C_i(U_i, V_i)$,

$$\Phi_i(\mathcal{M}_i) = M_i = \{(u_i, v_i) \mid u_i \in U_i, v_i = \sigma_i(u_i)\}, \tag{1.30}$$

where $C_i(U_i, V_i)$ is a space of bounded Lipschitz continuous functions from U_i to V_i. The situation is sketched in Figure 1.5 where U_i and V_i are drawn as 1-dimensional sets. The existence of an invariant manifold is shown using in each chart a graph

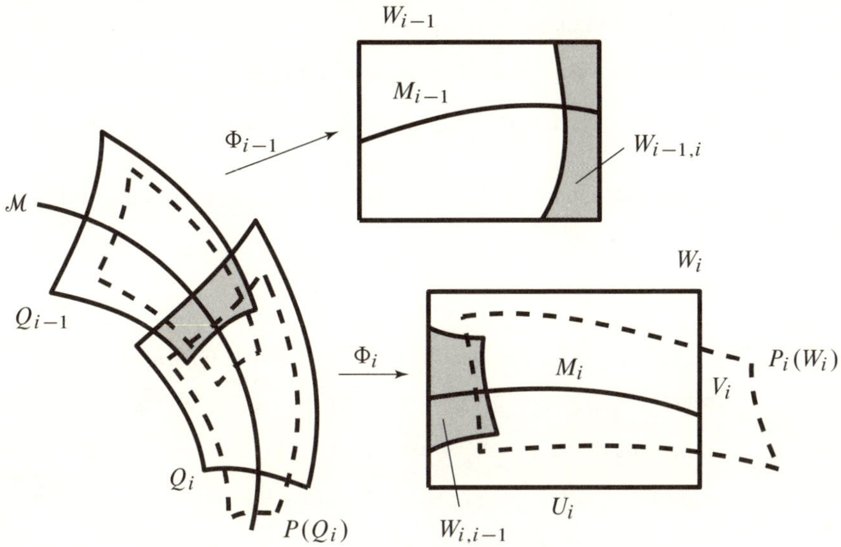

Figure 1.5. Sketch of the charts Φ_{i-1} and Φ_i.

transform result of the type of Theorem 1.11 in Subsection 1.4.2. For adjacent charts Φ_{i-1}, Φ_i, $i = 2, \dots, \kappa$, we introduce the sets $Q_{i-1,i} := Q_{i-1} \cap Q_i$, $W_{i-1,i} := \Phi_{i-1}(Q_{i-1,i}) \subset W_{i-1}$ and $W_{i,i-1} := \Phi_i(Q_{i-1,i}) \subset W_i$, cf Figure 1.5. In the set $Q_{i-1,i}$ a manifold $\mathcal{M}$ has a description in Φ_{i-1} and in Φ_i, respectively, with functions σ_{i-1} and σ_i, respectively. For $(u_{i-1}, \sigma_{i-1}(u_{i-1})) \in W_{i-1,i}$ we have

$$\Phi_i \circ \Phi_{i-1}^{-1} \begin{pmatrix} u_{i-1} \\ \sigma_{i-1}(u_{i-1}) \end{pmatrix} = \begin{pmatrix} u_i \\ \sigma_i(u_i) \end{pmatrix} \in W_{i,i-1}. \tag{1.31}$$

We introduce the linear function space $\Sigma^* = \{\sigma = (\sigma_1, \dots, \sigma_\kappa) \mid \sigma_i \in C_b^{\mathrm{Lip}}(U_i, \mathbb{R}^n)$, $i = 1, \dots, \kappa\}$ where $C_b^{\mathrm{Lip}}(U_i, \mathbb{R}^n)$ denotes the space of bounded Lipschitz continuous functions from U_i to $\mathbb{R}^n$. We equip the space Σ^* with the norm

$$\|\sigma\| = \max_{i=1,\dots,\kappa} |\sigma_i|_{U_i}, \tag{1.32}$$

where the norm $|\cdot|_{U_i}$ in $C_b^{\mathrm{Lip}}(U_i, \mathbb{R}^n)$ is given as

$$|\sigma_i|_{U_i} = N_i \sup_{u_i \in U_i} |\sigma_i(u_i)| \tag{1.33}$$

with a positive norm factor N_i. Restricting the functions σ_i to the subspaces $C_i(U_i, V_i)$ of $C_b^{\mathrm{Lip}}(U_i, \mathbb{R}^n)$ and requiring (1.31) for $i = 2, \dots, \kappa$ we obtain a metric space Σ where the metric is induced by the norm $\|\cdot\|$.

We describe a manifold $\mathcal{M}$ by a function $\sigma \in \Sigma$. Applying the map P to $\mathcal{M}$ we get a set $P(\mathcal{M})$. Under appropriate assumptions and under an appropriate choice of the spaces $C_i(U_i, V_i)$ there is a set $\bar{\mathcal{M}} \subset P(\mathcal{M})$ which is again described by some function $\bar{\sigma} \in \Sigma$, i.e., the map P induces an operator

$$\mathcal{F} : \Sigma \longrightarrow \Sigma, \quad \sigma \longmapsto \mathcal{F}(\sigma) := \bar{\sigma} \tag{1.34}$$

where the function $\bar{\sigma}$ is given by functions $\bar{\sigma}_i$ defined on the whole of U_i for every chart Φ_i.

In order to state our manifold result we make several assumptions, cf. Figures 1.5 and 1.6.

Hypothesis HMAK

For $i = 1, \dots, \kappa$ there are sets $Q_i \subset \widetilde{Q}_i \subset \mathbb{R}^\ell$ with $\widetilde{Q}_i \supset P(Q_i)$, sets $\check{U}_i \subset \hat{U}_i \subset U_i \subset \widetilde{U}_i \subset \mathbb{R}^m$ with $\hat{U}_1 = U_1$ and sets $V_i \subset \mathbb{R}^n$ and charts Φ_i with

$$\Phi_i(Q_i) = W_i := U_i \times V_i \quad \text{and} \quad \Phi_i(\widetilde{Q}_i) = \widetilde{W}_i := \widetilde{U}_i \times V_i$$

such that the following holds.

a) *In the chart Φ_i, $i = 1, \dots, \kappa$, the map P induces a map*

$$P_i := \Phi_i \circ P \circ \Phi_i^{-1} \colon W_i = U_i \times V_i \longrightarrow \widetilde{W}_i = \widetilde{U}_i \times V_i,$$
$$\begin{pmatrix} u_i \\ v_i \end{pmatrix} \longmapsto \begin{pmatrix} \bar{u}_i \\ \bar{v}_i \end{pmatrix} = \begin{pmatrix} F_i(u_i, v_i) \\ G_i(u_i, v_i) \end{pmatrix}.$$

b) *For $i = 1, \dots, \kappa$ there is a complete function space $C_i(U_i, V_i)$ of bounded Lipschitz continuous functions such that for $\sigma_i \in C_i(U_i, V_i)$ the map P_i maps the set $M_{\sigma_i} := \{(u_i, v_i) \mid u_i \in U_i, v_i = \sigma_i(u_i)\}$ to the set $P_i(M_{\sigma_i})$ which is the graph of some function $\tilde{\sigma}_i$. The restriction $\hat{\sigma}_i := \tilde{\sigma}_i|_{\hat{U}_i}$ lies in $C_i(\hat{U}_i, V_i)$, i.e., the induced operator $\mathcal{F}_i$ takes $\sigma_i \in C_i(U_i, V_i)$ to $\hat{\sigma}_i \in C_i(\hat{U}_i, V_i)$.*

c) *For every $\check{\sigma}_{i-1} \in C_{i-1}(\check{U}_{i-1}, V_{i-1})$, $i = 2, \dots, \kappa$, there is*

$$\bar{\sigma}_{i,i-1} \in C_i(U_i \setminus \hat{U}_i, V_i)$$

such that for every $u_i \in U_i \setminus \hat{U}_i$ there is $u_{i-1} \in \check{U}_{i-1}$ with

$$\Phi_i \circ \Phi_{i-1}^{-1}(u_{i-1}, \check{\sigma}_{i-1}(u_{i-1})) = (u_i, \bar{\sigma}_{i,i-1}(u_i)),$$

i.e., the induced operator $\mathcal{T}_{i,i-1}$ transforms $\check{\sigma}_{i-1} \in C_{i-1}(\check{U}_{i-1}, V_{i-1})$ to $\bar{\sigma}_{i,i-1} \in C_i(U_i \setminus \hat{U}_i, V_i)$.

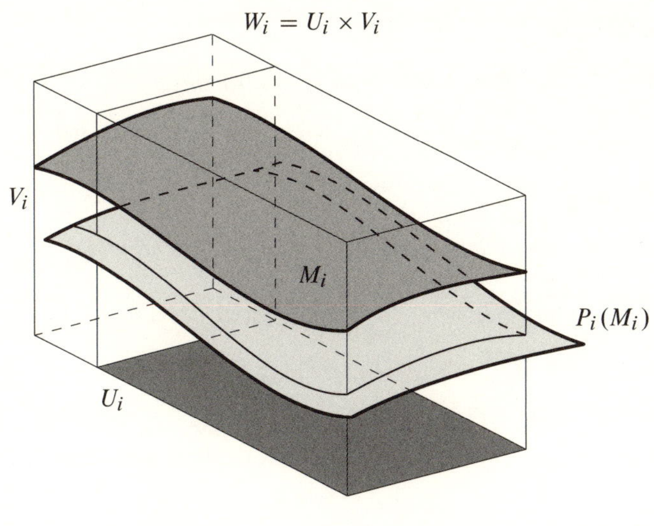

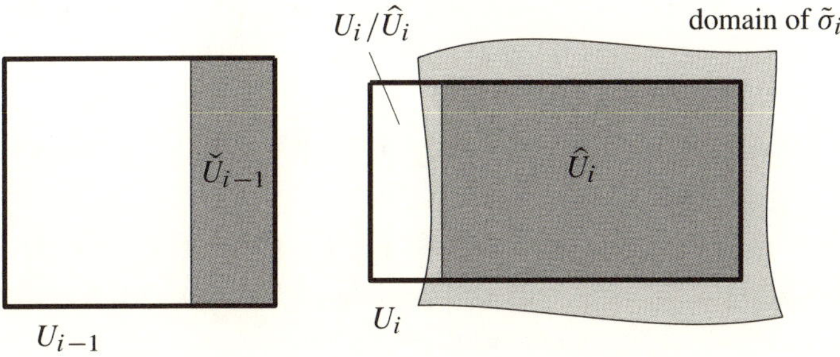

Figure 1.6. Above: M_i and $P_i(M_i)$ are the graphs of σ_i and $\tilde{\sigma}_i$. Below: the domains of the functions σ_{i-1}, $\check{\sigma}_{i-1}$, σ_i, $\tilde{\sigma}_i$, $\hat{\sigma}_i$, $\bar{\sigma}_{i,i-1}$ and $\bar{\sigma}_i$, respectively, are U_{i-1}, $\check{U}_{i-1}$, U_i, $\mathrm{dom}(\tilde{\sigma}_i)$, $\widehat{U}_i$, $U_i \setminus \widehat{U}_i$ and U_i, respectively.

d) *The function space*

$$\Sigma := \big\{\sigma = (\sigma_1, \dots, \sigma_\kappa) \mid \sigma_i \in C_i(U_i, V_i),\ i = 1, \dots, \kappa,\ \textit{and} \\ \mathcal{T}_{i,i-1}(\sigma_{i-1}|_{\check{U}_{i-1}}) = \sigma_i|_{U_i \setminus \widehat{U}_i},\ i = 2, \dots, \kappa\big\}$$

is nonempty and the operator $\mathcal{F}$ taking $\sigma = (\sigma_1, \dots, \sigma_\kappa) \in \Sigma$ to $\bar{\sigma} = (\bar{\sigma}_1, \dots, \bar{\sigma}_\kappa)$ with

$$\bar{\sigma}_i(u_i) := \begin{cases} \bar{\sigma}_{i,i-1}(u_i) := \mathcal{T}_{i,i-1}(\mathcal{F}_{i-1}(\sigma_{i-1})|_{\check{U}_{i-1}})(u_i) & \textit{for } u_i \in U_i \setminus \widehat{U}_i, \\ & i = 2, \dots, \kappa, \\ \hat{\sigma}_i(u_i) = \mathcal{F}_i(\sigma_i)(u_i) & \textit{for } u_i \in \widehat{U}_i, \\ & i = 1, \dots, \kappa, \end{cases}$$

maps Σ into itself, i.e., $\mathcal{F} : \Sigma \to \Sigma$.

Condition CMAK

a) *For* $i = 1, \dots, \kappa$ *the operator* $\mathcal{F}_i$ *is* χ_i*-contracting, i.e., for* $\sigma_i^{(1)}, \sigma_i^{(2)} \in C_i(U_i, V_i)$

$$|\mathcal{F}_i(\sigma_i^{(1)}) - \mathcal{F}_i(\sigma_i^{(2)})|_{\hat{U}_i} \le \chi_i \, |\sigma_i^{(1)} - \sigma_i^{(2)}|_{U_i}$$

holds with $\chi_i < 1$.

b) *For* $i = 2, \dots, \kappa$ *the operator* $\mathcal{T}_{i,i-1}$ *is nonaugmenting, i.e.,*

$$|\mathcal{T}_{i,i-1}(\check{\sigma}_{i-1}^{(1)}) - \mathcal{T}_{i,i-1}(\check{\sigma}_{i-1}^{(2)})|_{U_i \setminus \hat{U}_i} \le |\check{\sigma}_{i-1}^{(1)} - \check{\sigma}_{i-1}^{(2)}|_{\check{U}_{i-1}}$$

holds for $\check{\sigma}_{i-1}^{(1)}, \check{\sigma}_{i-1}^{(2)} \in C_{i-1}(\check{U}_{i-1}, V_{i-1})$.

Remark 1.9. Hypothesis HMAK means that the map P is flowing from chart Φ_{i-1} to chart Φ_i, $i = 2, \dots, \kappa$. Hypothesis HMAK d) states how the image $\bar{\sigma} = \mathcal{F}(\sigma)$ is obtained.

Theorem 1.10. *Assume that the map* P *of* (1.28) *is described in charts* Φ_i, $i = 1, \dots, \kappa$, *and let Hypothesis HMAK and Condition CMAK be satisfied.*

Then there is a function $\sigma = (\sigma_1, \dots, \sigma_\kappa) \in \Sigma$ *with* $\mathcal{F}(\sigma) = \sigma$ *such that the following assertions hold.*

i) *For* $M_i = \{(u_i, v_i) \mid u_i \in U_i, v_i = \sigma_i(u_i)\}$, $i = 1, \dots, \kappa$, *the set*

$$\mathcal{M} := \bigcup_{i=1}^{\kappa} \Phi_i^{-1}(M_i)$$

is a negatively invariant manifold of the map P.

ii) *The manifold* $\mathcal{M}$ *is* χ*-attractive with* $\chi := \max_{i=1,\dots,\kappa} \chi_i$, *i.e., for any manifold* $\tilde{\mathcal{M}}$ *described by a function* $\tau \in \Sigma$,

$$\|\mathcal{F}(\tau) - \sigma\| \le \chi \|\tau - \sigma\|.$$

Proof. By Hypothesis HMAK the map P maps any manifold $\mathcal{M}$ described by a function $\sigma \in \Sigma$, i.e., $\mathcal{M} := \bigcup_{i=1}^{\kappa} \Phi_i^{-1}(M_i)$ with $M_i = \{(u_i, v_i) \mid u_i \in U_i, v_i = \sigma_i(u_i)\}$ to $P(\mathcal{M}) \supset \bar{\mathcal{M}}$, where the manifold $\bar{\mathcal{M}}$ is described by the function $\bar{\sigma} \in \Sigma$ with $\mathcal{F}(\sigma) = \bar{\sigma}$. It suffices to show that the operator $\mathcal{F}$ is a contraction. Let $\sigma^{(1)}, \sigma^{(2)} \in \Sigma$. We estimate $|\bar{\sigma}_i^{(1)} - \bar{\sigma}_i^{(2)}|_{U_i}$ as follows: By Condition CMAK a) we have

$$\big|\bar{\sigma}_i^{(1)}|_{\hat{U}_i} - \bar{\sigma}_i^{(2)}|_{\hat{U}_i}\big|_{\hat{U}_i} = |\hat{\sigma}_i^{(1)} - \hat{\sigma}_i^{(2)}|_{\hat{U}_i} \le \chi_i |\sigma_i^{(1)} - \sigma_i^{(2)}|_{U_i} \le \chi \|\sigma^{(1)} - \sigma^{(2)}\|$$

and by Condition CMAK b) and a)

$$\begin{aligned} \big|\bar{\sigma}_i^{(1)}|_{U_i \setminus \hat{U}_i} - \bar{\sigma}_i^{(2)}|_{U_i \setminus \hat{U}_i}\big|_{U_i \setminus \hat{U}_i} &= |\bar{\sigma}_{i,i-1}^{(1)} - \bar{\sigma}_{i,i-1}^{(2)}|_{U_i \setminus \hat{U}_i} \\ &= \big|\mathcal{T}_{i,i-1}(\mathcal{F}_{i-1}(\sigma_{i-1}^{(1)})|_{\check{U}_{i-1}}) \\ &\qquad - \mathcal{T}_{i,i-1}(\mathcal{F}_{i-1}(\sigma_{i-1}^{(2)})|_{\check{U}_{i-1}})\big| \end{aligned}$$

$$\leq \left|\mathcal{F}_{i-1}(\sigma_{i-1}^{(1)})|_{\breve{U}_{i-1}} - \mathcal{F}_{i-1}(\sigma_{i-1}^{(2)})|_{\breve{U}_{i-1}}\right|_{\breve{U}_{i-1}}$$
$$\leq \chi_{i-1}\left|\sigma_{i-1}^{(1)} - \sigma_{i-1}^{(2)}\right|_{\breve{U}_{i-1}} \leq \chi\|\sigma^{(1)} - \sigma^{(2)}\|,$$

leading to

$$\left|\bar{\sigma}_i^{(1)} - \bar{\sigma}_i^{(2)}\right|_{U_i} = \max\left\{\left|\bar{\sigma}_i^{(1)}|_{\hat{U}_i} - \bar{\sigma}_i^{(2)}|_{\hat{U}_i}\right|_{\hat{U}_i}, \left|\bar{\sigma}_i^{(1)}|_{U_i\setminus\hat{U}_i} - \bar{\sigma}_i^{(2)}|_{U_i\setminus\hat{U}_i}\right|_{U_i\setminus\hat{U}_i}\right\}$$
$$\leq \chi\left\|\sigma^{(1)} - \sigma^{(2)}\right\|.$$

It follows that

$$\left\|\mathcal{F}(\sigma^{(1)}) - \mathcal{F}(\sigma^{(2)})\right\| = \max_{i=1,\dots,\kappa}\left\{|\bar{\sigma}_i^{(1)} - \bar{\sigma}_i^{(2)}|_{U_i}\right\} \leq \chi\left\|\sigma^{(1)} - \sigma^{(2)}\right\|. \qquad \square$$

1.4.2 Tools for Chapter 13

In this subsection we derive tools for verifying Hypothesis HMAK and Condition CMAK in applications. In every chart Φ_i we apply such a tool to the map P_i given in Hypothesis HMAK a). For simplicity, we omit the index i.

First tool. The invariant manifold result of Theorem 1.5 strongly relies on Hypothesis HM b) assuming that the map P_ϑ is outflowing with respect to X. In the situation of Section 1.4 this assumption is not satisfied. Instead, we assume that P_ϑ is flowing from one chart to the next one. The first tool corresponds to the standard situation where the map P takes (x, y) to $(\bar{x}, \bar{y})$ and depends on a parameter ϑ. The variables x, y, ϑ correspond to the variables $u = (x, \vartheta)$, $v = y$ in Theorem 1.10. Let

$$P_\vartheta\colon X \times Y \ni \begin{pmatrix} x \\ y \end{pmatrix} \longmapsto \begin{pmatrix} \bar{x} \\ \bar{y} \end{pmatrix} = \begin{pmatrix} F(x, y, \vartheta) \\ G(x, y, \vartheta) \end{pmatrix} \in \mathbb{R}^m \times \mathbb{R}^n. \tag{1.35}$$

We replace Hypothesis HM by

Hypothesis HMAG

The functions $F \in C^0(X \times Y \times E, \mathcal{B}_x)$, $G \in C^0(X \times Y \times E, \mathcal{B}_y)$ have the following properties.

a) *Hypothesis HM* a).

b) *There is $\widehat{X} \subset \mathcal{B}_x$ such that P_ϑ is flowing from X to $\widehat{X}$, i.e., for every $\bar{x} \in \widehat{X}$, $y \in Y$, $\vartheta \in E$ there is $x \in X$ such that $F(x, y, \vartheta) = \bar{x}$.*

c) *Hypothesis HM* c).

We introduce the space $C_{\alpha,\beta}(X \times E, Y)$ of bounded functions $\sigma\colon X \times E \to Y$ which are α-Lipschitz continuous with respect to x and β-Lipschitz continuous with respect to ϑ.

Let

$$\alpha_{\min}, \alpha_{\max} := \frac{2L_{21}}{\Gamma_{11} - L_{22} \pm \sqrt{(\Gamma_{11} - L_{22})^2 - 4L_{12}L_{21}}}, \tag{1.36}$$

and let for some $\alpha \in [\alpha_{\min}, \alpha_{\max}]$ the following condition hold.

Condition CMAG

$$L_{22} + L_{12}\alpha < 1.$$

Theorem 1.11. *Let the map* P_ϑ *of* (1.35) *satisfy Hypothesis HMAG and let Conditions CM, CMAG hold for some* $\alpha \in [\alpha_{\min}, \alpha_{\max}]$, *cf.* (1.36).

Then the following assertions hold.

i) *For every* $\sigma \in C_{\alpha,\beta}(X \times E, Y)$ *with*

$$\beta \geq \beta_{\min} = \frac{L_{23} + L_{13}\alpha}{1 - L_{22} - L_{12}\alpha},$$

there is a function $\hat{\sigma}: \hat{X} \times E \to Y$ *such that the image under* P_ϑ *of the graph* $M_{\vartheta,\sigma} = \{(x, y) \mid x \in X, y = \sigma(x, \vartheta)\}$ *of* σ *contains the graph* $M_{\vartheta,\hat{\sigma}}$ *of* $\hat{\sigma}$, *i.e.,* $P_\vartheta(M_{\vartheta,\sigma}) \supset M_{\vartheta,\hat{\sigma}}$.

ii) *The operator* $\mathcal{F}: \sigma \mapsto \hat{\sigma}$ *induced by the map* P_ϑ *has the following properties.*

a) $\mathcal{F}: C_{\alpha,\beta}(X \times E, Y) \to C_{\alpha,\beta}(\hat{X} \times E, Y)$.

b) $\mathcal{F}$ *is a contraction with contractivity constant* $\chi_A := L_{22} + L_{12}\alpha < 1$, *i.e.,*

$$|\hat{\sigma}^{(1)} - \hat{\sigma}^{(2)}|_{\hat{X}} \leq \chi_A\, |\sigma^{(1)} - \sigma^{(2)}|_X$$

for all $\sigma^{(i)} \in C_{\alpha,\beta}$ *and for* $\hat{\sigma}^{(i)} := \mathcal{F}(\sigma^{(i)})$, $i = 1, 2$.

Proof. The proof is almost identical to the proof of Theorem 1.5. □

Second Tool. The second tool is tailored to be applied to the special situations in Subsections 13.4.2 and 13.4.5. The maps considered are of the form

$$P: \quad \begin{aligned} \bar{h} &= h/\mu(r, \varepsilon, z, h), \\ \bar{r} &= r/\mu(r, \varepsilon, z, h), \\ \bar{\varepsilon} &= \varepsilon\mu(r, \varepsilon, z, h), \\ \bar{z} &= \bigl[1 + hB(r, \varepsilon, z, h)\bigr]z + h^2 G(r, \varepsilon, z, h), \end{aligned} \tag{1.37}$$

where $\mu(r, \varepsilon, z, h) = 1 + hA(r, \varepsilon, z, h)$. In Subsection 13.4.2 the domain $U \subset (0, R] \times (0, E] \times (H_0, H]$ of (r, ε, h) and $\hat{U} \subset U$ are depicted in Figure 13.4 on page 167. The domain V of z is $[-\eta, \eta]$ and $A > 0$ in $U \times V$. In Subsection 13.4.5 the domain

$U \subset (0, R] \times (0, E] \times (H_0, H]$ of (r, ε, h) and $\hat{U} \subset U$ are depicted in Figure 13.7 on page 180. The domain V of z is $[-\eta, \eta]$ and $A < 0$ in $U \times V$.

In this second tool the variable h essentially plays the role of a small parameter which is varying, however, and remains small. This situation is not covered by the first tool or the manifold result of Theorem 1.5.

Hypothesis HMAV

a) *The functions A, B and G are bounded in $U \times V$.*

b) *The functions A, B and G are Lipschitz continuous and satisfy*

$$\begin{aligned} |A(r_1, \varepsilon_1, z_1, h_1) - A(r_2, \varepsilon_2, z_2, h_2)| &\le \ell_{11}(|r_1 - r_2| + |\varepsilon_1 - \varepsilon_2|) \\ &\quad + \ell_{12}|z_1 - z_2| + \ell_{10}|h_1 - h_2|, \\ |B(r_1, \varepsilon_1, z_1, h_1) - B(r_2, \varepsilon_2, z_2, h_2)| &\le \ell_{21}(|r_1 - r_2| + |\varepsilon_1 - \varepsilon_2|) \\ &\quad + \ell_{22}|z_1 - z_2| + \ell_{20}|h_1 - h_2|, \\ |G(r_1, \varepsilon_1, z_1, h_1) - G(r_2, \varepsilon_2, z_2, h_2)| &\le L_{21}(|r_1 - r_2| + |\varepsilon_1 - \varepsilon_2|) \\ &\quad + L_{22}|z_1 - z_2| + L_{20}|h_1 - h_2|. \end{aligned}$$

Condition CMAV

There are constants a and b with $b > 2a > 0$ such that in $U \times V$ the estimates $|1 + hB| < 1 - hb$ and $|A| < a$ hold.

Definition. Let $C_{\alpha,\gamma}(U, V)$ denote the set of bounded functions $\sigma \colon U \to V$ which are α-Lipschitz continuous with respect to r and ε and γ-Lipschitz continuous with respect to h.

Theorem 1.12. *Let the map P of* (1.37) *satisfy Hypothesis HMAV and let Condition CMAV hold.*

Then there is a positive constant K such that for η and H small enough the following assertions hold.

i) *For every $\sigma \in C_{\alpha,\gamma}(U, V)$ with*

$$\alpha = \frac{2\ell_{21}\eta + KH}{b - a}, \quad \gamma = \frac{2(b\eta + \alpha) + KH}{b - 2a}$$

there is a function $\hat{\sigma} : \hat{U} \to V$ such that the image under P of the graph $M_\sigma = \{(r, \varepsilon, z, h) \mid (r, \varepsilon, h) \in U,\ z = \sigma(r, \varepsilon, h)\}$ of σ contains the graph $M_{\hat{\sigma}}$ of $\hat{\sigma}$, i.e., $P(M_\sigma) \supset M_{\hat{\sigma}}$.

ii) *The operator $\mathcal{F} : \sigma \mapsto \hat{\sigma}$ induced by the map P has the following properties.*

a) $\mathcal{F} : C_{\alpha,\gamma}(U, V) \to C_{\alpha,\gamma}(\hat{U}, V)$.

b) $\mathcal{F}$ *is a contraction with contractivity constant* $\chi_A := 1 - H_0 b/2 < 1$, *i.e.,*

$$|\hat{\sigma}^{(1)} - \hat{\sigma}^{(2)}|_{\hat{U}} \leq \chi_A \, |\sigma^{(1)} - \sigma^{(2)}|_U$$

for all $\sigma^{(i)} \in C_{\alpha,\gamma}$ *and for* $\hat{\sigma}^{(i)} := \mathcal{F}(\sigma^{(i)})$, $i = 1, 2$.

Proof. The proof is in the spirit of the proof of Theorem 1.5. □

Chapter 2
Perturbation and approximation

In this chapter we show that a perturbation of a map leads to a perturbation of the corresponding manifold. Moreover, we show that if a function approximatively satisfies the invariance equation of a manifold then its graph approximates the manifold.

2.1 Attractive negatively invariant manifolds

Let us consider the maps, cf. Section 1.2,

$$\begin{aligned} P_\vartheta \colon X \times Y \ni \begin{pmatrix} x \\ y \end{pmatrix} &\longmapsto \begin{pmatrix} F(x,y,\vartheta) \\ G(x,y,\vartheta) \end{pmatrix} \in \mathcal{B}_X \times \mathcal{B}_Y, \\ \bar{P}_{\bar{\vartheta}} \colon X \times Y \ni \begin{pmatrix} x \\ y \end{pmatrix} &\longmapsto \begin{pmatrix} \bar{F}(x,y,\bar{\vartheta}) \\ \bar{G}(x,y,\bar{\vartheta}) \end{pmatrix} \in \mathcal{B}_X \times \mathcal{B}_Y, \end{aligned} \tag{2.1}$$

and for fixed $\vartheta, \bar{\vartheta} \in E$ let us define

$$\begin{aligned} \Delta_1 &:= \sup_{(x,y)\in X\times Y} |\bar{F}(x,y,\bar{\vartheta}) - F(x,y,\vartheta)|, \\ \Delta_2 &:= \sup_{(x,y)\in X\times Y} |\bar{G}(x,y,\bar{\vartheta}) - G(x,y,\vartheta)|. \end{aligned} \tag{2.2}$$

We assume that P_ϑ and $\bar{P}_{\bar{\vartheta}}$ satisfy the assumptions of Theorem 1.5 with constants Γ_{11}, L_{12}, L_{13}, L_{21}, L_{22}, L_{23} and $\bar{\Gamma}_{11}$, $\bar{L}_{12}$, $\bar{L}_{13}$, $\bar{L}_{21}$, $\bar{L}_{22}$, $\bar{L}_{23}$, respectively. Then there are functions s_A and $\bar{s}_A$, uniformly λ_A-Lipschitz and $\bar{\lambda}_A$-Lipschitz, respectively, such that the manifolds M_ϑ described by the function s_A and $\bar{M}_{\bar{\vartheta}}$ described by $\bar{s}_A$ are invariant under P_ϑ and $\bar{P}_{\bar{\vartheta}}$, respectively, with all the properties stated in Theorem 1.5.

Theorem 2.1 (Perturbation). *Fix $\vartheta, \bar{\vartheta} \in E$ and let the maps P_ϑ and $\bar{P}_{\bar{\vartheta}}$ defined in* (2.1) *satisfy Hypotheses HM and HMA and assume that the constants Γ_{11}, L_{12}, L_{21}, L_{22} and $\bar{\Gamma}_{11}$, $\bar{L}_{12}$, $\bar{L}_{21}$, $\bar{L}_{22}$, respectively, satisfy Conditions CM and CMA. Let Δ_1, Δ_2 be defined as in equation* (2.2).

Then for the functions s_A and $\bar{s}_A$ of Theorem 1.5 *defining the negatively invariant manifolds M_ϑ and $\bar{M}_{\bar{\vartheta}}$, respectively, the estimate*

$$|\bar{s}_A(x,\bar{\vartheta}) - s_A(x,\vartheta)| \le \frac{1}{1-\chi_A}\,(\lambda_A\,\Delta_1 + \Delta_2) \tag{2.3}$$

holds for

$$\lambda_A = \frac{2L_{21}}{\Gamma_{11} - L_{22} + \sqrt{(\Gamma_{11} - L_{22})^2 - 4L_{12}L_{21}}}, \quad \chi_A = L_{22} + L_{12}\lambda_A,$$

as defined in Theorem 1.5.

Remark 2.2. In some applications λ_A might be much smaller than $\bar{\lambda}_A$ or vice versa. In equation (2.3) the "better" constants may be used.

Proof. The functions s_A and $\bar{s}_A$ satisfy the invariance equations (cf. equation (1.11))

$$\begin{aligned} G(x, s_A(x,\vartheta), \vartheta) &= s_A(F(x, s_A(x,\vartheta), \vartheta), \vartheta), \\ \bar{G}(x, \bar{s}_A(x,\bar{\vartheta}), \bar{\vartheta}) &= \bar{s}_A(\bar{F}(x, \bar{s}_A(x,\bar{\vartheta}), \bar{\vartheta}), \bar{\vartheta}). \end{aligned} \tag{2.4}$$

Defining

$$\begin{aligned} x_1 &:= F(x, s_A(x,\vartheta), \vartheta), \\ \bar{x}_1 &:= \bar{F}(x, \bar{s}_A(x,\bar{\vartheta}), \bar{\vartheta}) \end{aligned}$$

and taking differences leads to

$$\begin{aligned} |x_1 - \bar{x}_1| &\le |F(x, s_A(x,\vartheta), \vartheta) - F(x, \bar{s}_A(x,\bar{\vartheta}), \vartheta)| \\ &\quad + |F(x, \bar{s}_A(x,\bar{\vartheta}), \vartheta) - \bar{F}(x, \bar{s}_A(x,\bar{\vartheta}), \bar{\vartheta})| \\ &\le L_{12}\, |s_A(x,\vartheta) - \bar{s}_A(x,\bar{\vartheta})| + \Delta_1. \end{aligned} \tag{2.5}$$

Using (2.4) we obtain

$$\begin{aligned} |s_A(\bar{x}_1,\vartheta) - \bar{s}_A(\bar{x}_1,\bar{\vartheta})| &\le |s_A(\bar{x}_1,\vartheta) - s_A(x_1,\vartheta)| \\ &\quad + |G(x, s_A(x,\vartheta), \vartheta) - G(x, \bar{s}_A(x,\bar{\vartheta}), \vartheta)| \\ &\quad + |G(x, \bar{s}_A(x,\bar{\vartheta}), \vartheta) - \bar{G}(x, \bar{s}_A(x,\bar{\vartheta}), \bar{\vartheta})| \\ &\le \lambda_A\, |x_1 - \bar{x}_1| + L_{22}\, |s_A(x,\vartheta) - \bar{s}_A(x,\bar{\vartheta})| + \Delta_2. \end{aligned}$$

Together with (2.5) we get

$$|s_A(\bar{x}_1,\vartheta) - \bar{s}_A(\bar{x}_1,\bar{\vartheta})| \le \lambda_A\, \Delta_1 + \Delta_2 + (L_{22} + \lambda_A\, L_{12})\, |s_A(x,\vartheta) - \bar{s}_A(x,\bar{\vartheta})|.$$

Taking the supremum over x first on the right-hand side and then on the left-hand side implies

$$(1 - \chi_A) \sup_x |s_A(x,\vartheta) - \bar{s}_A(x,\bar{\vartheta})| \le \lambda_A\, \Delta_1 + \Delta_2. \qquad \square$$

As another application of Theorem 1.5 we state an approximation result.

Theorem 2.3 (Approximation). *Let the map P_ϑ of* (2.1) *satisfy Hypotheses HM and HMA and assume that the constants Γ_{11}, L_{12}, L_{21}, L_{22} satisfy Conditions CM and CMA. Let $s_A \in C_{\lambda_A,\nu_A}$ be the function describing the attractive negatively invariant manifold $M_\vartheta = \{(x,y) \mid x \in X,\ y = s_A(x,\vartheta)\}$ established in Theorem* 1.5. *Moreover, assume that the bounded function $\sigma \in C^0(X \times E, Y)$ satisfies*

$$G(x,\sigma(x,\vartheta),\vartheta) = \sigma(F(x,\sigma(x,\vartheta),\vartheta),\vartheta) + \rho(x,\vartheta)$$

for some bounded function $\rho \in C^0(X \times E, Y)$.

Then

$$|\sigma - s_A| \le \frac{1}{1-\chi_A}\,|\rho|$$

holds for

$$\chi_A = L_{22} + L_{12}\lambda_A, \quad \lambda_A = \frac{2L_{21}}{\Gamma_{11} - L_{22} + \sqrt{(\Gamma_{11}-L_{22})^2 - 4L_{12}L_{21}}}.$$

Proof. Taking into account the invariance equation (1.11) for the function s_A we obtain

$$\begin{aligned}
&|\sigma(F(x,\sigma(x,\vartheta),\vartheta),\vartheta) - s_A(F(x,\sigma(x,\vartheta),\vartheta),\vartheta)| \\
&\quad \le |G(x,\sigma(x,\vartheta),\vartheta) - G(x,s_A(x,\vartheta),\vartheta)| + |\rho(x,\vartheta)| \\
&\qquad + |s_A(F(x,s_A(x,\vartheta),\vartheta),\vartheta) - s_A(F(x,\sigma(x,\vartheta),\vartheta),\vartheta)| \\
&\quad \le L_{22}\,|\sigma(x,\vartheta) - s_A(x,\vartheta)| + |\rho(x,\vartheta)| + \lambda_A\,L_{12}\,|s_A(x,\vartheta) - \sigma(x,\vartheta)|.
\end{aligned}$$

Taking the supremum over $(x,\vartheta) \in X \times E$ first on the right-hand side and then on the left-hand side we thus have

$$[1 - (L_{22} + L_{12}\,\lambda_A)]\,|\sigma - s| \le |\rho|. \qquad \square$$

2.2 Repulsive positively invariant manifolds

For repulsive positively invariant manifolds we state results analogous to those of Section 2.1 omitting the proofs since they are almost identical.

We again consider the maps of equation (2.1) and we use the definitions (2.2).

Theorem 2.4 (Perturbation)). *Fix $\vartheta, \bar\vartheta \in E$ and let the maps P_ϑ and $\bar P_{\bar\vartheta}$ defined in* (2.1) *satisfy Hypotheses HM and HMR and assume that the constants Γ_{11}, L_{12}, L_{21}, L_{22} and $\bar\Gamma_{11}$, $\bar L_{12}$, $\bar L_{21}$, $\bar L_{22}$, respectively, satisfy Conditions CM and CMR. Let Δ_1, Δ_2 be defined as in equation* (2.2).

Then for the functions s_R and $\bar s_R$ of Theorem 1.3 *defining the positively invariant manifolds N_ϑ and $\bar N_{\bar\vartheta}$, respectively, the estimate*

$$|\bar s_R(x,\bar\vartheta) - s_R(x,\vartheta)| \le \frac{1}{\chi_R - 1}\,(\Delta_1 + \lambda_R\Delta_2)$$

holds for

$$\lambda_R = \frac{2L_{12}}{\Gamma_{11} - L_{22} + \sqrt{(\Gamma_{11} - L_{22})^2 - 4L_{12}L_{21}}}, \qquad \chi_R = \Gamma_{11} - L_{21}\lambda_R,$$

as defined in Theorem 1.3.

We state the approximation result for repulsive manifolds.

Theorem 2.5 (Approximation). *Let the map P_ϑ of* (2.1) *satisfy Hypotheses HM and HMR and assume that the constants Γ_{11}, L_{12}, L_{21}, L_{22} satisfy Conditions CM and CMR. Let $s_R \in C_{\lambda_R,\nu_R}$ be the function describing the repulsive positively invariant manifold $N_\vartheta = \{(x, y) \mid y \in Y, x = s_R(y, \vartheta)\}$ established in Theorem* 1.3. *Moreover, assume that the bounded function $\sigma \in C^0(Y \times E, X)$ satisfies*

$$F(\sigma(y, \vartheta), y, \vartheta) = \sigma\big(G(\sigma(y, \vartheta), y, \vartheta), \vartheta\big) + \rho(y, \vartheta)$$

for some bounded function $\rho \in C^0(Y \times E, X)$.

Then

$$|\sigma - s_R| \leq \frac{1}{\chi_R - 1}\,|\rho|$$

holds for

$$\chi_R = \Gamma_{11} - L_{21}\lambda_R, \quad \lambda_R = \frac{2L_{12}}{\Gamma_{11} - L_{22} + \sqrt{(\Gamma_{11} - L_{22})^2 - 4L_{12}L_{21}}}.$$

Chapter 3
Smoothness

The manifolds derived in Chapter 1 are Lipschitz continuous. In this chapter we show that if the map is k-times continuously differentiable and if some additional conditions are satisfied then the manifold is k-times continuously differentiable as well. We prove this result for repulsive manifolds in Section 3.1. In Section 3.2 we state the corresponding result for attractive manifolds without proof.

3.1 Repulsive positively invariant manifolds

Let $\mathcal{B}_x$, $\mathcal{B}_y$, $\mathcal{B}_\vartheta$ be Banach spaces and let $X \subset \mathcal{B}_x$, $Y \subset \mathcal{B}_y$, $E \subset \mathcal{B}_\vartheta$ be open sets. We consider the family of maps P_ϑ for $\vartheta \in E$ of the form

$$P_\vartheta : X \times Y \ni \begin{pmatrix} x \\ y \end{pmatrix} \longmapsto \begin{pmatrix} \bar{x} \\ \bar{y} \end{pmatrix} = \begin{pmatrix} F(x, y, \vartheta) \\ G(x, y, \vartheta) \end{pmatrix} \in \mathcal{B}_x \times \mathcal{B}_y. \tag{3.1}$$

We assume that the assumptions of Theorem 1.3 are satisfied and that the functions F, G are of class C_b^k, i.e., F and G are k-times continuously differentiable with bounded derivatives. If the additional condition

Condition CMR(k)

$$(L_{22} + \Delta)^k < \Gamma_{11} - \Delta$$

is satisfied then the invariant manifold N_ϑ established in Theorem 1.3 is also of class C_b^k. We state the precise result as

Theorem 3.1. *Let the map P_ϑ of equation* (3.1) *satisfy Hypotheses HM and HMR and assume that the constants Γ_{11}, L_{12}, L_{21}, L_{22} satisfy Conditions CM, CMR and CMR(k), $k \geq 1$. Let F and G be of class C_b^k.*

Then all assertions of Theorem 1.3 *hold and the function s_R describing the positively invariant manifold $N_\vartheta = \{(x, y) \mid y \in Y,\ x = s_R(y, \vartheta)\}$ is of class C_b^k.*

In the remaining part of this section we prove Theorem 3.1. In order to keep the proof as simple as possible we show in Subsections 3.1.1 and 3.1.2 in detail that the function s_R describing the manifold N_ϑ is differentiable with respect to y. In Subsection 3.1.1 the first derivative is treated. Subsection 3.1.2 deals with the higher derivatives. The reason for dividing the proof in two parts is that the first derivative has a structure different from the higher derivatives. For simplicity we suppress the dependence on ϑ. In Subsection 3.1.3 we indicate how to prove that s_R is of class C_b^k with respect to y *and* ϑ.

3.1.1 The first derivative

Let C_{λ_R} be the space

$$C_{\lambda_R} = \{\sigma \in C^0(Y, X) \mid \sigma \text{ is bounded and uniformly } \lambda_R\text{-Lipschitz continuous}\},$$

where λ_R is defined in Theorem 1.3. As in Section 1.1 we define the operator $\mathcal{K}: C_{\lambda_R} \to C_{\lambda_R}, \sigma \mapsto \bar{\sigma}$, by

$$F\big(\bar{\sigma}(y), y\big) = \sigma\big(G\big(\sigma(y), y\big)\big). \tag{3.2}$$

We show that if σ is differentiable then $\bar{\sigma} = \mathcal{K}\sigma$ is differentiable as well.

In the following we use the notation

$$F_1(x, y) = \frac{\partial}{\partial x} F(x, y), \quad F_2(x, y) = \frac{\partial}{\partial y} F(x, y), \text{ etc.},$$

for the partial derivatives of F, G. Moreover, $\sigma^{(1)}$ and $\bar{\sigma}^{(1)}$, respectively, denote the formal derivative of σ and $\bar{\sigma}$, respectively. We formally differentiate equation (3.2). If we assume that $F_1(x, y)$ has a bounded inverse for all (x, y) with $y \in Y$ and $F(x, y) \in X$ we obtain

$$\begin{aligned}\bar{\sigma}^{(1)}(y) = F_1(\bar{\sigma}(y), y)^{-1} \big\{\sigma^{(1)}\big(G(\sigma(y), y)\big)\big[G_1(\sigma(y), y)\,\sigma^{(1)}(y) \\ + G_2(\sigma(y), y)\big] - F_2(\bar{\sigma}(y), y)\big\}.\end{aligned} \tag{3.3}$$

We show that $F_1(x, y)$ has a bounded inverse.

Lemma 3.2. *Let Hypothesis HM be satisfied, assume $F \in C_b^1$ and let $(x, y) \in X \times Y$ such that $\bar{x} = F(x, y) \in X$.*

Then the derivative $F_1(x, y)$ has a bounded inverse and

$$|F_1(x, y)^{-1}| \le \frac{1}{\Gamma_{11}} \tag{3.4}$$

holds.

Proof. Let $u \in \mathcal{B}_x$ be given. From Hypothesis HM c) we have for small $\varepsilon > 0$,

$$|F(x + \varepsilon u, y) - F(x, y)| = |\varepsilon F_1(x, y)u + o(\varepsilon)| \ge \varepsilon \Gamma_{11} |u|.$$

We divide by ε and let ε tend to zero. We conclude that

$$|F_1(x, y)u| \ge \Gamma_{11} |u|. \tag{3.5}$$

It follows that $F_1(x, y)$ is injective. We show that $F_1(x, y)$ is surjective. Let $v \in \mathcal{B}_x$ be given. We show that there is $w \in \mathcal{B}_x$ such that $F_1(x, y)w = v$. Let ρ be small enough such that the ball $B_\rho(\bar{x}) \subset X$. Let $(\varepsilon_n)_{n \ge 0}$ with $|\varepsilon_0 v| < \rho$ and with $\varepsilon_n \downarrow 0$ for $n \to \infty$ be a monotonically decreasing sequence. By Hypothesis HM b) there is

$x_n \in X$ such that $F(x_n, y) = \bar{x} + \varepsilon_n v$ for all $n \geq 0$. We define $w_n := (x_n - x)/\varepsilon_n$. By Hypothesis HM c), we have

$$|F(x_n, y) - F(x, y)| \geq \Gamma_{11}|x_n - x|.$$

Dividing by ε_n we find

$$|v| \geq \Gamma_{11}|w_n|.$$

Hence, the sequence (w_n) is bounded. Since $x_n = x + \varepsilon_n w_n$ we find

$$\varepsilon_n v = F(x_n, y) - F(x, y) = \varepsilon_n F_1(x, y) w_n + o(\varepsilon_n)$$

and after dividing by ε_n,

$$F_1(x, y) w_n = v + o(1), \quad n \to \infty. \tag{3.6}$$

For $m, n \in \mathbb{N}$ we conclude from equation (3.5) that

$$\Gamma_{11}|w_n - w_m| \leq |F_1(x, y)(w_n - w_m)| = o(1), \quad m, n \to \infty.$$

Hence, (w_n) is a Cauchy sequence with some limit w. From equation (3.6) we conclude that $F_1(x, y)w = v$ implying that $F_1(x, y)$ is surjective. Therefore, $F_1(x, y)$ is invertible and from equation (3.5) we have

$$|F_1(x, y)^{-1}| \leq \frac{1}{\Gamma_{11}}. \qquad \square$$

By means of equations (3.2), (3.3) we may define the operator $\mathcal{K}^{(1)}\colon (\sigma, \sigma^{(1)}) \in C_{\lambda_R} \times C_{\lambda_R}^{(1)} \mapsto \bar{\sigma}^{(1)}$, where $C_{\lambda_R}^{(1)}$ is the space of functions

$$\sigma^{(1)}\colon Y \longrightarrow \mathcal{L}(\mathcal{B}_y, \mathcal{B}_x)$$

with $|\sigma^{(1)}| \leq \lambda_R$. Note that we do not assume that $\sigma^{(1)}$ is the derivative of σ. However,

if σ is differentiable and if $\sigma^{(1)}$ is its derivative then $\bar{\sigma}$ is differentiable as well and has derivative $\bar{\sigma}^{(1)}$. (3.7)

Lemma 3.3. *Under the assumptions of Theorem* 1.3 *and under condition CMR*(1) *the following statements hold.*

i) $\mathcal{K}^{(1)}\colon C_{\lambda_R} \times C_{\lambda_R}^{(1)} \to C_{\lambda_R}^{(1)}$.

ii) *For $\sigma_1^{(1)}, \sigma_2^{(1)} \in C_{\lambda_R}^{(1)}$ the estimate*

$$|\mathcal{K}^{(1)}(\sigma, \sigma_1^{(1)}) - \mathcal{K}^{(1)}(\sigma, \sigma_2^{(1)})| \leq \nu|\sigma_1^{(1)} - \sigma_2^{(1)}|$$

holds with $\nu := (L_{22} + 2\Delta)/\Gamma_{11} < 1$.

Proof. i) Equation (3.3) implies $\bar{\sigma}^{(1)} = \mathcal{K}^{(1)}(\sigma, \sigma^{(1)}) \in \mathcal{L}(\mathcal{B}_y, \mathcal{B}_x)$. With (3.4) we estimate

$$|\bar{\sigma}^{(1)}| \le \frac{1}{\Gamma_{11}} \left[\lambda_R(L_{21}\lambda_R + L_{22}) + L_{12}\right].$$

Since by definition λ_R satisfies

$$L_{21}{\lambda_R}^2 - (\Gamma_{11} - L_{22})\lambda_R + L_{12} = 0$$

we obtain $|\bar{\sigma}^{(1)}| \le \lambda_R$.

ii) Using equations (3.3) and (3.4) we estimate

$$\begin{aligned} |\bar{\sigma}_1^{(1)}(y) - \bar{\sigma}_2^{(1)}(y)| &\le \frac{1}{\Gamma_{11}} \Big| \sigma_1^{(1)}(G)G_1\big[\sigma_1^{(1)}(y) - \sigma_2^{(1)}(y)\big] \\ &\qquad + \big[\sigma_1^{(1)}(G) - \sigma_2^{(1)}(G)\big]\big[G_1\sigma_2^{(1)}(y) + G_2\big]\Big| \\ &\le \frac{1}{\Gamma_{11}} (2\Delta + L_{22})\,|\sigma_1^{(1)}(y) - \sigma_2^{(1)}(y)| \\ &= \nu|\sigma_1^{(1)}(y) - \sigma_2^{(1)}(y)| \end{aligned}$$

with $\Delta = L_{21}\lambda_R$. Condition CMR(1) implies $\nu < 1$. □

We introduce the operator $\mathcal{H}^{(1)}\colon C_{\lambda_R} \times C^{(1)}_{\lambda_R} \to C_{\lambda_R} \times C^{(1)}_{\lambda_R}$ taking $(\sigma, \sigma^{(1)})$ to $(\mathcal{K}(\sigma), \mathcal{K}^{(1)}(\sigma, \sigma^{(1)}))$. In order to show that $\mathcal{H}^{(1)}$ has a unique fixed point we need the following lemma known as Fiber Contraction Theorem, cf. Hirsch, Pugh, Shub [55].

Lemma 3.4. *Let U and V be complete metric spaces, let $f\colon U \to U$ be continuous and have a globally attractive fixed point u^*. Let $g\colon U \times V \to V$ be continuous and let for all $u \in U$ the function $g(u,\cdot)$ satisfy* $\operatorname{Lip} g(u,\cdot) \le \nu$ *for some constant $\nu < 1$. Let v^* be the uniquely determined fixed point of $g(u^*,\cdot)$.*

Then (u^*, v^*) is a globally attractive fixed point of the map

$$h\colon (u,v) \in U \times V \longmapsto (\bar{u}, \bar{v}) = \big(f(u), g(u,v)\big) \in U \times V.$$

For completeness, we present the proof.

Proof. For $(u_0, v_0) \in U \times V$ we set $(u_{n+1}, v_{n+1}) = h(u_n, v_n)$, $n \ge 0$. It suffices to show that $\operatorname{dist}(v_n, v^*) \to 0$ as $n \to \infty$. Define $\delta_n := \operatorname{dist}(g(u_n, v^*), v^*)$. The continuity of g and the fact that $u_n \to u^*$ as $n \to \infty$ implies

$$g(u_n, v^*) \longrightarrow g(u^*, v^*) = v^* \quad \text{for } n \to \infty$$

and therefore $\delta_n \to 0$ for $n \to \infty$. Define $v_0^* := v^*$ and $v_n^* := g(u_{n-1}, v_{n-1}^*)$, $n > 0$. Repeated use of the triangle inequality and the Lipschitz continuity of $g(u, \cdot)$ yields

$$\begin{aligned}
\operatorname{dist}(v_n, v^*) &\le \operatorname{dist}(v_n, v_n^*) + \operatorname{dist}(v_n^*, v^*) \\
&\le \nu \operatorname{dist}(v_{n-1}, v_{n-1}^*) + \operatorname{dist}\big(v_n^*, g(u_{n-1}, v^*)\big) + \operatorname{dist}\big(g(u_{n-1}, v^*), v^*\big) \\
&\le \nu\big[\operatorname{dist}(v_{n-1}, v_{n-1}^*) + \operatorname{dist}(v_{n-1}^*, v^*)\big] + \delta_{n-1} \\
&\ \ \vdots \\
&\le \nu^n \operatorname{dist}(v_0, v^*) + \nu^{n-1}\,\delta_0 + \nu^{n-2}\,\delta_1 + \cdots + \delta_{n-1}.
\end{aligned} \tag{3.8}$$

For $0 < \ell < n$ we estimate

$$\begin{aligned}
\sum_{j=0}^{n-1} \nu^{n-1-j}\,\delta_j &= \sum_{j=0}^{\ell-1} \nu^{n-1-j}\,\delta_j + \sum_{j=\ell}^{n-1} \nu^{n-1-j}\,\delta_j \\
&\le \nu^{n-\ell}(1 + \nu + \cdots + \nu^{\ell-1}) \max_{j\ge 0} \delta_j \\
&\qquad + (1 + \nu + \cdots + \nu^{n-\ell-1}) \max_{j\ge\ell} \delta_j \\
&\le \frac{1}{1-\nu}\,\big[\nu^{n-\ell} \max_{j\ge 0} \delta_j + \max_{j\ge\ell} \delta_j\big].
\end{aligned}$$

Setting $\ell = [\frac{n}{2}]$ one gets $\sum_{j=0}^{n-1} \nu^{n-1-j}\,\delta_j \to 0$ as $n \to \infty$. It follows from equation (3.8) that $\operatorname{dist}(v_n, v^*) \to 0$ for $n \to \infty$. □

We now construct a sequence of differentiable functions (σ_n) together with their derivatives $\sigma_n^{(1)}$ by setting

$$\sigma_0(y) \equiv 0, \quad \sigma_0^{(1)}(y) \equiv 0, \quad y \in Y,$$

$$(\sigma_{n+1}, \sigma_{n+1}^{(1)}) := \mathcal{H}^{(1)}(\sigma_n, \sigma_n^{(1)}) = \big(\mathcal{K}(\sigma_n), \mathcal{K}^{(1)}(\sigma_n, \sigma_n^{(1)})\big).$$

By (3.7) the functions σ_n are differentiable and have derivative $\sigma_n^{(1)}$. By Lemma 3.4 using Lemma 3.3 the sequence $(\sigma_n, \sigma_n^{(1)})$ converges to a limit $(s_R, s_R^{(1)})$. In other words, (σ_n) is a convergent sequence of differentiable functions and their derivatives converge uniformly to $s_R^{(1)}$. It follows that s_R is differentiable and has derivative $s_R' = s_R^{(1)}$.

3.1.2 The higher derivatives

In order to treat the higher derivatives we introduce the spaces $C_{\alpha_j}^{(j)}$, $j = 2, \ldots, k$, of functions

$$\sigma^{(j)} \colon X \longrightarrow \mathcal{L}([\mathcal{B}_y]^j, \mathcal{B}_x)$$

with $|\sigma^{(j)}| \le \alpha_j$. Note that $\alpha_1 = \lambda_R$ holds by Lemma 3.3.

We rewrite equation (3.3) in the form

$$\begin{aligned} &F_1(\bar{\sigma}(y), y)\,\bar{\sigma}^{(1)}(y) \\ &\quad = \sigma^{(1)}\big(G(\sigma(y), y)\big)\big[G_1(\sigma(y), y)\sigma^{(1)}(y) + G_2(\sigma(y), y)\big] - F_2\big(\bar{\sigma}(y), y\big), \end{aligned}$$

differentiate formally with respect to y and write $\sigma^{(2)}$, $\bar{\sigma}^{(2)}$ for the formal second derivatives of σ and $\bar{\sigma}$, respectively. We get

$$\begin{aligned} F_1(\bar{\sigma}(y), y)\,\bar{\sigma}^{(2)}(y) = {} & \sigma^{(2)}\big(G(\sigma(y), y)\big)\big[G_1(\sigma(y), y)\,\sigma^{(1)}(y) + G_2(\sigma(y), y)\big]^2 \\ & + \sigma^{(1)}\big(G(\sigma(y), y)\big)\, G_1(\sigma(y), y)\,\sigma^{(2)}(y) + T_2, \end{aligned} \tag{3.9}$$

where the term T_2 only depends on y, σ, $\bar{\sigma}$, $\sigma^{(1)}$ and $\bar{\sigma}^{(1)}$. Equation (3.9) allows to introduce the operators

$$\mathcal{K}^{(2)}\colon C_{\lambda_R} \times C^{(1)}_{\alpha_1} \times C^{(2)}_{\alpha_2} \ni (\sigma, \sigma^{(1)}, \sigma^{(2)}) \longmapsto \bar{\sigma}^{(2)}$$

and

$$\mathcal{H}^{(2)}\colon C_{\lambda_R} \times C^{(1)}_{\alpha_1} \times C^{(2)}_{\alpha_2} \ni (\sigma, \sigma^{(1)}, \sigma^{(2)}) \longmapsto (\bar{\sigma}, \bar{\sigma}^{(1)}, \bar{\sigma}^{(2)})$$

with $(\bar{\sigma}, \bar{\sigma}^{(1)}, \bar{\sigma}^{(2)}) := \big(\mathcal{H}^{(1)}(\sigma, \sigma^{(1)}), \mathcal{K}^{(2)}(\sigma, \sigma^{(1)}, \sigma^{(2)})\big)$ and where α_2 will be determined later such that $\mathcal{H}^{(2)}$ maps $C_{\lambda_R} \times C^{(1)}_{\alpha_1} \times C^{(2)}_{\alpha_2}$ into itself. Note that the definition of $\mathcal{H}^{(2)}$ is done such that if σ is of class C^2_b and if $\sigma^{(1)} = \sigma'$ and $\sigma^{(2)} = \sigma''$ then $\bar{\sigma}$ is also of class C^2_b and $\bar{\sigma}' = \bar{\sigma}^{(1)}$, $\bar{\sigma}'' = \bar{\sigma}^{(2)}$.

For derivatives of order $j > 2$ we proceed in the same way. Let $\sigma^{(j)}$ be functions in $C^{(j)}_{\alpha_j}$, $j = 2, \dots, k$. We again get by successive formal differentiation of equation (3.9) that

$$\begin{aligned} F_1(\bar{\sigma}(y), y)\,\bar{\sigma}^{(j)}(y) = \sigma^{(j)}\big(G(\sigma(y), y)\big)\big[G_1(\sigma(y), y)\,\sigma^{(1)}(y) + G_2(\sigma(y), y)\big]^j \\ + \sigma^{(1)}\big(G(\sigma(y), y)\big)\, G_1(\sigma(y), y)\,\sigma^{(j)}(y) + T_j. \end{aligned} \tag{3.10}$$

Note that T_j only depends on $y, \sigma, \bar{\sigma}$ and on $\sigma^{(i)}, \bar{\sigma}^{(i)}$ with $i < j$. In a similar way as $\mathcal{K}^{(2)}$ and $\mathcal{H}^{(2)}$ we inductively define operators

$$\mathcal{K}^{(j)}\colon C_{\lambda_R} \times C^{(1)}_{\alpha_1} \times \dots \times C^{(j)}_{\alpha_j} \ni (\sigma, \sigma^{(1)}, \dots, \sigma^{(j)}) \longmapsto \bar{\sigma}^{(j)}$$

by means of equation (3.10) and

$$\mathcal{H}^{(j)}\colon (\sigma, \sigma^{(1)}, \dots \sigma^{(j)}) \longmapsto \big(\mathcal{H}^{(j-1)}(\sigma, \sigma^{(1)}, \dots, \sigma^{(j-1)}), \mathcal{K}^{(j)}(\sigma, \dots, \sigma^{(j)})\big).$$

Lemma 3.5. *Under the assumptions of Theorem* 1.3 *and condition CMR*(k)*,* $k \ge 1$*, the following statements hold for* $j = 1, \dots, k$*.*

i) *There are positive numbers* $\nu_j < 1, \alpha_j$ *such that the operator* $\mathcal{K}^{(j)}$ *is well defined and maps* $C_{\lambda_R} \times C^{(1)}_{\alpha_1} \times \cdots \times C^{(j)}_{\alpha_j}$ *to* $C^{(j)}_{\alpha_j}$. *Moreover,*

$$\begin{aligned}&|\mathcal{K}^{(j)}(\sigma, \sigma^{(1)}, \ldots, \sigma^{(j-1)}, \sigma^{(j)}_1) - \mathcal{K}^{(j)}(\sigma, \sigma^{(1)}, \ldots, \sigma^{(j-1)}, \sigma^{(j)}_2)| \\ &\le \nu_j |\sigma^{(j)}_1 - \sigma^{(j)}_2|.\end{aligned}$$

ii) *The operator* $\mathcal{H}^{(j)}$ *is well defined and maps* $C_{\lambda_R} \times C^{(1)}_{\alpha_1} \times \cdots \times C^{(j)}_{\alpha_j}$ *into itself and has a globally attractive fixed point* $(s_R, s^{(1)}_R, \ldots, s^{(j)}_R)$.

Proof. The proof is by induction with respect to j. Assume that the assertion holds for $j-1$. Note that we have already proved the statement for $j = 1$, cf. Lemma 3.3.

i) We have to show that there is $\alpha_j > 0$, such that $\mathcal{K}^{(j)}$ maps $C_{\lambda_R} \times C^{(1)}_{\alpha_1} \times \cdots \times C^{(j)}_{\alpha_j}$ into $C^{(j)}_{\alpha_j}$ and that $\mathcal{K}^{(j)}$ is contracting with respect to the last argument. Using the estimate (3.4) we get from equation (3.10), with $\Delta = L_{21}\lambda_R$, that

$$|\bar{\sigma}^{(j)}| \le \frac{1}{\Gamma_{11}} \left[\alpha_j (\Delta + L_{22})^j + \Delta\, \alpha_j + |T_j|\right].$$

It follows that $\bar{\sigma}^{(j)} \in C^{(j)}_{\alpha_j}$ if α_j is taken such that

$$\alpha_j \ge \frac{|T_j|}{\Gamma_{11} - \Delta - (L_{22} + \Delta)^j}.$$

Note that $\Gamma_{11} - \Delta - (L_{22} + \Delta)^j > 0$ by Conditions CMR and CMR(k). $\mathcal{K}^{(j)}$ is contracting with respect to its last argument since by equation (3.10)

$$\begin{aligned}|\bar{\sigma}^{(j)}_1 - \bar{\sigma}^{(j)}_2| &\le \frac{1}{\Gamma_{11}} \left[(L_{22} + \Delta)^j + \Delta\right] |\sigma^{(j)}_1 - \sigma^{(j)}_2| \\ &=: \nu_j\, |\sigma^{(j)}_1 - \sigma^{(j)}_2|\end{aligned}$$

with $\nu_j < 1$ by Conditions CMR and CMR(k).

ii) By definition, $\mathcal{H}^{(j)}$ is well defined and maps $C_{\lambda_R} \times C^{(1)}_{\alpha_1} \times \cdots \times C^{(j)}_{\alpha_j}$ into itself. The statement follows immediately from Lemma 3.4 with $f = \mathcal{H}^{(j-1)}$ and $g = \mathcal{K}^{(j)}$. □

By construction, the operator $\mathcal{H}^{(k)}$ has the following property. If σ is of class C^k_b and if $\sigma^{(j)}$, $j = 1, \ldots, k$, are the derivatives of σ then $\bar{\sigma}$ is also of class C^k_b and $\bar{\sigma}^{(j)}$, $j = 1, \ldots, k$, are the derivatives of $\bar{\sigma}$.

We define a sequence $\left((\sigma_n, \sigma^{(1)}_n, \ldots, \sigma^{(k)}_n)\right)_{n \ge 0}$ of differentiable functions σ_n together with their derivatives $\sigma^{(j)}_n$, $j = 1, \ldots, k$, in the following way. Let

$$(\sigma_0, \sigma^{(1)}_0, \ldots, \sigma^{(k)}_0) \equiv (0, 0, \ldots, 0)$$

and

$$(\sigma_{n+1}, \sigma_{n+1}^{(1)}, \dots, \sigma_{n+1}^{(k)}) := \mathcal{H}^{(k)}(\sigma_n, \sigma_n^{(1)}, \dots, \sigma_n^{(k)}).$$

By means of Lemma 3.5 the sequence $(\sigma_n, \sigma_n^{(1)}, \dots, \sigma_n^{(k)})$ converges uniformly to the globally attractive fixed point $(s_R, s_R^{(1)}, \dots, s_R^{(k)})$. Since the convergence is uniform it follows that s_R is of class C_b^k with respect to y with derivatives $s_R^{(j)}$, $j = 1, \dots, k$.

3.1.3 The differentiability with respect to parameters

We outline how to show that the function s_R is of class C_b^k with respect to (y, ϑ). We need to show that all partial derivatives

$$s_R^{(j,\ell)}(y, \vartheta) = \frac{\partial^j}{\partial y^j} \frac{\partial^\ell}{\partial \vartheta^\ell} s_R(y, \vartheta), \quad j + \ell \le k,$$

exist and are bounded continuous functions. The assertion has already been proved for $\ell = 0$. The proof is done by induction with respect to ℓ. For fixed ℓ the proof is done by induction with respect to j, $j = 0, 1, \dots, k - \ell$. By formal differentiation of equation (3.2) with respect to ϑ one finds

$$\begin{aligned} &F_1(\bar\sigma(y,\vartheta), y, \vartheta)\, \bar\sigma^{(0,1)}(y,\vartheta) \\ &\quad = \sigma^{(1,0)}\big(G(\sigma(y,\vartheta), y, \vartheta), \vartheta\big) G_1(\sigma(y,\vartheta), y, \vartheta)\, \sigma^{(0,1)}(y,\vartheta) \\ &\qquad + \sigma^{(0,1)}\big(G(\sigma(y,\vartheta), y, \vartheta), \vartheta\big) + T_{0,1}. \end{aligned}$$

Further differentiation with respect to y and ϑ leads to

$$\begin{aligned} &F_1(\bar\sigma(y,\vartheta), y, \vartheta)\, \bar\sigma^{(j,\ell)}(y,\vartheta) \\ &\quad = \sigma^{(j,\ell)}(G(\sigma(y,\vartheta), y, \vartheta), \vartheta)\big[G_1(\sigma(y,\vartheta), y, \vartheta)\, \sigma^{(1,0)}(y,\vartheta) \\ &\qquad\qquad + G_2(\sigma(y,\vartheta), y, \vartheta)\big]^j \\ &\qquad + \sigma^{(1,0)}(G(\sigma(y,\vartheta), y, \vartheta), \vartheta)\, G_1(\sigma(y,\vartheta), y, \vartheta)\, \sigma^{(j,\ell)}(y,\vartheta) + T_{j,\ell} \end{aligned} \tag{3.11}$$

for all $\ell = 1, \dots, k$ and $j = 0, \dots, k - \ell$. The terms $T_{j,\ell}$ contain only derivatives $\sigma^{(j',\ell')}$ of order $j' + \ell' < j + \ell$ and of order $j' + \ell' = j + \ell$ with $\ell' < \ell$, respectively.

For the induction step $\ell \to \ell + 1$ we proceed as in the proof that s_R is differentiable with respect to y up to order k. We again use Lemma 3.4. The estimate in Lemma 3.5 is replaced by the estimate

$$|\bar\sigma_1^{(j,\ell)} - \bar\sigma_2^{(j,\ell)}| \le \nu_j\, |\sigma_1^{(j,\ell)} - \sigma_2^{(j,\ell)}|, \quad \nu_j = \frac{1}{\Gamma_{11}}\,[(L_{22} + \Delta)^j + \Delta]$$

obtained from equation (3.11). The estimate $\nu_j < 1$ follows from Conditions CMR and CMR(k). This completes the proof of Theorem 3.1.

3.2 Attractive negatively invariant manifolds

We assume that the map P_ϑ in equation (3.1) satisfies the assumptions of Theorem 1.5 and that the functions F and G are of class C_b^k. If the additional condition

Condition CMA(k)

$$L_{22} + \Delta < (\Gamma_{11} - \Delta)^k$$

is satisfied the following result holds.

Theorem 3.6. *Let the map P_ϑ of equation* (3.1) *satisfy the Hypotheses HM and HMA and assume that the constants Γ_{11}, L_{12}, L_{21}, L_{22} satisfy Conditions CM, CMA and CMA(k), $k \geq 1$. Let F and G be of class C_b^k.*

Then all assertions of Theorem 1.5 *hold and the function s_A describing the negatively invariant manifold $M_\vartheta = \{(x, y) \mid x \in X,\ y = s_A(x, \vartheta)\}$ is of class C_b^k.*

We omit the proof of Theorem 3.6 since it is similar to the proof of Theorem 3.1.

Chapter 4
Foliation

The existence of an invariant manifold of a map P gives rise to a foliation of the adjacent space. In the case of a χ_A-attractive negatively invariant manifold containing a positively invariant set there exists a stable fiber through every point z of this set. The stable fiber through z consists of all points whose orbits under the map P approach the z-orbit exponentially with rate χ_A. The stable fibers are described by some smooth function and they form a positively invariant family under the map P. We prove this result in Section 4.1. The corresponding results for the unstable foliation and the hyperbolic case are given in Section 4.2 and 4.3, respectively.

4.1 The stable foliation

As in Section 1.2 we consider a family of maps

$$P_\vartheta \colon X \times Y \ni \begin{pmatrix} x \\ y \end{pmatrix} \longmapsto \begin{pmatrix} \bar{x} \\ \bar{y} \end{pmatrix} = \begin{pmatrix} F(x, y, \vartheta) \\ G(x, y, \vartheta) \end{pmatrix} \in \mathcal{B}_x \times \mathcal{B}_y, \quad \vartheta \in E \subset \mathcal{B}_\vartheta, \tag{4.1}$$

satisfying Hypotheses HM and HMA and Conditions CM and CMA. From Theorem 1.5 we know that P_ϑ admits an attractive negatively invariant manifold M_ϑ.

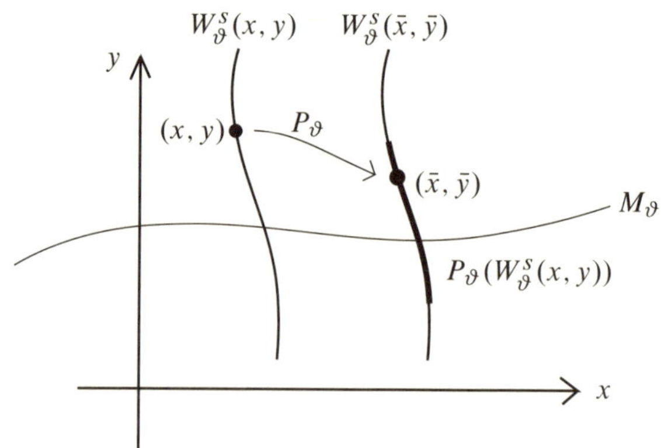

Figure 4.1. The stable fibers form a positively invariant family.

We assume that there is a positively invariant set $\Omega_\vartheta \subset X \times Y$, i.e., $P_\vartheta(\Omega_\vartheta) \subset \Omega_\vartheta$. We show that there is a stable fiber through every point $(x, y) \in \Omega_\vartheta$. We denote it by

$W^s_\vartheta(x,y)$. The stable fibers are Lipschitz manifolds and they form a positively invariant family under the map P_ϑ, i.e., for $(x,y) \in \Omega_\vartheta$ and for $(\bar{x},\bar{y}) := P_\vartheta(x,y) \in \Omega_\vartheta$ the inclusion $P_\vartheta\big(W^s_\vartheta(x,y)\big) \subset W^s_\vartheta(\bar{x},\bar{y})$ holds, cf. Figure 4.1.

For $(x,y) \in \Omega_\vartheta$ the stable fiber $W^s_\vartheta(x,y)$ may be described as the graph of some function from Y to X. We show that for $\vartheta \in E$ there is a function

$$w^s_\vartheta \colon \Omega_\vartheta \times Y \longrightarrow X, \quad (x,y,\eta) \longmapsto w^s_\vartheta(x,y,\eta),$$

such that for every $(x,y) \in \Omega_\vartheta$ the stable fiber $W^s_\vartheta(x,y)$ through the point (x,y) is the graph of $w^s_\vartheta(x,y,\cdot)$, cf. Figure 4.2.

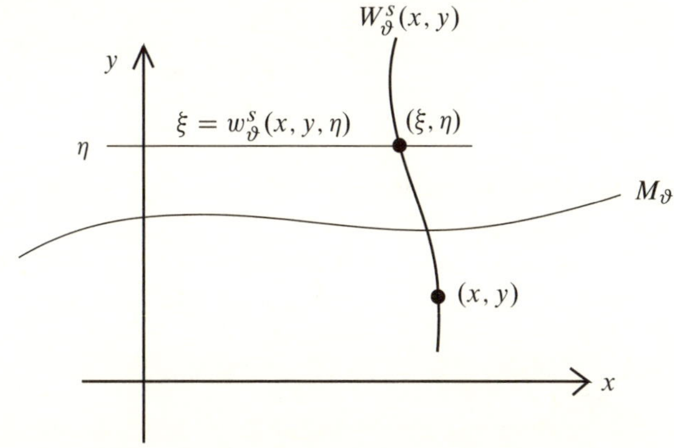

Figure 4.2. The function w^s_ϑ describing the stable fibers.

Theorem 4.1. *Let $\vartheta \in E$ and let the map P_ϑ of* (4.1) *satisfy Hypotheses HM and HMA and let the constants Γ_{11}, L_{12}, L_{21}, L_{22} satisfy Conditions CM and CMA. Let $\Omega_\vartheta \subset X \times Y$ be a positively invariant set of P_ϑ, i.e., $P_\vartheta(\Omega_\vartheta) \subset \Omega_\vartheta$.*

Then there is a continuous function $w^s_\vartheta \colon \Omega_\vartheta \times Y \to X$ such that the following assertions hold.

i) $w^s_\vartheta(x,y,y) = x$ *for all* $(x,y) \in \Omega_\vartheta$.

ii) *For all $(x,y) \in \Omega_\vartheta$ the function $w^s_\vartheta(x,y,\cdot)$ is uniformly λ_R-Lipschitz with*

$$\lambda_R = \frac{2L_{12}}{\Gamma_{11} - L_{22} + \sqrt{(\Gamma_{11} - L_{22})^2 - 4L_{12}L_{21}}}.$$

iii) *The stable fibers $W^s_\vartheta(x,y) := \{(\xi,\eta) \mid \eta \in Y,\ \xi = w^s_\vartheta(x,y,\eta)\}$, $(x,y) \in \Omega_\vartheta$, form a positively invariant family, i.e.,*

$$P_\vartheta(W^s_\vartheta(x,y)) \subset W^s_\vartheta(P_\vartheta(x,y)).$$

iv) *The function* w^s_ϑ *satisfies the invariance equation*

$$F(w^s_\vartheta(x,y,\eta),\eta,\vartheta) = w^s_\vartheta(F(x,y,\vartheta),G(x,y,\vartheta),G(w^s_\vartheta(x,y,\eta),\eta,\vartheta)). \tag{4.2}$$

v) *The stable fibers are disjoint, i.e., for* $(x_i, y_i) \in \Omega_\vartheta$, $i = 1,2$, *either* $W^s_\vartheta(x_1,y_1) \cap W^s_\vartheta(x_2,y_2) = \emptyset$ *or* $W^s_\vartheta(x_1,y_1) = W^s_\vartheta(x_2,y_2)$.

vi) *The map* P_ϑ *is contracting along the stable fibers: For* $(x,y) \in \Omega_\vartheta$ *let* $(x_0,y_0) \in W^s_\vartheta(x,y)$ *and* $(\tilde{x}_0,\tilde{y}_0) \in W^s_\vartheta(x,y)$ *and define* $(x_j,y_j) := P^j_\vartheta(x_0,y_0)$ *and* $(\tilde{x}_j,\tilde{y}_j) := P^j_\vartheta(\tilde{x}_0,\tilde{y}_0)$, $j \in \mathbb{N}_0$. *Then the estimates*

$$|x_j - \tilde{x}_j| \le \lambda_R\, {\chi_A}^j\, |y_0 - \tilde{y}_0|,$$
$$|y_j - \tilde{y}_j| \le {\chi_A}^j\, |y_0 - \tilde{y}_0|$$

hold with $\chi_A = L_{22} + \Delta < 1$, $\Delta = L_{21}\lambda_R$.

vii) *The negatively invariant manifold* $M_\vartheta = \{(x,y) \mid x \in X,\ y = s_A(x,\vartheta)\}$ *of Theorem* 1.5 *has the property of asymptotic phase: For every* $(\xi_0,\eta_0) \in W^s_\vartheta(\Omega_\vartheta) := \bigcup_{(x,y)\in\Omega_\vartheta} W^s_\vartheta(x,y)$ *there is* $(x_0,y_0) \in M_\vartheta$ *such that for* $(\xi_j,\eta_j) := P^j_\vartheta(\xi_0,\eta_0)$ *and* $(x_j,y_j) := P^j_\vartheta(x_0,y_0)$, $j \in \mathbb{N}_0$, *the estimates*

$$|\xi_j - x_j| \le q_A\, {\chi_A}^j\, |\eta_0 - s_A(\xi_0,\vartheta)|,$$
$$|\eta_j - y_j| \le (1 + \lambda_A q_A)\, {\chi_A}^j\, |\eta_0 - s_A(\xi_0,\vartheta)|$$

hold with

$$q_A = \frac{L_{12}}{\sqrt{(\Gamma_{11} - L_{22})^2 - 4L_{12}L_{21}}}.$$

viii) *For* $(x,y) \in \Omega_\vartheta$ *the stable fiber* $W^s_\vartheta(x,y)$ *contains all points which under iteration of the map* P_ϑ *exponentially tend to the iterates of* (x,y) *with rate* ${\chi_A}^j$, *i.e., for* $(x,y) \in \Omega_\vartheta$ *every point* $(\xi,\eta) \in X \times Y$, *for which there is a constant* c *such that for* $j \in \mathbb{N}$,

$$|P^j_\vartheta(\xi,\eta) - P^j_\vartheta(x,y)| \le c\, {\chi_A}^j \tag{4.3}$$

holds, belongs to $W^s_\vartheta(x,y)$ *and hence* $\xi = w^s_\vartheta(x,y,\eta)$.

ix) *If there is a map* $\kappa\colon X \to X$ *such that* F *is* κ*-equivariant and* G *is* κ*-invariant and such that* Ω_ϑ *is invariant with respect to* $(x,y) \mapsto (\kappa x, y)$ *then* w^s_ϑ *is* κ*-equivariant, i.e., if for all* $(u,v) \in X \times Y$ *and all* $(x,y) \in \Omega_\vartheta$,

$$F(\kappa u, v, \vartheta) = \kappa F(u,v,\vartheta),$$
$$G(\kappa u, v, \vartheta) = G(u,v,\vartheta),$$
$$(\kappa x, y) \in \Omega_\vartheta,$$

then $w^s_\vartheta(\kappa x, y, \cdot) = \kappa\, w^s_\vartheta(x,y,\cdot)$ *holds.*

x) *If F and G are of class C_b^k, $k \geq 1$ and if Condition CMA(k) holds then for $(x, y) \in \Omega_\vartheta$ the function $w_\vartheta^s(x, y, \cdot)$ is of class C_b^k.*

Remark 4.2. (1) In the case $X = \mathcal{B}_X$ one may choose $\Omega_\vartheta = X \times Y$. It follows that the whole space $X \times Y$ is foliated by the fibers $W_\vartheta^s\big(x, s_A(x, \vartheta)\big)$, $x \in X$, i.e., by the fibers through the base points on the manifold M_ϑ.

(2) Assertion x) states that under appropriate smoothness conditions the function $w_\vartheta^s\colon (x, y, \eta) \mapsto w_\vartheta^s(x, y, \eta)$ is of class C_b^k with respect to η. In Chapter 5 we show that under additional conditions the function w_ϑ^s is of class C_b^{k-1}.

Proof. To simplify notation we suppress the parameter ϑ in the functions F and G and write w, Ω instead of w_ϑ^s, Ω_ϑ, for short.

i), ii), iv) Let C_{λ_R} be the space

$$C_{\lambda_R} := \big\{w \in C^0(\Omega \times Y \to X) \mid w(x, y, y) = x,\ w(x, y, \cdot) \text{ is } \lambda_R\text{-Lipschitz}\big\}.$$

Equipped with the metric

$$\operatorname{dist}(w_1, w_2) := \sup_{\substack{x,y,\eta \\ \text{with } \eta \neq y}} \frac{|w_1(x, y, \eta) - w_2(x, y, \eta)|}{|y - \eta|},$$

C_{λ_R} becomes a complete metric space with

$$\begin{aligned}\operatorname{dist}(w_1, w_2) &\leq \sup_{\substack{x,y,\eta \\ \text{with } \eta \neq y}} \frac{|w_1(x, y, \eta) - w_1(x, y, y)| + |w_2(x, y, y) - w_2(x, y, \eta)|}{|y - \eta|} \\ &\leq 2\lambda_R.\end{aligned}$$

We define the operator $\mathcal{F} : w \to \underline{w}$ acting on C_{λ_R} by

$$F(\underline{w}(x, y, \eta), \eta) = w(F(x, y), G(x, y), G(w(x, y, \eta), \eta)). \tag{4.4}$$

By Hypothesis HM the operator $\mathcal{F}$ is well defined.

Assertion 4.1.1. $\mathcal{F} : C_{\lambda_R} \to C_{\lambda_R}$.

The function $\underline{w}$ is continuous, since F, G and w are continuous and since $F(\cdot, \eta)$ is a homeomorphism by Hypothesis HM c). We set $\eta = y$ in equation (4.4) and use that $w(x, y, y) = x$ for all $y \in Y$ to get

$$F(\underline{w}(x, y, y), y) = w(F(x, y), G(x, y), G(x, y)) = F(x, y).$$

We conclude that $\underline{w}(x, y, y) = x$ since $F(\cdot, y)$ is a homeomorphism.

We show that $\underline{w}(x, y, \eta)$ is λ_R-Lipschitz with respect to η. From equation (4.4) and Hypothesis HM c) we obtain for $\eta_1, \eta_2 \in Y$,

$$\begin{aligned}
\Gamma_{11}|\underline{w}(x, y, \eta_1) - \underline{w}(x, y, \eta_2)| - L_{12}|\eta_1 - \eta_2| \\
&\leq |F(\underline{w}(x, y, \eta_1), \eta_1) - F(\underline{w}(x, y, \eta_2), \eta_2)| \\
&= |w(F(x, y),\, G(x, y),\, G(w(x, y, \eta_1), \eta_1)) \\
&\qquad - w(F(x, y),\, G(x, y),\, G(w(x, y, \eta_2), \eta_2))| \\
&\leq \lambda_R(L_{21}\lambda_R + L_{22})\,|\eta_1 - \eta_2|
\end{aligned}$$

leading to

$$|\underline{w}(x, y, \eta_1) - \underline{w}(x, y, \eta_2)| \leq \frac{L_{21}\lambda_R^2 + L_{22}\lambda_R + L_{12}}{\Gamma_{11}}\,|\eta_1 - n_2|.$$

We conclude that $\underline{w}$ is λ_R-Lipschitz with respect to η since λ_R satisfies

$$L_{21}\lambda_R^2 - (\Gamma_{11} - L_{22})\lambda_R + L_{12} = 0.$$

This proves Assertion 4.1.1.

Assertion 4.1.2. *$\mathcal{F}$ is a contraction in C_{λ_R}.*

From Hypothesis HM c) and equation (4.4) we obtain for $w_1, w_2 \in C_{\lambda_R}$,

$$\begin{aligned}
\Gamma_{11}|\underline{w}_1(x, y, \eta) - \underline{w}_2(x, y, \eta)| \\
&\leq |F(\underline{w}_1(x, y, \eta), \eta) - F(\underline{w}_2(x, y, \eta), \eta)| \\
&= |w_1(F(x, y),\, G(x, y),\, G(w_1(x, y, \eta), \eta)) \\
&\qquad - w_2(F(x, y),\, G(x, y),\, G(w_2(x, y, \eta), \eta))| \\
&\leq |w_1(\bar{x}, \bar{y}, \bar{\eta}_1) - w_2(\bar{x}, \bar{y}, \bar{\eta}_1)| + |w_2(\bar{x}, \bar{y}, \bar{\eta}_1) - w_2(\bar{x}, \bar{y}, \bar{\eta}_2)|
\end{aligned}$$

where we have set $\bar{x} = F(x, y)$, $\bar{y} = G(x, y)$, $\bar{\eta}_i = G(w_i(x, y, \eta), \eta)$, $i = 1, 2$. Dividing this inequality by $|y - \eta|$ we get

$$\begin{aligned}
\Gamma_{11}\frac{|\underline{w}_1(x, y, \eta) - \underline{w}_2(x, y, \eta)|}{|y - \eta|} \\
&\leq \frac{|w_1(\bar{x}, \bar{y}, \bar{\eta}_1) - w_2(\bar{x}, \bar{y}, \bar{\eta}_1)|}{|\bar{y} - \bar{\eta}_1|}\,\frac{|\bar{y} - \bar{\eta}_1|}{|y - \eta|} + \lambda_R\,\frac{|\bar{\eta}_1 - \bar{\eta}_2|}{|y - \eta|}.
\end{aligned}$$

Using $|\bar{y}-\bar{\eta}_1| \leq (L_{21}\lambda_R+L_{22})|y-\eta|$ and $|\bar{\eta}_1-\bar{\eta}_2| \leq L_{21}|w_1(x, y, \eta)-w_2(x, y, \eta)|$ and taking the supremum first on the right-hand side and then on the left-hand side yields

$$\operatorname{dist}(\underline{w}_1, \underline{w}_2) \leq \frac{L_{22} + 2L_{21}\lambda_R}{\Gamma_{11}}\,\operatorname{dist}(w_1, w_2).$$

Since by the definition of λ_R,

$$2L_{21}\lambda_R = \Gamma_{11} - L_{22} - \sqrt{(\Gamma_{11} - L_{22})^2 - 4L_{12}L_{21}}$$

holds, it follows that $\mathcal{F}$ is a contraction.

Assertions 4.1.1 and 4.1.2 imply that $\mathcal{F}$ has a unique fixed point w which satisfies (4.2). This proves the assertions i), ii) and iv) of the theorem.

iii) Let $(\xi, \eta) \in W^s_\vartheta(x, y)$ and $(\bar{x}, \bar{y}) := P_\vartheta(x, y)$, $(\bar{\xi}, \bar{\eta}) := P_\vartheta(\xi, \eta)$. By definition, $\xi = w(x, y, \eta)$ holds and by (4.2) one has $\bar{\xi} = F(\xi, \eta) = w(\bar{x}, \bar{y}, \bar{\eta})$ implying $(\bar{\xi}, \bar{\eta}) \in W^s_\vartheta(\bar{x}, \bar{y})$.

v) We show that if the stable fibers $W^s_\vartheta(x_1, y_1)$ and $W^s_\vartheta(x_2, y_2)$ have one common point, then they are identical. We show that $w(x_1, y_1, \eta) = w(x_2, y_2, \eta)$ implies $w(x_1, y_1, \zeta) = w(x_2, y_2, \zeta)$ for all $\zeta \in Y$. For given ζ we define for $j \in \mathbb{N}_0$ (cf. Figure 4.3)

$$
\begin{aligned}
(p_j, q_j) &:= P^j_\vartheta(w(x_1, y_1, \zeta), \zeta), \\
(u_j, v_j) &:= P^j_\vartheta(w(x_2, y_2, \zeta), \zeta), \\
(\xi_j, \eta_j) &:= P^j_\vartheta(w(x_1, y_1, \eta), \eta) = P^j_\vartheta(w(x_2, y_2, \eta), \eta).
\end{aligned}
$$

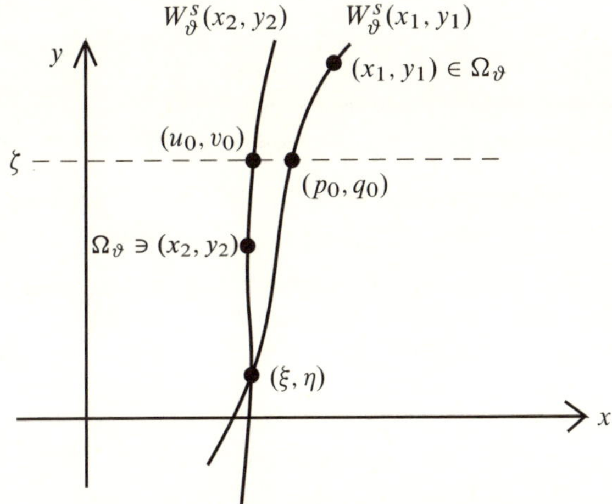

Figure 4.3. Illustration for the proof of Assertion v).

The points (p_0, q_0) and (u_0, v_0) lie on the manifold given by $y = \zeta$. Since by Assertion 1.5.1 a λ_A-Lipschitz manifold is mapped to a λ_A-Lipschitz manifold it holds that $|q_j - v_j| \leq \lambda_A |p_j - u_j|$. We estimate

$$
|p_{j+1} - u_{j+1}| = |F(p_j, q_j) - F(u_j, v_j)| \geq (\Gamma_{11} - L_{12}\lambda_A)\,|p_j - u_j| \qquad (4.5)
$$

and conclude that

$$|p_j - u_j| \geq (\Gamma_{11} - \Delta)^j \, |p_0 - u_0| \tag{4.6}$$

with $\Delta = L_{12}\lambda_A = L_{21}\lambda_R$. Since the points (p_j, q_j) and (ξ_j, η_j) lie on the same fiber (which is the graph of a λ_R-Lipschitz function) we have

$$|p_j - \xi_j| \leq \lambda_R \, |q_j - \eta_j|$$

for all $j \in \mathbb{N}_0$. We find

$$\begin{aligned} |p_j - \xi_j| &\leq \lambda_R \, |q_j - \eta_j| = \lambda_R \, |G(p_{j-1}, q_{j-1}) - G(\xi_{j-1}, \eta_{j-1})| \\ &\leq \lambda_R (L_{22} + L_{21}\, \lambda_R) \, |q_{j-1} - \eta_{j-1}| \\ &\leq \lambda_R (L_{22} + \Delta)^j \, |q_0 - \eta_0|. \end{aligned} \tag{4.7}$$

Analogously, we have

$$|u_j - \xi_j| \leq \lambda_R \, (L_{22} + \Delta)^j \, |v_0 - \eta_0| = \lambda_R \, (L_{22} + \Delta)^j \, |q_0 - \eta_0|$$

since $v_0 = q_0 = \zeta$ and we obtain

$$|p_j - u_j| \leq |p_j - \xi_j| + |\xi_j - u_j| \leq 2\lambda_R \, (L_{22} + \Delta)^j \, |q_0 - \eta_0|.$$

Combining this result with equation (4.6) we get

$$|p_0 - u_0| \leq 2\lambda_R \left(\frac{L_{22} + \Delta}{\Gamma_{11} - \Delta} \right)^j \, |q_0 - \eta_0|.$$

We know that $L_{22} + \Delta < \Gamma_{11} - \Delta$ from equation (1.16). Letting $j \to \infty$ yields $|p_0 - u_0| = 0$ and therefore $w(x_1, y_1, \zeta) = w(x_2, y_2, \zeta)$.

vi) Assertion vi) follows immediately from equation (4.7) with $|p_j - \xi_j|$ replaced by $|x_j - \tilde{x}_j|$ and $|q_j - \eta_j|$ replaced by $|y_j - \tilde{y}_j|$.

vii) The stable fiber $W^s_\vartheta(\xi_0, \eta_0)$ has a unique intersection point (x_0, y_0) with the manifold M_ϑ. This follows from the fact that $W^s_\vartheta(x, y)$ is λ_R-Lipschitz and M_ϑ is λ_A-Lipschitz with $\lambda_R \lambda_A < 4L_{12}L_{21}/(\Gamma_{11} - L_{22})^2 < 1$ due to Condition CM. From vi) we know that

$$\begin{aligned} |\xi_j - x_j| &\leq \lambda_R \, {\chi_A}^j \, |\eta_0 - s_A(x_0)|, \\ |\eta_j - y_j| &\leq {\chi_A}^j \, |\eta_0 - s_A(x_0)|. \end{aligned}$$

Since s_A is λ_A-Lipschitz we estimate

$$|\eta_0 - s_A(x_0)| \leq |\eta_0 - s_A(\xi_0)| + \lambda_A |\xi_0 - x_0|. \tag{4.8}$$

It remains to estimate $|\xi_0 - x_0|$. Since $(x_0, s_A(x_0))$ lies on the fiber through (ξ_0, η_0) which is the graph of the λ_R-function $w(\xi_0, \eta_0, \cdot)$ we get

$$|\eta_0 - s_A(x_0)| \geq \frac{1}{\lambda_R} |\xi_0 - x_0|.$$

Combining this estimate with (4.8) we find

$$|\eta_0 - s_A(\xi_0)| \geq \left(\frac{1}{\lambda_R} - \lambda_A\right)|\xi_0 - x_0| = \frac{1}{q_A}|\xi_0 - x_0|.$$

By equation (4.8) we obtain

$$|\eta_0 - s_A(x_0)| \leq (1 + \lambda_A q_A)|\eta_0 - s_A(\xi_0)|.$$

The fact that $\lambda_R(1 + \lambda_A q_A) = q_A$ concludes the proof of vii).

viii) Let (ξ, η) be a point satisfying (4.3). We define $u := w(x, y, \eta)$, $v := \eta$ and $(x_j, y_j) := P_\vartheta^j(x, y)$, $(\xi_j, \eta_j) := P_\vartheta^j(\xi, \eta)$, $(u_j, v_j) := P_\vartheta^j(u, v)$, $j \geq 0$ (cf. Figure 4.4). Since the points (u, v) and (ξ, η) lie on the manifold given by $y = \eta$ Assertion 1.5.1 implies

$$|\eta_j - v_j| \leq \lambda_A|\xi_j - u_j|.$$

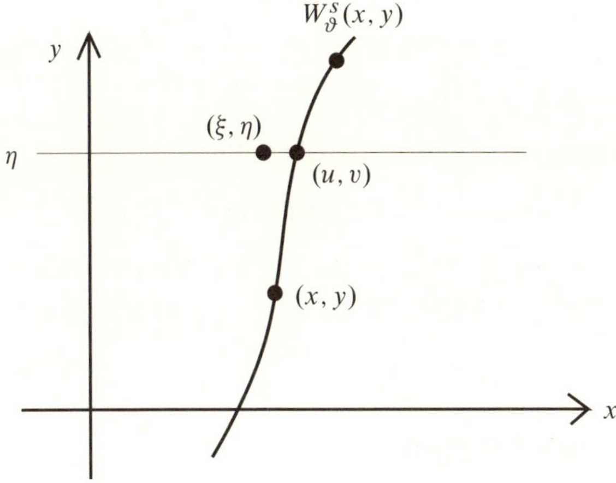

Figure 4.4. Illustration for the proof of Assertion viii).

We estimate for $j > 0$,

$$\begin{aligned}
|\xi_j - u_j| &= |F(\xi_{j-1}, \eta_{j-1}) - F(u_{j-1}, v_{j-1})| \\
&\geq (\Gamma_{11} - L_{12}\lambda_A)\,|\xi_{j-1} - u_{j-1}| \\
&\geq (\Gamma_{11} - \Delta)^j\,|\xi - u|
\end{aligned}$$

implying

$$
\begin{aligned}
|\xi - u| &\leq \frac{1}{(\Gamma_{11} - \Delta)^j} \, |\xi_j - u_j| \\
&\leq \frac{1}{(\Gamma_{11} - \Delta)^j} \, (|\xi_j - x_j| + |x_j - u_j|).
\end{aligned}
$$

By assumption we know that $|\xi_j - x_j| \leq c(L_{22} + \Delta)^j$. Assertion vi) yields $|x_j - u_j| \leq \lambda_R (L_{22} + \Delta)^j \, |y - \eta|$. We thus have

$$
|\xi - u| \leq \left(\frac{L_{22} + \Delta}{\Gamma_{11} - \Delta} \right)^j \, (c + \lambda_R |y - \eta|).
$$

From equation (1.16) we know that $L_{22} + \Delta < \Gamma_{11} - \Delta$ implying $\xi = u$.

ix) Let $C_{\lambda_R}|_\kappa$ be the set of κ-equivariant functions $w \in C_{\lambda_R}$, i.e., $C_{\lambda_R}|_\kappa = \{w \in C_{\lambda_R} \mid w(\kappa x, y, \eta) = \kappa w(x, y, \eta)\}$. We show that $\mathcal{F} : C_{\lambda_R}|_\kappa \to C_{\lambda_R}|_\kappa$. For $w \in C_{\lambda_R}|_\kappa$ we have using equation (4.4)

$$
\begin{aligned}
F(\underline{w}(\kappa x, y, \eta), \eta) &= w(F(\kappa x, y), G(\kappa x, y), G(w(\kappa x, y, \eta), \eta)) \\
&= w(\kappa F(x, y), G(x, y), G(\kappa w(x, y, \eta), \eta)) \\
&= \kappa w(F(x, y), G(x, y), G(w(x, y, \eta), \eta)) \\
&= \kappa F(\underline{w}(x, y, \eta), \eta) \\
&= F(\kappa \underline{w}(x, y, \eta), \eta).
\end{aligned}
$$

Since $F(\cdot, \eta)$ is a homeomorphism this implies $\underline{w} \in C_{\lambda_R}|_\kappa$.

x) The smoothness of the function $w(x, y, \cdot)$ is shown in an analogous way as the smoothness of $s_A(\cdot, \vartheta)$ in the proof of Theorem 3.1 and is therefore omitted here. □

4.2 The unstable foliation

As in Section 1.1 we consider a family of maps

$$
P_\vartheta : X \times Y \ni \begin{pmatrix} x \\ y \end{pmatrix} \longmapsto \begin{pmatrix} \bar{x} \\ \bar{y} \end{pmatrix} = \begin{pmatrix} F(x, y, \vartheta) \\ G(x, y, \vartheta) \end{pmatrix} \in \mathcal{B}_x \times \mathcal{B}_y, \quad \vartheta \in E \subset \mathcal{B}_\vartheta, \tag{4.9}
$$

satisfying Hypothesis HM and HMR and Conditions CM and CMR. From Theorem 1.3 we know that P_ϑ admits a repulsive positively invariant manifold N_ϑ.

We assume that there is a negatively invariant set $\Omega_\vartheta \subset X \times Y$, i.e., $\Omega_\vartheta \subset P_\vartheta(\Omega_\vartheta)$. In order to show that there is an unstable fiber $W_\vartheta^u(x, y)$ through every point $(x, y) \in \Omega_\vartheta$ we need an additional assumption.

Hypothesis HMRF

There is a set $\Omega_\vartheta \subset X \times Y$ which is negatively invariant under P_ϑ. The map P_ϑ is invertible on Ω_ϑ, i.e., for every $(x, y) \in \Omega_\vartheta$ there is a unique $(\underline{x}, \underline{y}) \in \Omega_\vartheta$ such that $P_\vartheta(\underline{x}, \underline{y}) = (x, y)$.

We state the result on unstable fibers without proof since the proof is similar to the one of Theorem 4.1.

Theorem 4.3. *Let $\vartheta \in E$ and let the map P_ϑ of (4.9) satisfy Hypotheses HM, HMR and HMRF and let the constants Γ_{11}, L_{12}, L_{21}, L_{22} satisfy Conditions CM and CMR.*

Then there is a continuous function $w^u_\vartheta \colon \Omega_\vartheta \times X \to Y$ such that the following assertions hold

i) $w^u_\vartheta(x, y, x) = y$ *for all* $(x, y) \in \Omega_\vartheta$.

ii) *For all $(x, y) \in \Omega_\vartheta$ the function $w^u_\vartheta(x, y, \cdot)$ is uniformly λ_A-Lipschitz with*

$$\lambda_A = \frac{2L_{21}}{\Gamma_{11} - L_{22} + \sqrt{(\Gamma_{11} - L_{22})^2 - 4L_{12}L_{21}}}.$$

iii) *The unstable fibers $W^u_\vartheta(x, y) := \{(\xi, \eta) \mid \xi \in X,\ \eta = w^u_\vartheta(x, y, \xi)\}$, $(x, y) \in \Omega_\vartheta$, form a negatively invariant family, i.e.,*

$$W^u_\vartheta(x, y) \subset P_\vartheta\big(W^u_\vartheta(\underline{x}, \underline{y})\big)$$

with $(\underline{x}, \underline{y}) := P_\vartheta^{-1}(x, y)$.

iv) *The function w^u_ϑ satisfies the invariance equation*

$$G(\xi, w^u_\vartheta(x, y, \xi), \vartheta) = w^u_\vartheta(F(x, y, \vartheta), G(x, y, \vartheta), F(\xi, w^u_\vartheta(x, y, \xi), \vartheta)).$$

v) *The unstable fibers are disjoint, i.e., for $(x_i, y_i) \in \Omega_\vartheta$, $i = 1, 2$, one has either $W^u_\vartheta(x_1, y_1) \cap W^u_\vartheta(x_2, y_2) = \emptyset$ or $W^u_\vartheta(x_1, y_1) = W^u_\vartheta(x_2, y_2)$.*

vi) *The map P_ϑ is invertible on the set $W^u_\vartheta(\Omega_\vartheta) := \bigcup_{(x,y)\in\Omega_\vartheta} W^u_\vartheta(x, y)$, i.e., for every $(\xi, \eta) \in W^u_\vartheta(\Omega_\vartheta)$ there is a unique $(\underline{\xi}, \underline{\eta}) \in W^u_\vartheta(\Omega_\vartheta)$ with $P_\vartheta(\underline{\xi}, \underline{\eta}) = (\xi, \eta)$. The map P_ϑ^{-1} is contracting along the unstable fibers: For $(x, y) \in \Omega_\vartheta$ let $(x_0, y_0) \in W^u_\vartheta(x, y)$ and $(\tilde{x}_0, \tilde{y}_0) \in W^u_\vartheta(x, y)$ and define $(x_{-j}, y_{-j}) := P_\vartheta^{-j}(x_0, y_0)$ and $(\tilde{x}_{-j}, \tilde{y}_{-j}) := P_\vartheta^{-j}(\tilde{x}_0, \tilde{y}_0)$, $j \in \mathbb{N}_0$. Then the estimates*

$$|x_{-j} - \tilde{x}_{-j}| \le \chi_R{}^{-j} |x_0 - \tilde{x}_0|,$$
$$|y_{-j} - \tilde{y}_{-j}| \le \lambda_A\, \chi_R{}^{-j} |x_0 - \tilde{x}_0|$$

hold with $\chi_R := \Gamma_{11} - \Delta > 1$, $\Delta = L_{12}\lambda_A$.

vii) *The positively invariant manifold* $N_\vartheta = \{(x, y) \mid y \in Y,\ x = s_R(y, \vartheta)\}$ *of Theorem* 1.3 *has the property of asymptotic phase:*

For every $(\xi_0, \eta_0) \in W^u_\vartheta(\Omega_\vartheta) := \bigcup_{(x,y)\in\Omega_\vartheta} W^u_\vartheta(x, y)$ *there is* $(x_0, y_0) \in N_\vartheta$ *such that for* $(\xi_{-j}, \eta_{-j}) := P_\vartheta^{-j}(\xi_0, \eta_0)$ *and* $(x_{-j}, y_{-j}) := P_\vartheta^{-j}(x_0, y_0)$, $j \in \mathbb{N}_0$, *the estimates*

$$|\xi_{-j} - x_{-j}| \le (1 + \lambda_R q_R)\, {\chi_R}^{-j}\, |\xi_0 - s_R(\eta_0, \vartheta)|,$$
$$|\eta_{-j} - y_{-j}| \le q_R\, {\chi_R}^{-j}\, |\xi_0 - s_R(\eta_0, \vartheta)|$$

hold with

$$q_R = \frac{L_{21}}{\sqrt{(\Gamma_{11} - L_{22})^2 - 4L_{12}L_{21}}}.$$

viii) *For* $(x, y) \in \Omega_\vartheta$ *the unstable fiber* $W^u_\vartheta(x, y)$ *contains all points which under iteration of* P_ϑ^{-1} *exponentially tend to the iterates of* (x, y) *with rate* ${\chi_R}^{-j}$, *i.e., for* $(x, y) \in \Omega_\vartheta$ *every point* $(\xi, \eta) \in X \times Y$, *for which there is a constant* c *such that for* $j \in \mathbb{N}$

$$|P_\vartheta^{-j}(\xi, \eta) - P_\vartheta^{-j}(x, y)| \le c\, {\chi_R}^{-j}$$

holds, belongs to $W^u_\vartheta(x, y)$ *and hence* $\eta = w^u_\vartheta(x, y, \xi)$.

ix) *If there is a map* $\kappa\colon Y \to Y$ *such that* F *is* κ*-invariant and* G *is* κ*-equivariant and such that* Ω_ϑ *is invariant with respect to* $(x, y) \mapsto (x, \kappa y)$ *then* w^u_ϑ *is* κ*-equivariant, i.e., if for all* $(u, v) \in X \times Y$, $(x, y) \in \Omega_\vartheta$,

$$F(u, \kappa v, \vartheta) = F(u, v, \vartheta),$$
$$G(u, \kappa v, \vartheta) = \kappa G(u, v, \vartheta),$$
$$(x, \kappa y) \in \Omega_\vartheta,$$

then $w^u_\vartheta(x, \kappa y, \cdot) = \kappa\, w^u_\vartheta(x, y, \cdot)$ *holds.*

x) *If* F *and* G *are of class* C^k_b, $k \ge 1$, *and if Condition CMR(*k*) holds then for* $(x, y) \in \Omega_\vartheta$ *the function* $w^u_\vartheta(x, y, \cdot)$ *is of class* C^k_b.

Remark 4.4. In the case $Y = \mathcal{B}_Y$ one may choose $\Omega_\vartheta = X \times Y$. It follows that the whole space is foliated by the fibers $W^u_\vartheta(s_R(y, \vartheta), y)$, $y \in Y$, i.e., by the fibers through the base points on the manifold N_ϑ.

4.3 The hyperbolic case

As in Section 1.3 we consider a map

$$P\colon X \times Z \times Y \ni \begin{pmatrix} x \\ z \\ y \end{pmatrix} \longmapsto \begin{pmatrix} \bar{x} \\ \bar{z} \\ \bar{y} \end{pmatrix} = \begin{pmatrix} F(x, z, y) \\ H(x, z, y) \\ G(x, z, y) \end{pmatrix}$$

where we again suppress a possible dependence on parameters. We consider the two forms P_A and P_R of the map P as introduced in (1.23) and (1.24). We assume that all assumptions of Theorem 1.7 hold. It follows that the map P admits the attractive negatively invariant manifold

$$M = \{(x, z, y) \mid (x, z) \in X \times Z,\ y = s_A(x, z)\},$$

the repulsive positively invariant manifold

$$N = \{(x, z, y) \mid (z, y) \in Z \times Y,\ x = s_R(z, y)\}$$

and the hyperbolic invariant manifold

$$K = M \cap N = \{(x, z, y) \mid z \in Z,\ x = r_x(z),\ y = r_y(z)\}.$$

Since K is invariant under the map P we may apply Theorems 4.1 and 4.3, respectively, with $\Omega := K$, yielding functions w^s_ϑ and w^u_ϑ, respectively. Thus, there is a positively invariant family of stable fibers

$$W^s(x, z, y) = \{(\xi, \zeta, \eta) \mid \eta \in Y,\ (\xi, \zeta) = w^s_\vartheta(x, z, y, \eta)\}, \quad (x, z, y) \in K, \tag{4.10}$$

and a negatively invariant family of unstable fibers

$$W^u(x, y, z) = \{(\xi, \zeta, \eta) \mid \xi \in X,\ (\zeta, \eta) = w^u_\vartheta(x, y, z, \xi)\}, \quad (x, y, z) \in K. \tag{4.11}$$

We show that the repulsive manifold N is foliated by the family of stable fibers and the attractive manifold M is foliated by the family of unstable fibers (cf. Figure 4.5).

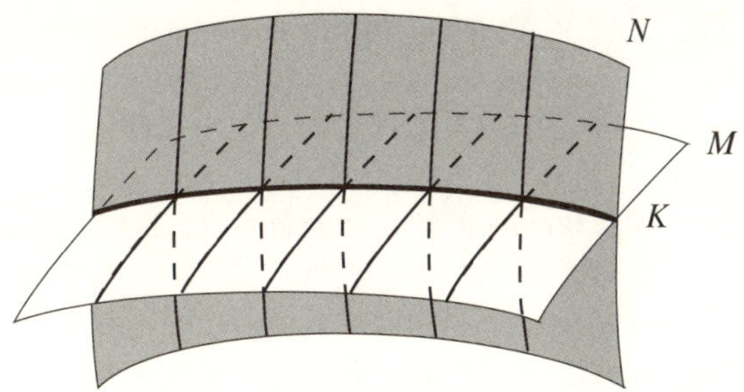

Figure 4.5. The hyperbolic manifold K and the foliation of the unstable manifold M and of the stable manifold N.

Theorem 4.5. *Assume that the maps P_A and P_R defined in* (1.23) *and* (1.24), *respectively, satisfy Hypothesis HM with constants Γ^A_{11}, L^A_{12}, L^A_{21}, L^A_{22} and Γ^R_{11}, L^R_{12}, L^R_{21}, L^R_{22}, respectively. Let P_A satisfy Hypothesis HMA and assume that the constants Γ^A_{11}, L^A_{12}, L^A_{21}, L^A_{22} satisfy Conditions CM and CMA with*

$$\lambda_A = 2L^A_{21} \Big/ \Big(\Gamma^A_{11} - L^A_{22} + \sqrt{(\Gamma^A_{11} - L^A_{22})^2 - 4L^A_{12} L^A_{21}}\Big)$$

and $\Delta_A = L^A_{12}\lambda_A$. *Let* P_R *satisfy Hypothesis HMR and assume that the constants* Γ^R_{11}, L^R_{12}, L^R_{21}, L^R_{22} *satisfy Conditions CM and CMR with*

$$\lambda_R = 2L^R_{12}\Big/\Big(\Gamma^R_{11} - L^R_{22} + \sqrt{(\Gamma^R_{11} - L^R_{22})^2 - 4L^R_{12}L^R_{21}}\Big)$$

and $\Delta_R = L^R_{21}\lambda_R$. *Moreover, assume that* $\lambda_A\lambda_R < 1$.

Then for the attractive manifold M, *the repulsive manifold* N *and the hyperbolic manifold* $K = M \cap N$ *the following assertions hold.*

i) $M = W^u(K) = \bigcup_{(x,z,y)\in K} W^u(x,z,y)$,

ii) $N = W^s(K) = \bigcup_{(x,z,y)\in K} W^s(x,z,y)$,

where $W^s(x,z,y)$ *and* $W^u(x,z,y)$, *respectively, are defined in equations* (4.10) *and* (4.11), *respectively.*

Proof. We apply Theorem 4.1 to the map P_A and Theorem 4.3 to the map P_R. Note that Hypothesis HMRF is satisfied since $K \subset M$ and since $P_A|_M$ is invertible due to Theorem 1.5 vii).

i) We show that $M \subset W^u(K)$. Let $(\xi,\zeta,\eta) \in M$. Since $M \subset P(M)$ and $P|_M$ is invertible (cf. Theorem 1.5 vii)) we may apply Theorem 4.3 with $\Omega = M$. We define the sequence $(\xi_{-j},\zeta_{-j},\eta_{-j}) := P^{-j}(\xi,\zeta,\eta) \in M$, $j \geq 0$. Through every point $(\xi_{-j},\zeta_{-j},\eta_{-j})$ there is an unstable fiber $W^u(\xi_{-j},\zeta_{-j},\eta_{-j})$. Since $W^u(\xi_{-j},\zeta_{-j},\eta_{-j})$ are λ_A-Lipschitz manifolds and since N is a λ_R-Lipschitz manifold and since $\lambda_A\lambda_R < 1$ there are unique intersection points $(\tilde{x}_{-j},\tilde{z}_{-j},\tilde{y}_{-j}) := W^u(\xi_{-j},\zeta_{-j},\eta_{-j}) \cap N$. The set $\Lambda := \bigcup_{j\geq 0}(\tilde{x}_{-j},\tilde{z}_{-j},\tilde{y}_{-j})$ is negatively invariant, i.e., $\Lambda \subset P(\Lambda)$. By Theorem 4.3 vi) we have

$$|\tilde{y}_{-j} - \eta_{-j}| \leq \lambda_A\, \chi_R{}^{-j}\, |\tilde{x}_0 - \xi_0|$$

and hence $|\tilde{y}_{-j}| \leq |\tilde{y}_{-j} - \eta_{-j}| + |\eta_{-j}| \leq \lambda_A|\tilde{x}_0 - \xi_0| + |s_A|$, $j \in \mathbb{N}$, are uniformly bounded. Therefore, Λ is bounded with respect to y. It follows from Theorem 1.5 iv) that $\Lambda \subset M$ and hence $(\tilde{x}_0,\tilde{z}_0,\tilde{y}_0) \in M \cap N = K$. Theorem 4.3 v) implies $W^u(\xi,\zeta,\eta) = W^u(\tilde{x}_0,\tilde{z}_0,\tilde{y}_0)$ and therefore $(\xi,\zeta,\eta) \in W^u(\tilde{x}_0,\tilde{z}_0,\tilde{y}_0) \subset W^u(K)$.

We now show that $W^u(K) \subset M$. Let $(\xi,\zeta,\eta) \in W^u(K)$. By definition of $W^u(K)$ there is a base point $(\tilde{x},\tilde{z},\tilde{y}) \in K$ such that $(\xi,\zeta,\eta) \in W^u(\tilde{x},\tilde{z},\tilde{y})$. We define the points $(\tilde{x}_{-j},\tilde{z}_{-j},\tilde{y}_{-j}) := P^{-j}(\tilde{x},\tilde{z},\tilde{y}) \in M$ and $(\xi_{-j},\zeta_{-j},\eta_{-j}) := P^{-j}(\xi,\zeta,\eta)$, $j \geq 0$. The points $(\xi_{-j},\zeta_{-j},\eta_{-j})$ are well defined since P is invertible on $W^u(K)$, cf. Theorem 4.3 vi). The set $\Lambda := \bigcup_{j\geq 0}(\xi_{-j},\zeta_{-j},\eta_{-j})$ is negatively invariant, i.e., $\Lambda \subset P(\Lambda)$. Since by Theorem 4.3 vi)

$$|\tilde{y}_{-j} - \eta_{-j}| \leq \lambda_A\, \chi_R{}^{-j}\, |\tilde{x}_0 - \xi_0|$$

and since $\tilde{y}_{-j} = s_A(\tilde{x}_{-j},\tilde{z}_{-j})$, $j \in \mathbb{N}$, are bounded we conclude that $|\eta_{-j}|$, $j \in \mathbb{N}$, are uniformly bounded. Therefore, the set Λ is bounded with respect to y. Theorem 1.5 iv) implies that $\Lambda \subset M$ and therefore $(\xi,\zeta,\eta) \in M$.

ii) The proof is similar to the proof of Assertion i). □

Chapter 5

Smoothness of the foliation with respect to the base point

Under the assumption of Theorem 4.1 the map P_ϑ of (5.1) below (which is the map (4.1)) admits a negatively invariant manifold M_ϑ and stable fibers through every point of Ω_ϑ. The fiber through $(x,y)\in\Omega_\vartheta$ is described by a continuous function $w^s_\vartheta(x,y,\eta)$, $\eta\in Y$, which is smooth with respect to η. The manifold M_ϑ intersects each fiber in one point $(x,s_A(x,\vartheta))$, the so-called base point. In Section 5.1 we show that under additional conditions the function $w^s_\vartheta(x,s_A(x,\vartheta),\eta)$ is smooth with respect to all arguments, i.e., with respect to x, η, ϑ. In Section 5.2 we state the analogous result for the unstable foliation.

5.1 The stable foliation

As in Section 4.1 we consider a family of maps P_ϑ of the form

$$P_\vartheta\colon X\times Y\ni\begin{pmatrix}x\\ y\end{pmatrix}\longmapsto\begin{pmatrix}\bar{x}\\ \bar{y}\end{pmatrix}=\begin{pmatrix}F(x,y,\vartheta)\\ G(x,y,\vartheta)\end{pmatrix}\in\mathcal{B}_x\times\mathcal{B}_y,\quad \vartheta\in E\subset\mathcal{B}_\vartheta,\tag{5.1}$$

where $\mathcal{B}_x$, $\mathcal{B}_y$, $\mathcal{B}_\vartheta$ are Banach spaces and $X\subset\mathcal{B}_x$, $Y\subset\mathcal{B}_y$, $E\subset\mathcal{B}_\vartheta$ are open subspaces. We consider the setting of Sections 1.2 and 3.2 where the map (5.1) admits an attractive negatively invariant manifold M_ϑ. In this section we make the assumption that M_ϑ is invariant. Then with the choice $\Omega_\vartheta=M_\vartheta$ Theorem 4.1 yields for $x\in X$ the existence of stable fibers with base points $(x,s_A(x,\vartheta))$ in M_ϑ. Every stable fiber is given as $W^s_\vartheta(x,s_A(x,\vartheta))=\{(\xi,\eta)\mid \eta\in Y,\ \xi=w^s_\vartheta(x,s_A(x,\vartheta),\eta)\}$. We show that under appropriate conditions the function $w\colon (x,\eta,\vartheta)\mapsto w(x,\eta,\vartheta):=w^s_\vartheta(x,s_A(x,\vartheta),\eta)$ is of class C^{k-1}_b provided the map P_ϑ is of class C^k_b. More precisely, we show that w is of the form

$$w(x,\eta,\vartheta)=x+R(x,\eta,\vartheta)\big[\eta-s_A(x,\vartheta)\big]$$

with $R\in C^{k-1}_b\big(X\times Y\times E,\mathcal{L}_b(\mathcal{B}_y,\mathcal{B}_x)\big)$ where $\mathcal{L}_b$ denotes the space of bounded linear operators.

We make the following assumptions.

Hypothesis HMB

a) *The sets X and Y are convex.*

b) *There is a constant* Γ_{22} *such that for* $x \in X$, $y_1, y_2 \in Y, \vartheta \in E$ *the function* G *satisfies*

$$|G(x, y_1, \vartheta) - G(x, y_2, \vartheta)| \geq \Gamma_{22}\, |y_1 - y_2|.$$

Hypothesis HMAB

There is a constant L_{11} *such that for* $x_1, x_2 \in X$, $y \in Y$, $\vartheta \in E$ *the function* F *satisfies*

$$|F(x_1, y, \vartheta) - F(x_2, y, \vartheta)| \leq L_{11}\, |x_1 - x_2|.$$

Condition CMB

$$\Gamma_{22} - \Delta > 0,$$

where

$$\Delta = \frac{2L_{12}L_{21}}{\Gamma_{11} - L_{22} + \sqrt{(\Gamma_{11} - L_{22})^2 - 4L_{12}L_{21}}}.$$

Condition CMAB($k-1$)

$$\frac{\Gamma_{11} - \Delta}{L_{22} + \Delta} > (L_{11} + \Delta)^{k-1}.$$

We state the smoothness result for the stable foliation.

Theorem 5.1. *Let the map* P_ϑ *of* (5.1) *satisfy Hypotheses HM, HMA, HMB, HMAB and Conditions CM, CMA, CMA*(k), $k > 1$, *and CMB and CMAB*$(k-1)$. *Let* F *and* G *be of class* C_b^k. *Assume that the manifold* $M_\vartheta = \{(x, y) \mid x \in X,\ y = s_A(x, \vartheta)\}$ *established in Theorem* 1.5 *is invariant under* P_ϑ.

Then there is a function w *describing stable fibers* $W_\vartheta^s\big(x, s_A(x, \vartheta)\big) = \{(\xi, \eta) \mid \eta \in Y,\ \xi = w(x, \eta, \vartheta)\}$ *with base point* $\big(x, s_A(x, \vartheta)\big) \in M_\vartheta$. *The function* w *is of class* C_b^{k-1} *and is of the form*

$$w(x, \eta, \vartheta) = x + R(x, \eta, \vartheta)\big[\eta - s_A(x, \vartheta)\big] \tag{5.2}$$

with

$$R \in C_b^{k-1}\big(X \times Y \times E,\ \mathcal{L}_b(\mathcal{B}_y, \mathcal{B}_x)\big).$$

Proof. Setting $\Omega_\vartheta = M_\vartheta$ we know from Theorem 4.1 that the map P_ϑ acts as follows:

$$\begin{pmatrix} x \\ y \\ \xi \\ \eta \end{pmatrix} = \begin{pmatrix} x \\ s_A(x, \vartheta) \\ w(x, \eta, \vartheta) \\ \eta \end{pmatrix} \longmapsto \begin{pmatrix} \bar{x} \\ \bar{y} \\ \bar{\xi} \\ \bar{\eta} \end{pmatrix} = \begin{pmatrix} F(x, s_A(x, \vartheta), \vartheta) \\ G(x, s_A(x, \vartheta), \vartheta) \\ F(w(x, \eta, \vartheta), \eta, \vartheta) \\ G(w(x, \eta, \vartheta), \eta, \vartheta) \end{pmatrix} = \begin{pmatrix} \bar{x} \\ s_A(\bar{x}, \vartheta) \\ w(\bar{x}, \bar{\eta}, \vartheta) \\ \bar{\eta} \end{pmatrix}.$$

Assuming that the function w has the form (5.2) leads to

$$\begin{pmatrix} x \\ s_A(x,\vartheta) \\ x + R(x,\eta,\vartheta)\big[\eta - s_A(x,\vartheta)\big] \\ \eta \end{pmatrix} \longmapsto \begin{pmatrix} \bar{x} \\ s_A(\bar{x},\vartheta) \\ \bar{x} + R(\bar{x},\bar{\eta},\vartheta)\big[\bar{\eta} - s_A(\bar{x},\vartheta)\big] \\ \bar{\eta} \end{pmatrix}.$$

Hence, if there is a function R satisfying (5.2) then it has to satisfy the invariance equation

$$\begin{aligned} &R(\bar{x},\bar{\eta},\vartheta)\big[G\big(x + R(x,\eta,\vartheta)\big[\eta - s_A(x,\vartheta)\big],\eta,\vartheta\big) - G(x,s_A(x,\vartheta),\vartheta)\big] \\ &\quad = F\big(x + R(x,\eta,\vartheta)\big[\eta - s_A(x,\vartheta)\big],\eta,\vartheta\big) - F\big(x,s_A(x,\vartheta),\vartheta\big). \end{aligned} \tag{5.3}$$

For $u \in C_b^{k-1}(X \times Y \times E, \mathcal{L}_b(\mathcal{B}_y,\mathcal{B}_x))$ we differentiate

$$F\big(x + \tau u[\eta - s_A(x,\vartheta)], s_A(x,\vartheta) + \tau(\eta - s_A(x,\vartheta)),\vartheta\big)$$

with respect to τ and integrate from 0 to 1 and get

$$F\big(x + u[\eta - s(x,\vartheta)],\eta,\vartheta\big) - F\big(x,s_A(x,\vartheta),\vartheta\big) = \widetilde{F}(u,x,\eta,\vartheta)\big[\eta - s_A(x,\vartheta)\big],$$

where

$$\begin{aligned} \widetilde{F}(u,x,\eta,\vartheta) := \int_0^1 &\{F_1\big(x + \tau u[\eta - s_A(x,\vartheta)], s_A(x,\vartheta) + \tau[\eta - s_A(x,\vartheta)],\vartheta\big)u \\ &+ F_2\big(x + \tau u[\eta - s_A(x,\vartheta)]\big), s_A(x,\vartheta) + \tau[\eta - s_A(x,\vartheta)],\vartheta\big)\}d\tau. \end{aligned} \tag{5.4}$$

Analogously we have

$$G\big(x + u[\eta - s(x,\vartheta)],\eta,\vartheta\big) - G\big(x,s_A(x,\vartheta),\vartheta\big) = \widetilde{G}(u,x,\eta,\vartheta)\big[\eta - s_A(x,\vartheta)\big]$$

where

$$\begin{aligned} \widetilde{G}(u,x,\eta,\vartheta) := \int_0^1 &\{G_1\big(x + \tau u[\eta - s_A(x,\vartheta)], s_A(x,\vartheta) + \tau[\eta - s_A(x,\vartheta)],\vartheta\big)u \\ &+ G_2\big(x + \tau u[\eta - s_A(x,\vartheta)]\big), s_A(x,\vartheta) + \tau[\eta - s_A(x,\vartheta)],\vartheta\big)\}d\tau. \end{aligned} \tag{5.5}$$

The invariance equation (5.3) for R takes the form

$$\begin{aligned} &R(\bar{x},\bar{\eta},\vartheta)\,\widetilde{G}\big(R(x,\eta,\vartheta),x,\eta,\vartheta\big)\big[\eta - s_A(x,\vartheta)\big] \\ &\quad = \widetilde{F}\big(R(x,\eta,\vartheta),x,\eta,\vartheta\big)\big[\eta - s_A(x,\vartheta)\big] \end{aligned} \tag{5.6}$$

where

$$\begin{aligned} \bar{x} &= F(x,s_A(x,\vartheta),\vartheta) \\ \bar{\eta} &= G\big(x + R(x,\eta,\vartheta)[\eta - s_A(x,\vartheta)]\big),\eta,\vartheta\big). \end{aligned}$$

Instead of (5.6) we make the following stronger assumption on R:

$$R(\bar{x},\bar{\eta},\vartheta)\,\widetilde{G}\big(R(x,\eta,\vartheta),x,\eta,\vartheta\big)=\widetilde{F}\big(R(x,\eta,\vartheta),x,\eta,\vartheta\big). \tag{5.7}$$

In order to show that such a function R exists we introduce a certain map $\widehat{P}_\vartheta$ admitting an invariant manifold. We define the spaces

$$\mathcal{B}_u := \mathcal{L}_b(\mathcal{B}_y,\mathcal{B}_x) \quad \text{and} \quad \mathcal{B}_v := \mathcal{B}_x\times\mathcal{B}_y.$$

Equipped with the norms

$$|u| = \sup_{\eta\in\mathcal{B}_y\setminus\{0\}}\frac{|u\eta|}{|\eta|}, \qquad u\in\mathcal{B}_u,$$

$$|v| = \max\{\alpha|x|,\ |\eta|\}, \quad v\in\mathcal{B}_v,$$

the spaces $\mathcal{B}_u$ and $\mathcal{B}_v$ become Banach spaces (α will be determined later). Moreover, we define the sets

$$\begin{aligned} U &:= \big\{u\in\mathcal{B}_u \mid |u|\le\tilde{\lambda}_R\big\},\\ V_\varepsilon &:= \big\{v=(x,\eta)\in\mathcal{B}_v \mid x\in X,\ \eta\in Y,\ |\eta-s_A(x,\vartheta)|<\varepsilon\big\}, \end{aligned}$$

where the positive constants $\tilde{\lambda}_R$ and ε will be fixed later. We consider the map

$$\widehat{P}_\vartheta\colon U\times V_\varepsilon \ni \begin{pmatrix}u\\v\end{pmatrix}=\begin{pmatrix}u\\x\\\eta\end{pmatrix}\longmapsto\begin{pmatrix}\bar{u}\\\bar{v}\end{pmatrix}=\begin{pmatrix}\bar{u}\\\bar{x}\\\bar{\eta}\end{pmatrix}=:\begin{pmatrix}\widehat{F}(u,v,\vartheta)\\\widehat{G}(u,v,\vartheta)\end{pmatrix}\in\mathcal{B}_u\times\mathcal{B}_v$$

defined by

$$\begin{aligned} \bar{u}\,\widetilde{G}(u,x,\eta,\vartheta) &= \widetilde{F}(u,x,\eta,\vartheta),\\ \bar{x} &= F(x,s_A(x,\vartheta),\vartheta),\\ \bar{\eta} &= G\big(x+u\big[\eta-s_A(x,\vartheta)\big],\eta,\vartheta\big), \end{aligned} \tag{5.8}$$

where $\widetilde{F}$ and $\widetilde{G}$ are given in (5.4), (5.5).

We show that the map $\widehat{P}_\vartheta$ is well defined and that it admits a repulsive positively invariant manifold $\{(u,v)\mid v\in V_\varepsilon,\ u=R(v,\vartheta)\}$ described by a function R of class C_b^{k-1}. Then this function $R\colon (v,\vartheta)=(x,\eta,\vartheta)\mapsto R(x,\eta,\vartheta)$ satisfies the invariance equation (5.7). It follows that the function w of the form (5.2) satisfies the invariance equation $F(w(x,\eta,\vartheta),\eta,\vartheta)=w(\bar{x},\bar{\eta},\vartheta)$ and hence describes the stable fibers of the map P_ϑ.

We determine $\tilde{\lambda}_R$ defining the set U such that the map $\widehat{P}_\vartheta$ is well defined for ε sufficiently small. Note that the Lipschitz constant λ_R of the stable fibers W^s_ϑ (cf. Theorem 4.1) is the smaller root of the quadratic equation $L_{21}\,p^2-(\Gamma_{11}-L_{22})\,p+L_{12}=0$. We choose $\tilde{\lambda}_R>\lambda_R$ such that

$$L_{21}\,\tilde{\lambda}_R^2-(\Gamma_{11}-L_{22})\,\tilde{\lambda}_R+L_{12}<0 \tag{5.9}$$

and such that Conditions CMA, CMA(k), CMB and CMAB($k-1$) hold if $\Delta = L_{21}\,\lambda_R$ is replaced by $\tilde{\Delta} = L_{21}\,\tilde{\lambda}_R > \Delta$.

In what follows we mostly suppress the parameter ϑ in the functions F, G, etc. Due to Hypothesis HMB we have

$$\left|G_2\big(x, s_A(x)\big)^{-1}\right| \le \frac{1}{\Gamma_{22}}.$$

We write $\widetilde{G}$ defined in (5.5) as follows:

$$\begin{aligned}
&\widetilde{G}(u, x, \eta) \\
&\quad = G_2\big(x, s_A(x)\big)\,\Big[\textstyle\int_0^1 G_2\big(x, s_A(x)\big)^{-1}\, G_2\big(x + \tau u[\eta - s_A(x)], s_A(x) \\
&\qquad\qquad\qquad\qquad\qquad\qquad + \tau[\eta - s_A(x)]\big)\,d\tau \\
&\qquad\qquad + G_2\big(x, s_A(x)\big)^{-1} \textstyle\int_0^1 G_1\big(x + \tau u[\eta - s_A(x)], s_A(x) \\
&\qquad\qquad\qquad\qquad\qquad\qquad + \tau[\eta - s_A(x)]\big)u\,d\tau\Big] \\
&\quad = G_2\big(x, s_A(x)\big)\,\big[I + G_2\big(x, s_A(x)\big)^{-1}\, G_1\big(x, s_A(x)\big)u + O(\varepsilon)\big].
\end{aligned}$$

We have $\left|G_2\big(x, s_A(x)\big)^{-1}\, G_1\big(x, s_A(x)\big)u\right| \le L_{21}\,\tilde{\lambda}_R/\Gamma_{22} = \tilde{\Delta}/\Gamma_{22} < 1$ since Condition CMB also holds for $\tilde{\Delta}$. Hence, $\widetilde{G}(u, x, \eta)$ is invertible for sufficiently small ε and the inverse may be estimated as

$$|\widetilde{G}(u, x, \eta)^{-1}| \le \frac{1}{\Gamma_{22} - L_{21}\tilde{\lambda}_R} + O(\varepsilon). \tag{5.10}$$

It follows that the map $\widehat{P}_\vartheta$ is well defined in $U \times V_\varepsilon$ for ε small enough.

We show that for sufficiently small ε the map $\widehat{P}_\vartheta$ satisfies the assumptions of Theorem 3.1 and hence admits a smooth repulsive positively invariant manifold of class C_b^{k-1} given as the graph $\widehat{N}_\vartheta = \{(u, v) \mid v = (x, \eta) \in V_\varepsilon,\ u = R(v)\}$ of some function R.

We first show that the map $\widehat{P}_\vartheta$ satisfies Hypothesis HM. To verify HM a) we have to show that $\bar{v} = \widehat{G}(u, v) \in V_\varepsilon$ holds for all $(u, v) \in U \times V_\varepsilon$. Obviously, we have

$$\bar{x} = F\big(x, s_A(x)\big) \in X.$$

Since P_ϑ of (5.1) satisfies Hypothesis HM we find

$$\bar{\eta} = G\big(x + u[\eta - s_A(x)], \eta\big) \in Y.$$

Moreover, we have

$$|\bar{\eta} - s_A(\bar{x})| = \left|G\big(x + u[\eta - s_A(x)], \eta\big) - G\big(x, s_A(x)\big)\right| \le (L_{22} + L_{21}\,\tilde{\lambda}_R)\,|\,\eta - s_A(x)|.$$

Since Condition CMA also holds for $\widetilde{\Delta} = L_{21}\tilde{\lambda}_R$ we have $|\bar{\eta} - s_A(\bar{x})| < \varepsilon$. Therefore $\bar{v} \in V_\varepsilon$ holds.

For HM b) to hold we have to verify that for every $\bar{u} \in U$, $v \in V_\varepsilon$ there is $u \in U$ such that $\widehat{F}(u,v) = \bar{u}$, i.e., we have to show that $\bar{u}\,\widetilde{G}(u,v) = \widetilde{F}(u,v)$ may be solved for $u \in U$. Writing $\widetilde{F}(u,v) = \widetilde{F}^1(u,v)\,u + \widetilde{F}^2(u,v)$, cf. (5.4), we have

$$\bar{u}\,\widetilde{G}(u,v) = \widetilde{F}^1(u,v)\,u + \widetilde{F}^2(u,v). \tag{5.11}$$

We solve this equation by iteration according to

$$\bar{u}\,\widetilde{G}(u^{(i)},v) = \widetilde{F}^1(u^{(i)},v)\,u^{(i+1)} + \widetilde{F}^2(u^{(i)},v). \tag{5.12}$$

For $x \in X$ the derivative $F_1\big(x, s_A(x)\big)$ is an isomorphism of $\mathcal{B}_x$ and satisfies

$$\big|F_1\big(x, s_A(x)\big)^{-1}\big| \le 1/\Gamma_{11},$$

cf. Lemma 3.2. Since $\widetilde{F}^1(u^{(i)},v) = F_1\big(x, s_A(x)\big) + O(\varepsilon)$ it follows that $\widetilde{F}^1(u^{(i)},v)$ is invertible for ε sufficiently small and that

$$|\widetilde{F}^1(u^{(i)},v)^{-1}| \le \frac{1}{\Gamma_{11}} + O(\varepsilon). \tag{5.13}$$

We show that the map $u^{(i)} \mapsto u^{(i+1)}$ takes U into itself if ε is sufficiently small. We show that $u^{(i)} \in U$ implies $u^{(i+1)} \in U$. Using the estimates

$$\big|\widetilde{F}^2(u^{(i)},v)\big| \le L_{12}, \quad |\widetilde{G}(u^{(i)},v)| \le L_{21}\,\tilde{\lambda}_R + L_{22}$$

and the estimate (5.13) we obtain from (5.12)

$$|u^{(i+1)}| \le \frac{L_{21}\,\tilde{\lambda}_R^2 + L_{22}\,\tilde{\lambda}_R + L_{12}}{\Gamma_{11}} + O(\varepsilon).$$

By our choice of $\tilde{\lambda}_R$, cf. (5.9), it follows that $|u^{(i+1)}| \le \tilde{\lambda}_R$ if ε is chosen sufficiently small. We show that the map $u^{(i)} \mapsto u^{(i+1)}$ is a contraction. From (5.12) we get

$$\begin{aligned}\widetilde{F}^1(u_1^{(i)},v)(u_1^{(i+1)} - u_2^{(i+1)}) &= \bar{u}\big(\widetilde{G}(u_1^{(i)},v) - \widetilde{G}(u_2^{(i)},v)\big)\\ &\quad - \big(\widetilde{F}^2(u_1^{(i)},v) - \widetilde{F}^2(u_2^{(i)},v)\big)\\ &\quad - \big(\widetilde{F}^1(u_1^{(i)},v) - \widetilde{F}^1(u_2^{(i)},v)\big)\,u_2^{(i+1)}.\end{aligned}$$

Since

$$|\widetilde{G}(u_1^{(i)},v) - \widetilde{G}(u_2^{(i)},v)| \le \big(L_{21} + O(\varepsilon)\big)\big|u_1^{(i)} - u_2^{(i)}\big|$$

and

$$|\widetilde{F}^j(u_1^{(i)},v) - \widetilde{F}^j(u_2^{(i)},v)| = O(\varepsilon)\,|u_1^{(i)} - u_2^{(i)}|, \quad j = 1,2,$$

we find

$$|u_1^{(i+1)} - u_2^{(i+1)}| \leq \left(\frac{L_{21}\,\tilde{\lambda}_R}{\Gamma_{11}} + O(\varepsilon) \right) |u_1^{(i)} - u_2^{(i)}|.$$

Since Condition CMR also holds for $\widetilde{\Delta} = L_{21}\tilde{\lambda}_R$ it follows that $L_{21}\tilde{\lambda}_R/\Gamma_{11} < 1$ and that the map $u^{(i)} \mapsto u^{(i+1)}$ is a contraction if ε is sufficiently small. Hence equation (5.11) has a unique solution $u \in U$.

To verify Hypothesis HM c) we determine Lipschitz constants $\widehat{L}_{12}$, $\widehat{L}_{21}$, $\widehat{L}_{22}$ and the "lower Lipschitz constant" $\widehat{\Gamma}_{11}$ of the map $\widehat{P}_\vartheta$.

We first determine $\widehat{L}_{22}$ and $\widehat{L}_{12}$. For $u \in U$, $v_i \in V_\varepsilon$, $i = 1, 2$, we set $\bar{v}_i = (\bar{x}_i, \bar{\eta}_i) := \widehat{G}(u, v_i)$ and $\bar{u}_i := \widehat{F}(u, v_i)$. From (5.8) we get

$$|\bar{x}_1 - \bar{x}_2| \leq (L_{11} + L_{12}\,\lambda_A)\,|x_1 - x_2|,$$

$$|\bar{\eta}_1 - \bar{\eta}_2| \leq L_{21}(1 + \tilde{\lambda}_R\,\lambda_A)|x_1 - x_2| + (L_{21}\,\tilde{\lambda}_R + L_{22})\,|\eta_1 - \eta_2|.$$

With the norm $|v| = \max\{\alpha|x|, |\eta|\}$ in the space $\mathcal{B}_v$ we find the estimate

$$|\bar{v}_1 - \bar{v}_2| \leq \max\left\{L_{11} + L_{12}\,\lambda_A,\ \frac{1}{\alpha}\,L_{21}(1 + \lambda_A\,\tilde{\lambda}_R) + L_{21}\,\tilde{\lambda}_R + L_{22}\right\} |v_1 - v_2|.$$

Since $L_{11} > L_{22}$ and since $L_{12}\,\lambda_A = \Delta < \widetilde{\Delta} = L_{21}\,\tilde{\lambda}_R$ we may choose α large enough such that

$$\frac{1}{\alpha}L_{21}(1 + \lambda_A\tilde{\lambda}_R) + \widetilde{\Delta} + L_{22} < \widetilde{\Delta} + L_{11}$$

and hence set

$$\widehat{L}_{22} := L_{11} + \widetilde{\Delta}. \tag{5.14}$$

We show that $\widehat{L}_{12}$ may be taken as

$$\widehat{L}_{12} = \text{const}\,(|D^2 F| + |D^2 G|), \tag{5.15}$$

where D^2F, D^2G denote the second derivatives of F and G. From (5.8) we get

$$(\bar{u}_1 - \bar{u}_2)\,\widetilde{G}(u, v_1) = \widetilde{F}(u, v_1) - \widetilde{F}(u, v_2) - \bar{u}_2\big(\widetilde{G}(u, v_1) - \widetilde{G}(u, v_2)\big).$$

The right-hand side of this equation may be estimated as (cf. equations (5.4), (5.5))

$$|\text{r.h.s.}| \leq \text{const}\,(|D^2F| + |D^2G|)\,|v_1 - v_2|.$$

Equation (5.15) follows using the estimate (5.10).

Next, we determine $\widehat{\Gamma}_{11}$ and $\widehat{L}_{21}$. For $u_i \in U$, $i = 1, 2$, $v \in V_\varepsilon$ we set $\bar{u}_i = \widehat{F}(u_i, v)$ and $\bar{v}_i = (\bar{x}_i, \bar{\eta}_i) = \widehat{G}(u_i, v)$. We show that $\widehat{\Gamma}_{11}$ may be taken as

$$\widehat{\Gamma}_{11} = \frac{\Gamma_{11} - \widetilde{\Delta}}{L_{22} + \widetilde{\Delta}} + O(\varepsilon). \tag{5.16}$$

From (5.8) we obtain

$$(\bar{u}_1 - \bar{u}_2)\,\widetilde{G}(u_1, v) = \widetilde{F}(u_1, v) - \widetilde{F}(u_2, v) - \bar{u}_2\big(\widetilde{G}(u_1, v) - \widetilde{G}(u_2, v)\big).$$

Since $\widetilde{G}$ is an isomorphism of $\mathcal{B}_y$ (cf. (5.10)) we may estimate

$$\begin{aligned}
|\bar{u}_1 - \bar{u}_2| &= \sup_{\eta \in \mathcal{B}_y \setminus \{0\}} \frac{|(\bar{u}_1 - \bar{u}_2)\eta|}{|\eta|} \\
&= \sup_{\xi \in \mathcal{B}_y \setminus \{0\}} \frac{\big|\big(\widetilde{F}(u_1, v) - \widetilde{F}(u_2, v) - \bar{u}_2\big(\widetilde{G}(u_1, v) - \widetilde{G}(u_2, v)\big)\big)\,\xi\big|}{|\widetilde{G}(u_1, v)\,\xi|} \\
&\geq \frac{\Gamma_{11} - L_{21}\,\tilde{\lambda}_R + O(\varepsilon)}{L_{22} + L_{21}\,\tilde{\lambda}_R}\,|u_1 - u_2|
\end{aligned}$$

yielding (5.16).

In order to obtain

$$\widehat{L}_{21} = \varepsilon\, L_{21} \tag{5.17}$$

we use (5.8) to find

$$\begin{aligned}
&|\bar{x}_1 - \bar{x}_2| = 0, \\
&|\bar{\eta}_1 - \bar{\eta}_2| \leq L_{21}\,|\eta - s_A(x, \vartheta)|\,|u_1 - u_2| \leq \varepsilon\, L_{21}\,|u_1 - u_2|.
\end{aligned}$$

Next we show that the Lipschitz constants with respect to the parameter ϑ satisfy

$$\widehat{L}_{13} = \text{const}\,(|D^2 F| + |D^2 G|), \tag{5.18}$$

$$\widehat{L}_{23} = \max\{\alpha\,(L_{13} + L_{12}\,\nu_A),\ L_{23} + \widetilde{\Delta}\,\nu_A\}. \tag{5.19}$$

For given $u \in U$, $v \in V_\varepsilon$ and $\vartheta_1, \vartheta_2 \in E$, let $\bar{u}_i := \widehat{F}(u, v, \vartheta_i)$, $\bar{v}_i = (\bar{x}_i, \bar{\eta}_i) := \widehat{G}(u, v, \vartheta_i)$, $i = 1, 2$. From (5.8) we obtain

$$(\bar{u}_1 - \bar{u}_2)\,\widetilde{G}(u, v, \vartheta_1) = \widetilde{F}(u, v, \vartheta_1) - \widetilde{F}(u, v, \vartheta_2) - \bar{u}_2\big(\widetilde{G}(u, v, \vartheta_1) - \widetilde{G}(u, v, \vartheta_2)\big).$$

As in the estimate of $\widehat{L}_{12}$ one gets an expression involving second derivatives of F and G. This leads to equation (5.18). Again using (5.8) we find

$$\begin{aligned}
|\bar{x}_1 - \bar{x}_2| &\leq (L_{12}\,\nu_A + L_{13})\,|\vartheta_1 - \vartheta_2|, \\
|\bar{\eta}_1 - \bar{\eta}_2| &\leq (L_{21}\,\tilde{\lambda}_R\,\nu_A + L_{23})\,|\vartheta_1 - \vartheta_2|.
\end{aligned}$$

The norm chosen in $\mathcal{B}_v$ leads to equation (5.19).

With equations (5.14), (5.15), (5.16), (5.17), (5.18) and (5.19) we have verified Hypothesis HM c).

We verify that Hypothesis HMR is satisfied with $u^* = 0$. From (5.4), (5.5) we find

$$|\widetilde{F}(u^*, v)| = \left| \int_0^1 F_2\big(x, s_A(x) + \tau[\eta - s_A(x)]\big) d\tau \right| \le L_{12}.$$

Using (5.10) we get from (5.8)

$$|\widehat{F}(u^*, v)| \le \frac{L_{12}}{\Gamma_{22} - \widetilde{\Delta}} + O(\varepsilon).$$

Hence, Hypothesis HMR is verified.

In order to apply Theorem 3.1 we have to verify Conditions CM, CMR, CMR$(k-1)$.

Condition CM: $2\sqrt{\widehat{L}_{12}\,\widehat{L}_{21}} < \widehat{\Gamma}_{11} - \widehat{L}_{22}$.

Taking into account equations (5.14), (5.15), (5.16) and (5.17) Condition CM has the form

$$O(\varepsilon^{1/2}) < \frac{\Gamma_{11} - \widetilde{\Delta}}{L_{22} + \widetilde{\Delta}} - (L_{11} + \widetilde{\Delta}) + O(\varepsilon). \tag{5.20}$$

Since Condition CMA(1) also holds for $\widetilde{\Delta}$ it follows that

$$\frac{\Gamma_{11} - \widetilde{\Delta}}{L_{22} + \widetilde{\Delta}} > 1. \tag{5.21}$$

(Note that Conditions CMA and CMA(k) imply that Condition CMA(j) holds for $j = 0, \dots, k$). Since Condition CMAB$(k-1)$ also holds for $\widetilde{\Delta}$, condition (5.20) is satisfied for ε sufficiently small.

Condition CMR:

$$\widehat{\Gamma}_{11} - \widehat{\Delta} > 1, \quad \text{where } \widehat{\Delta} = \frac{2\widehat{L}_{12}\,\widehat{L}_{21}}{\widehat{\Gamma}_{11} - \widehat{L}_{22} + \sqrt{(\widehat{\Gamma}_{11} - \widehat{L}_{22})^2 - 4\widehat{L}_{12}\,\widehat{L}_{21}}}.$$

From $\widehat{L}_{21} = \varepsilon L_{21}$ (cf. (5.17)) we get $\widehat{\Delta} = O(\varepsilon)$. Using (5.16) Condition CMR has the form

$$\frac{\Gamma_{11} - \widetilde{\Delta}}{L_{22} + \widetilde{\Delta}} + O(\varepsilon) > 1$$

and is satisfied by (5.21) for ε sufficiently small.

Condition CMR$(k-1)$: $\big(\widehat{L}_{22} + \widehat{\Delta}\big)^{k-1} < \widehat{\Gamma}_{11} - \widehat{\Delta}$.

Using $\widehat{\Delta} = O(\varepsilon)$ and equations (5.14), (5.16) yields that Condition CMR$(k-1)$ is of the form

$$\big(L_{11} + \widetilde{\Delta} + O(\varepsilon)\big)^{k-1} < \frac{\Gamma_{11} - \widetilde{\Delta}}{L_{22} + \widetilde{\Delta}} + O(\varepsilon).$$

Since Condition CMAB$(k-1)$ also holds for $\widetilde{\Delta}$ this condition is satisfied for ε sufficiently small.

From the definition of the map $\widehat{P}_\vartheta$ (cf. equations (5.8), (5.4), (5.5)) it follows that $\widehat{F}$ and $\widehat{G}$ are of class C_b^{k-1}. Hence, we have verified all assumptions of Theorem 3.1 for the map $\widehat{P}_\vartheta$. It follows that the map $\widehat{P}_\vartheta$ admits a repulsive positively invariant manifold $\{(u,v) \mid v \in V_\varepsilon,\ u = R(v,\vartheta)\}$ described by a function R of class C_b^{k-1}. This terminates the proof of Theorem 5.1. □

5.2 The unstable foliation

We consider the setting of Sections 1.1, 3.1 and 4.2 where the map P_ϑ of (4.9) (which is the map of (1.1)) admits a repulsive positively invariant manifold N_ϑ. In this section we make the assumption that N_ϑ is invariant. Then with the choice $\Omega_\vartheta = N_\vartheta$ Theorem 4.3 yields for $y \in Y$ the existence of unstable fibers with base points $(s_R(y,\vartheta), y)$ in N_ϑ. Every unstable fiber is given as $W_\vartheta^u(s_R(y,\vartheta), y) = \{(\xi,\eta) \mid \xi \in X,\ \eta = w_\vartheta^u(s_R(y,\vartheta), y, \xi)\}$. Under appropriate conditions the function $w\colon (y,\xi,\vartheta) \mapsto w(y,\xi,\vartheta) := w_\vartheta^u(s_R(y,\vartheta), y, \xi)$ is of class C_b^{k-1} provided the map P_ϑ is of class C_b^k. More precisely, the function w is of the form

$$w(y,\xi,\vartheta) = y + Q(y,\xi,\vartheta)\big[\xi - s_R(y,\vartheta)\big]$$

with $Q \in C_b^{k-1}\big(Y \times X \times E,\ \mathcal{L}_b(\mathcal{B}_x, \mathcal{B}_y)\big)$ where $\mathcal{L}_b$ denotes the space of bounded linear operators.

We need an additional condition:

Condition CMRB$(k-1)$

$$\frac{L_{22} + \Delta}{\Gamma_{11} - \Delta} < (\Gamma_{22} - \Delta)^{k-1}.$$

We state the smoothness result with respect to the base point for the unstable fibers. We omit the proof since it is similar to the one of Theorem 5.1.

Theorem 5.2. *Let the map P_ϑ of* (4.9) *satisfy Hypotheses HM, HMR, HMB and Conditions CM, CMR, CMR(k), $k > 1$, and CMB and CMRB$(k-1)$. Let F and G be of class C_b^k. Assume that the manifold $N_\vartheta = \{(x,y) \mid y \in Y,\ x = s_R(y,\vartheta)\}$ established in Theorem* 1.3 *is invariant under P_ϑ.*

Then there is a function w describing unstable fibers $W_\vartheta^u(s_R(y,\vartheta), y) = \{(\xi,\eta) \mid \xi \in X,\ \eta = w(y,\xi,\vartheta)\}$ with base point $(s_R(y,\vartheta), y) \in N_\vartheta$. The function w is of class C_b^{k-1} and is of the form

$$w(y,\xi,\vartheta) = y + Q(y,\xi,\vartheta)\big[\xi - s_R(y,\vartheta)\big]$$

with

$$Q \in C_b^{k-1}\big(Y \times X \times E, \mathcal{L}_b(\mathcal{B}_x, \mathcal{B}_y)\big).$$

Part II

Continuous Dynamical Systems – ODEs

In Part II we investigate continuous dynamical systems defined by some autonomous ordinary differential equation (ODE). We give assumptions on the ODE such that the dynamical system admits an invariant manifold. Invariant manifold results for ODEs are considered, e.g., in Perron [105], Hirsch, Pugh, Shub [55], Kelley [63], Carr [23], Fenichel [38]–[42], Knobloch, Kappel [72], Knobloch [69], Stuart, Humphries [125], Yi [130].

The ODEs we consider are of the form

$$\begin{aligned} \dot{x} &= f(x, y, \vartheta), \\ \dot{y} &= g(x, y, \vartheta), \end{aligned} \qquad (x, y) \in X \times Y \subset \mathbb{R}^m \times \mathbb{R}^n, \quad \vartheta \in E \subset \mathbb{R}^\ell$$

where ϑ is a parameter and where we assume that the flow is inflowing with respect to Y and outflowing with respect to X. Our approach is to consider the time-T map of the ODE. If the time-T map satisfies the assumptions made in Part I this map admits an invariant manifold which is also invariant under the ODE. This result is established in Chapter 6. In Chapter 7 we consider the time-T map for small T. We derive conditions on the vector field (f, g) such that the time-T map for sufficiently small T satisfies the assumptions made in Part I. It then follows that the ODE admits an invariant manifold. Similar to Part I we make assumptions on the Lipschitz constants of f and g and in addition on the logarithmic norms of the derivatives $\partial f/\partial x$ and $\partial g/\partial y$.

Chapter 6
A general result for the time-T map

Let $X \subset X' \subset \mathbb{R}^m$, $Y \subset \mathbb{R}^n$, $E \subset \mathbb{R}^\ell$ be nonempty open sets and let $f\colon X' \times Y \times E \to \mathbb{R}^m$ and $g\colon X' \times Y \times E \to \mathbb{R}^n$ be of class C_b^k, $k \geq 1$, i.e., the space of bounded, k-times continuously differentiable functions. We consider the autonomous system of ODEs

$$\begin{aligned} \dot{x} &= f(x, y, \vartheta), \\ \dot{y} &= g(x, y, \vartheta), \end{aligned} \tag{6.1}$$

and denote by $(\varphi(t; x, y, \vartheta),\ \psi(t; x, y, \vartheta))$ the solution with $\varphi(0; x, y, \vartheta) = x$, $\psi(0; x, y, \vartheta) = y$. We aim at the situation where the flow of (6.1) is outflowing with respect to X and inflowing with respect to Y. We make the following assumptions.

Hypothesis HD0

Let the closure of X satisfy $\bar{X} \subset X'$. Assume that there is $T > 0$ such that for all $(x, y, \vartheta) \in X \times Y \times E$ the solution $(\varphi(t; x, y, \vartheta), \psi(t; x, y, \vartheta))$ exists and remains in $X' \times Y$ for all $t \in [0, T]$ and assume that if x is on the boundary ∂X of X then $\varphi(t; x, y, \vartheta) \notin X$ for $t \in (0, T]$.

Hypothesis HDA

There is $y^* \in Y$ such that $g(\cdot, y^*, \cdot)$ is bounded.

Hypothesis HDR

There is $x^* \in X$ such that $f(x^*, \cdot, \cdot)$ is bounded.

For $(x, y) \in X \times Y$ we consider the time-T map of the differential equation (6.1)

$$P_\vartheta^T \colon \begin{pmatrix} x \\ y \end{pmatrix} \longmapsto \begin{pmatrix} \bar{x} \\ \bar{y} \end{pmatrix} = \begin{pmatrix} F(x, y, \vartheta) \\ G(x, y, \vartheta) \end{pmatrix} := \begin{pmatrix} \varphi(T; x, y, \vartheta) \\ \psi(T; x, y, \vartheta) \end{pmatrix}. \tag{6.2}$$

Note that for fixed $\vartheta \in E$ it holds that $P_\vartheta^T \colon X \times Y \to X' \times Y$ and that P_ϑ^T is invertible where the inverse is given by P_ϑ^{-T} (a consequence of the group property of the flow of (6.1) and the uniqueness of the solutions). We show that if the time-T map (6.2) admits a negatively invariant manifold M_ϑ then M_ϑ is a negatively invariant manifold of the ODE (6.1). Analogously, if the time-T map (6.2) admits a positively invariant manifold N_ϑ then N_ϑ is a positively invariant manifold of the ODE (6.1).

Theorem 6.1. *Let f and g in the differential equation* (6.1) *be of class C_b^k, $k \geq 1$, and let Hypothesis HD*0 *be satisfied. Assume that the map P_ϑ^T given in* (6.2) *satisfies Hypothesis HM.*

a) *Let Hypothesis HDA hold and assume that the constants* Γ_{11}, L_{12}, L_{21}, L_{22} *in Hypothesis HM* c) *satisfy Conditions CM, CMA and CMA*(k). *Then there is a function* $s_A\colon X \times E \to Y$ *of class* C_b^k *such that the following assertions hold for* $\vartheta \in E$.

i) *The set* $M_\vartheta = \{(x, y) \mid x \in X,\ y = s_A(x, \vartheta)\}$ *is negatively invariant under the differential equation* (6.1), *i.e., for* $(x, y) \in M_\vartheta$ *the solution* $(\varphi(t; x, y, \vartheta), \psi(t; x, y, \vartheta))$ *remains in* M_ϑ *for all* $t \in (-\infty, t^+)$ *where* $(-\infty, t^+)$ *is the interval of existence with respect to the set* $X \times Y$.

The function s_A *satisfies the invariance equation*

$$\left[\frac{\partial}{\partial x} s_A(x, \vartheta)\right] f(x, s_A(x, \vartheta), \vartheta) = g(x, s_A(x, \vartheta), \vartheta)$$

for $x \in X$.

ii) *The manifold* M_ϑ *satisfies all assertions of Theorem* 1.5 *for the map* (6.2).

b) *Let Hypothesis HDR hold and assume that the constants* Γ_{11}, L_{12}, L_{21}, L_{22} *in Hypothesis HM* c) *satisfy Conditions CM, CMR and CMR*(k). *Then there is a function* $s_R\colon Y \times E \to X$ *of class* C_b^k *such that the following assertions hold for* $\vartheta \in E$.

i) *The set* $N_\vartheta = \{(x, y) \mid y \in Y,\ x = s_R(y, \vartheta)\}$ *is positively invariant under the differential equation* (6.1), *i.e., for* $(x, y) \in N_\vartheta$ *the solution* $(\varphi(t; x, y, \vartheta), \psi(t;\ x, y, \vartheta))$ *remains in* N_ϑ *for all* $t \in (t^-, \infty)$ *where* (t^-, ∞) *is the interval of existence with respect to the set* $X \times Y$.

The function s_R *satisfies the invariance equation*

$$\left[\frac{\partial}{\partial y} s_R(y, \vartheta)\right] g(s_R(y, \vartheta), y, \vartheta) = f(s_R(y, \vartheta), y, \vartheta)$$

for $y \in Y$.

ii) *The manifold* N_ϑ *satisfies all assertions of Theorem* 1.3 *for the map* (6.2).

Proof. a) In order to show that Theorem 3.6 may be applied we have to show that the map (6.2) satisfies Hypothesis HMA. We show that $\psi(t; x, y^*, \vartheta)$ is bounded for $t \in [0, T]$, $x \in X$, $\vartheta \in E$, implying that $G(\cdot, y^*, \cdot)$ is bounded. We have

$$\psi(t; x, y^*, \vartheta) = y^* + \textstyle\int_0^t g(\varphi(\sigma; x, y^*, \vartheta), \psi(\sigma; x, y^*, \vartheta), \vartheta)\, d\sigma$$

and estimate

$$|\psi(t; x, y^*, \vartheta) - y^*| \le t|g(\cdot, y^*, \cdot)| + |g_y| \textstyle\int_0^t |\psi(\sigma; x, y^*, \vartheta) - y^*|\, d\sigma. \tag{6.3}$$

Applying Gronwall's lemma it follows that ψ is bounded for $y = y^*$ and $t \in [0, T]$. Thus, Hypothesis HMA is satisfied. Theorem 3.6 implies that the map P_ϑ^T admits a negatively invariant manifold M_ϑ of class C_b^k and that all assertions of Theorem 1.5 hold.

We show that M_ϑ is negatively invariant under the differential equation (6.1). Let $(x, y) \in M_\vartheta$ and define the set $\Lambda := \{(\varphi(t; x, y, \vartheta),\ \psi(t; x, y, \vartheta) \mid t \in (t^-, t^+)\} \subset X \times Y$ where (t^-, t^+) is the interval of existence with respect to $X \times Y$. It suffices to verify that $t^- = -\infty$ and that $\Lambda \subset M_\vartheta$. Assume $t^- > -\infty$. Then $\varphi(t^-; x, y, \vartheta) \in \partial X$ and there is $\Delta t \in (0, T]$ with $\varphi(t^- + \Delta t; x, y, \vartheta) \in X$ contradicting Hypothesis HD0. We justify that $\Lambda \subset X \times Y_0$ for some bounded Y_0. The manifold M_ϑ is bounded with respect to y. Hence, for $(x_j, y_j) := (\varphi(jT; x, y, \vartheta), \psi(jT; x, y, \vartheta))$ it holds that y_j is uniformly bounded for $j < t^+/T$. We obtain a uniform bound for $|\psi(t; x_j, y_j, \vartheta)|$, $t \in [0, T]$, $j < t^+/T$, by estimating $|\psi(t; x_j, y_j, \vartheta) - y^*|$ as in (6.3). Since $t^- = -\infty$ we have $P_\vartheta^T(\Lambda) \supset \Lambda$. Theorem 1.5 iv) implies $\Lambda \subset M_\vartheta$.

A solution $(x(t), y(t))$ on the manifold M_ϑ satisfies $y(t) = s_A(x(t), \vartheta)$. Taking the derivative with respect to t yields the invariance equation.

b) The proof for the repulsive case is completely analogous. □

Chapter 7
Invariant manifold results

Let $X \subset X' \subset \mathbb{R}^m$, $Y \subset \mathbb{R}^n$, $E \subset \mathbb{R}^\ell$ be nonempty open convex sets and let $f\colon X' \times Y \times E \to \mathbb{R}^m$ and $g\colon X' \times Y \times E \to \mathbb{R}^n$ be of class C_b^k, $k \geq 1$, i.e., the space of bounded, k-times continuously differentiable functions. We consider the autonomous system of ODEs

$$\begin{aligned} \dot{x} &= f(x, y, \vartheta), \\ \dot{y} &= g(x, y, \vartheta), \end{aligned} \tag{7.1}$$

and denote by

$$\bigl(\varphi(t; x, y, \vartheta), \psi(t; x, y, \vartheta)\bigr)$$

the solution with $\varphi(0; x, y, \vartheta) = x$, $\psi(0; x, y, \vartheta) = y$. For $(x, y) \in X \times Y$ we consider the time-τ map of the differential equation (7.1)

$$P_\vartheta^\tau\colon \begin{pmatrix} x \\ y \end{pmatrix} \longmapsto \begin{pmatrix} \bar{x} \\ \bar{y} \end{pmatrix} = \begin{pmatrix} F(x, y, \vartheta) \\ G(x, y, \vartheta) \end{pmatrix} := \begin{pmatrix} \varphi(\tau; x, y, \vartheta) \\ \psi(\tau; x, y, \vartheta) \end{pmatrix}. \tag{7.2}$$

We state hypotheses on the functions f and g of the ODE (7.1) such that for sufficiently small τ the time-τ map satisfies the assumptions made in Part I.

In this chapter we use the *logarithmic norm* of a square matrix defined as

$$\mu(A) := \lim_{\varepsilon \to 0+} \frac{|I + \varepsilon A| - 1}{\varepsilon}.$$

Some basic properties of the logarithmic norm may be found, e.g., in Ström [124]. In Section 7.1 we derive relations of the Lipschitz constants, hypotheses and conditions of the vector field (f, g) with those of the time-τ map for small τ. In Section 7.2 we prove results for attractive negatively invariant manifolds and in Section 7.3 we state the corresponding results for repulsive positively invariant manifolds without proof.

7.1 Auxiliary results

In this section we state auxiliary results needed in the next two sections. Consider the following hypotheses.

Hypothesis HD

Let $X \subset X' \subset \mathbb{R}^m$, $Y \subset \mathbb{R}^n$, $E \subset \mathbb{R}^\ell$ be nonempty open convex sets and let $f\colon X' \times Y \times E \to \mathbb{R}^m$ and $g\colon X' \times Y \times E \to \mathbb{R}^n$ be of class C_b^k, $k \geq 1$. Let $X \subset X'$ be such *that the closure of X satisfies $\bar{X} \subset X'$ and assume that there is $T > 0$ such that for all $(x, y, \vartheta) \in X \times Y \times E$ the solution $(\varphi(t; x, y, \vartheta), \psi(t; x, y, \vartheta))$ of the differential equation* (7.1) *remains in $X' \times Y$ for $t \in [0, T]$.*

a) *Let the flow of the differential equation* (7.1) *be inflowing with respect to Y, i.e., if Y has a boundary ∂Y, then it is piecewise of class C^1 and $n_Y(y) \cdot g(x, y, \vartheta) < 0$ for all $(x, y, \vartheta) \in X \times \partial Y \times E$, n_Y being an outer normal with respect to Y.*

b) *The flow of the differential equation* (7.1) *is outflowing with respect to X, i.e., if X has a boundary ∂X, then it is piecewise of class C^1 and $n_X(x) \cdot f(x, y, \vartheta) > 0$ for all $(x, y, \vartheta) \in \partial X \times Y \times E$, n_X being an outer normal with respect to X.*

c) i) *There are nonnegative constants ℓ_{12}, ℓ_{13}, ℓ_{21}, ℓ_{23} such that on the set $X \times Y \times E$,*

$$\left|\frac{\partial f}{\partial y}\right| \le \ell_{12}, \quad \left|\frac{\partial f}{\partial \vartheta}\right| \le \ell_{13},$$

$$\left|\frac{\partial g}{\partial x}\right| \le \ell_{21}, \quad \left|\frac{\partial g}{\partial \vartheta}\right| \le \ell_{23}.$$

ii) *There are constants $\ell_{22} < 0$ and γ_{11} such that the logarithmic norms of the Jacobians $-\partial f/\partial x$ and $\partial g/\partial y$ satisfy*

$$\mu\Big(-\frac{\partial f}{\partial x}\Big) \le -\gamma_{11} \quad \text{and} \quad \mu\Big(\frac{\partial g}{\partial y}\Big) \le \ell_{22}$$

on the set $X \times Y \times E$.

Hypothesis HDA

There is y^* such that $g(\cdot, y^*, \cdot)$ is bounded.

Hypothesis HDR

There is x^* such that $f(x^*, \cdot, \cdot)$ is bounded.

Hypothesis HDRF

There is a set $\Omega_\vartheta \subset X \times Y$ which is negatively invariant under the flow of the differential equation (7.1), *i.e., for $x, y \in \Omega_\vartheta$ the solution satisfies $(\varphi(t; x, y, \vartheta), \psi(t; x, y, \vartheta)) \in \Omega_\vartheta$ for all $t \le 0$.*

Hypothesis HDB

There is a constant γ_{22} such that the logarithmic norm of the Jacobian $-\partial g/\partial y$ satisfies

$$\mu\Big(-\frac{\partial g}{\partial y}\Big) \le -\gamma_{22}.$$

Hypothesis HDAB

There is a constant ℓ_{11} such that the logarithmic norm of the Jacobian $\partial f/\partial x$ satisfies

$$\mu\Big(\frac{\partial f}{\partial x}\Big) \le \ell_{11}.$$

Remark 7.1. In applications it is important to allow the constants γ_{11}, γ_{22} and ℓ_{11}, $\ell_{12}, \ell_{13}, \ell_{21}, \ell_{22}, \ell_{23}$ to depend on ϑ.

In the following lemma we give conditions on f and g leading to estimates of the "upper and lower Lipschitz constants" of the time-τ map (7.2).

Lemma 7.2. *Let the functions f and g in the differential equation* (7.1) *satisfy Hypothesis HD and let $\tau < T$.*

Then for $\vartheta \in E$ the following assertions hold for the Lipschitz constants L_{11}, L_{12}, L_{13}, L_{21}, L_{22}, L_{23} and the "lower Lipschitz constants" Γ_{11}, Γ_{22} of the functions F, G in (7.2) *for $x \in X$, $y \in Y$ and $\tau \to 0^+$:*

$$\text{i)} \quad \begin{aligned} L_{12} &= \tau\,\ell_{12} + o(\tau),\\ L_{13} &= \tau\,\ell_{13} + o(\tau),\\ L_{21} &= \tau\,\ell_{21} + o(\tau),\\ L_{23} &= \tau\,\ell_{23} + o(\tau); \end{aligned}$$

$$\text{ii)} \quad \begin{aligned} \Gamma_{11} &= 1 + \tau\,\gamma_{11} + o(\tau),\\ L_{11} &= 1 + \tau\,\ell_{11} + o(\tau),\\ \Gamma_{22} &= 1 + \tau\,\gamma_{22} + o(\tau),\\ L_{22} &= 1 + \tau\,\ell_{22} + o(\tau). \end{aligned}$$

Proof. i) By the smooth dependence of the solutions on initial conditions and on parameters it follows that the functions F and G in equation (7.2) are of class C_b^k, $k \geq 1$. We determine the Lipschitz constants of F and G. The functions φ and ψ, cf. (7.2), satisfy the integral equations

$$\varphi(\tau; x, y, \vartheta) = x + \int_0^\tau f(\varphi(t; x, y, \vartheta), \psi(t; x, y, \vartheta), \vartheta)\, dt,$$

$$\psi(\tau; x, y, \vartheta) = y + \int_0^\tau g(\varphi(t; x, y, \vartheta), \psi(t; x, y, \vartheta), \vartheta)\, dt.$$

Using the notations $\varphi^i(t) := \varphi(t; x_i, y_i, \vartheta_i)$ for $(x_i, y_i, \vartheta_i) \in X \times Y \times E$, $i = 1, 2$, and analogously for $\psi^i(t)$, we have

$$\begin{aligned} \varphi^1(\tau) - \varphi^2(\tau) = x_1 - x_2 &+ \int_0^\tau [f(\varphi^1(t), \psi^1(t), \vartheta_1) - f(\varphi^2(t), \psi^1(t), \vartheta_1)]\, dt\\ &+ \int_0^\tau [f(\varphi^2(t), \psi^1(t), \vartheta_1) - f(\varphi^2(t), \psi^2(t), \vartheta_1)]\, dt \qquad (7.3)\\ &+ \int_0^\tau [f(\varphi^2(t), \psi^2(t), \vartheta_1) - f(\varphi^2(t), \psi^2(t), \vartheta_2)]\, dt \end{aligned}$$

and

$$\begin{aligned}\psi^1(\tau) - \psi^2(\tau) = y_1 - y_2 &+ \int_0^\tau [g(\varphi^1(t), \psi^1(t), \vartheta_1) - g(\varphi^2(t), \psi^1(t), \vartheta_1)]\, dt \\ &+ \int_0^\tau [g(\varphi^2(t), \psi^1(t), \vartheta_1) - g(\varphi^2(t), \psi^2(t), \vartheta_1)]\, dt \\ &+ \int_0^\tau [g(\varphi^2(t), \psi^2(t), \vartheta_1) - g(\varphi^2(t), \psi^2(t), \vartheta_2)]\, dt.\end{aligned} \tag{7.4}$$

Applying Gronwall's lemma we obtain for some constant K and small τ

$$\begin{aligned}&|\varphi^1(\tau) - \varphi^2(\tau)| + |\psi^1(\tau) - \psi^2(\tau)| \\ &\quad \le |x_1 - x_2| + |y_1 - y_2| + K\tau\big(|x_1 - x_2| + |y_1 - y_2| + |\vartheta_1 - \vartheta_2|\big).\end{aligned}$$

Using this estimate on the right-hand side of equations (7.3) and (7.4) we find for τ small enough that

$$\begin{aligned}\varphi^1(\tau) - \varphi^2(\tau) &= x_1 - x_2 + O\big(\tau(|x_1 - x_2| + |y_1 - y_2| + |\vartheta_1 - \vartheta_2|)\big), \\ \psi^1(\tau) - \psi^2(\tau) &= y_1 - y_2 + O\big(\tau(|x_1 - x_2| + |y_1 - y_2| + |\vartheta_1 - \vartheta_2|)\big).\end{aligned}$$

Again from equations (7.3), (7.4) we obtain for τ small and for some $\bar{K} > 0$ the estimates

$$\begin{aligned}|\varphi^1(\tau) - \varphi^2(\tau)| \le{}& \Big| I + \int_0^\tau \int_0^1 f_1(t, \sigma)\, d\sigma dt \Big|\, |x_1 - x_2| \\ &+ \tau \operatorname{Lip}_y(f)\, |y_1 - y_2| + \tau \operatorname{Lip}_\vartheta(f)\, |\vartheta_1 - \vartheta_2| \\ &+ \bar{K}\tau^2(|x_1 - x_2| + |y_1 - y_2| + |\vartheta_1 - \vartheta_2|), \\ |\psi^1(\tau) - \psi^2(\tau)| \le{}& \Big| I + \int_0^\tau \int_0^1 g_2(t, \sigma)\, d\sigma dt \Big|\, |y_1 - y_2| \\ &+ \tau \operatorname{Lip}_x(g)\, |x_1 - x_2| + \tau \operatorname{Lip}_\vartheta(g)\, |\vartheta_1 - \vartheta_2| \\ &+ \bar{K}\tau^2(|x_1 - x_2| + |y_1 - y_2| + |\vartheta_1 - \vartheta_2|),\end{aligned} \tag{7.5}$$

where

$$f_1(t, \sigma) := \frac{\partial}{\partial x}\, f(\varphi^2(t) + \sigma(\varphi^1(t) - \varphi^2(t)), \psi^1(t), \vartheta_1),$$

$$g_2(t, \sigma) := \frac{\partial}{\partial y}\, g(\varphi^2(t), \psi^2(t) + \sigma(\psi^1(t) - \psi^2(t)), \vartheta_1),$$

and where $\mathrm{Lip}_y(f)$ is the Lipschitz constant of f with respect to y, etc. From (7.5) we find for $x_1 = x_2, \vartheta_1 = \vartheta_2$ that $\mathrm{Lip}_y(F) = \tau \,\mathrm{Lip}_y(f) + o(\tau)$, $\tau \to 0$, which yields the first estimate of i) and similarly for the other three estimates.

ii) We estimate

$$
\begin{aligned}
\Big| I + \int_0^\tau \int_0^1 f_1(t,\sigma)\, d\sigma dt \Big| &= \Big| \frac{1}{\tau}\Big[\int_0^\tau \int_0^1 (I + \tau (f_1(t,\sigma))\, d\sigma\, dt \Big]\Big| \\
&\le 1 + \int_0^\tau \int_0^1 \frac{|I + \tau\, f_1(t,\sigma)| - 1}{\tau}\, d\sigma\, dt \qquad (7.6) \\
&= 1 + \int_0^\tau \int_0^1 (\ell_{11} + v(t,\sigma,\tau)) d\sigma\, dt,
\end{aligned}
$$

where the remainder $v(t,\sigma,\tau) \to 0$ for $\tau \to 0$. We find from equations (7.5) and (7.6) for $y_1 = y_2$, $\vartheta_1 = \vartheta_2$ that $\mathrm{Lip}_x(F) = 1 + \tau\, \ell_{11} + o(\tau)$ as $\tau \to 0$. In exactly the same way one finds $\mathrm{Lip}_y(G) = 1 + \tau\, \ell_{22} + o(\tau)$, $\tau \to 0$. This proves the second and the fourth estimate of ii).

Similarly as for (7.5) we obtain

$$
\begin{aligned}
&|\varphi(\tau; x_1, y, \vartheta) - \varphi(\tau; x_2, y, \vartheta)| \\
&\quad = \Big| \Big(I + \int_0^\tau \int_0^1 f_1(t,\sigma)\, d\sigma\, dt \Big)(x_1 - x_2) + R(\tau; x_1, x_2, y, \vartheta) \Big|
\end{aligned}
$$

with $R = O(\tau^2\, |x_1 - x_2|)$. Using the inequality $|(I + B)\, u + v| \ge (2 - |I - B|)|u| - |v|$ we find

$$
\begin{aligned}
&|\varphi(\tau; x_1, y, \vartheta) - \varphi(\tau; x_2, y, \vartheta)| \\
&\quad \ge \Big[2 - \frac{|\int_0^\tau \int_0^1 (I - \tau f_1(t,\sigma)) d\sigma dt|}{\tau} \Big] |x_1 - x_2| - |R| \\
&\quad = \Big[1 - \frac{\int_0^\tau \int_0^1 (|I - \tau f_1(t,\sigma)| - 1) d\sigma dt}{\tau} \Big] |x_1 - x_2| - |R| \\
&\quad = \Big[1 - \int_0^\tau \int_0^1 (-\gamma_{11} + w(t,\sigma,\tau)) d\sigma\, dt \Big] |x_1 - x_2| - |R| \\
&\quad \le [1 + \tau\gamma_{11} + o(\tau)] |x_1 - x_2|, \quad \tau \to 0.
\end{aligned}
$$

Hence, we have for the "lower Lipschitz constant" of F with respect to x that $\Gamma_{11} = 1 + \tau\gamma_{11} + o(\tau)$, $\tau \to 0$. Analogously, one finds for the "lower Lipschitz constant" of G with respect to y that $\Gamma_{22} = 1 + \tau\gamma_{22} + o(\tau)$, $\tau \to 0$. This implies the first and the third estimate of ii). □

In the next lemma we state that our hypotheses on the vector field (f, g) of the differential equation (7.1) imply the corresponding hypotheses on the time-τ map (7.2).

Lemma 7.3. *Let the functions f and g in (7.1) satisfy Hypothesis HD.*

Then there is $\tau_0 \in (0, T]$ such that for $\tau \leq \tau_0$ the following assertions hold for the functions F and G in the time-τ map P_ϑ^τ of (7.2).

i) *F and G satisfy Hypothesis HM.*

ii) *If f and g satisfy Hypothesis HDA, then F and G satisfy Hypothesis HMA.*

iii) *If f and g satisfy Hypothesis HDR, then F and G satisfy Hypothesis HMR.*

iv) *If f and g satisfy Hypothesis HDRF, then F and G satisfy Hypothesis HMRF.*

v) *If f and g satisfy Hypothesis HDB, then F and G satisfy Hypothesis HMB.*

vi) *If f and g satisfy Hypothesis HDAB, then F and G satisfy Hypothesis HMAB.*

Proof. In this proof we skip the dependence on the parameter ϑ, for short.

i) Hypothesis HM a) is satisfied since the flow of the differential equation (7.1) is inflowing with respect to Y by Hypothesis HD a). By hypothesis HD b) the flow of (7.1) is outflowing with respect to X. Therefore, since X has finite dimension, the inclusion $F(X, y) \supset X$ holds, implying Hypothesis HM b). Hypothesis HM c) is a consequence of Lemma 7.2.

ii) We have

$$\begin{aligned}\psi(\tau; x, y) = y + \int_0^\tau \Big\{ & g(\varphi(t; x, y), y) \\ & + \int_0^1 g_2(\varphi(t; x, y), y + \sigma[\psi(t; x, y) - y])\, d\sigma\, [\psi(t; x, y) - y]\Big\}\, dt.\end{aligned} \tag{7.7}$$

For $y = y^*$ Hypothesis HDA and $g \in C_b^k$ imply that for some positive constants K_0, K_1,

$$|\psi(\tau; x, y^*) - y^*| \leq K_0 \tau + K_1 \int_0^\tau |\psi(t; x, y^*) - y^*|\, dt.$$

Using Gronwall's lemma we conclude that $\psi(\tau; \cdot, y^*)$ is bounded for $\tau \leq T$ fixed.

iii) The proof is completely analogous to the proof of assertion ii).

iv) The map P_ϑ^τ of (7.2) is invertible and $(P_\vartheta^\tau)^{-1} = P_\vartheta^{-\tau}$. Hypothesis HDRF implies that $P_\vartheta^{-\tau}(\Omega_\vartheta) \subset \Omega_\vartheta$. This implies Hypothesis HMRF.

v), vi) Hypotheses HMB and HMAB are consequences of Hypotheses HDB, HDAB and of Lemma 7.2. □

In order to apply the invariant manifold results of Part I the Lipschitz constants of the time-τ map (7.2) have to satisfy appropriate conditions. We give conditions on the Lipschitz constants of the vector field (f, g) of the ODE (7.1) and on the logarithmic norms of the derivatives of f and g such that the corresponding conditions of the time-τ map hold.

Condition CD

$$2\sqrt{\ell_{12}\ell_{21}} < \gamma_{11} - \ell_{22}.$$

Condition CDA

$$\ell_{22} + \delta < 0,$$

where

$$\delta = \frac{2\ell_{12}\ell_{21}}{\gamma_{11} - \ell_{22} + \sqrt{(\gamma_{11} - \ell_{22})^2 - 4\ell_{12}\ell_{21}}}.$$

Condition CDA(k)

$$\ell_{22} + \delta < k(\gamma_{11} - \delta).$$

Condition CDR

$$0 < \gamma_{11} - \delta.$$

Condition CDR(k)

$$k(\ell_{22} + \delta) < \gamma_{11} - \delta.$$

Condition CDAB$(k-1)$

$$\gamma_{11} - \ell_{22} - 2\delta > (k-1)(\ell_{11} + \delta).$$

Lemma 7.4. *Let the functions f and g in (7.1) satisfy Hypothesis HD. Let L_{11}, L_{12}, L_{13}, L_{21}, L_{22}, L_{23} be the Lipschitz constants of the functions F and G of the time-τ map (7.2) and let Γ_{11}, Γ_{22} be the "lower Lipschitz constants".*

Then there is $\tau_1 \leq \tau_0$, with τ_0 of Lemma 7.3, such that the following assertions hold for $\tau \leq \tau_1$.

i) *Condition CD implies Condition CM.*

ii) *Condition CDA implies Condition CMA.*

iii) *Condition CDA(k) implies Condition CMA(k).*

iv) *Condition CDR implies Condition CMR*

v) *Condition CDR(k) implies Condition CMR(k).*

vi) *Condition CMB holds.*

vii) *Condition CDAB$(k-1)$ implies Condition CMAB$(k-1)$.*

Proof. The assertions are easily verified by means of Lemma 7.2. We give the details for assertion v) only. We first express the constant Δ of Condition CMR(k) in terms of the Lipschitz constants of the functions f and g. We have

$$\begin{aligned}
\Delta &= \frac{2L_{12}L_{21}}{\Gamma_{11} - L_{22} + \sqrt{(\Gamma_{11} - L_{22})^2 - 4L_{12}L_{21}}} \\
&= \frac{2(\tau\ell_{12} + o(\tau))(\tau\ell_{21} + o(\tau))}{\tau(\gamma_{11} - \ell_{22}) + o(\tau) + \sqrt{\tau^2(\gamma_{11} - \ell_{22})^2 + 4\tau^2\,\ell_{12}\,\ell_{21} + \tau\, o(\tau)}} \\
&= \tau\Big(\frac{2\ell_{12}\ell_{21}}{\gamma_{11} - \ell_{22} + \sqrt{(\gamma_{11} - \ell_{22})^2 + 4\ell_{12}\ell_{21}}} + o(1)\Big) \\
&= \tau(\delta + o(1))
\end{aligned}$$

for $\tau \to 0$. We now estimate

$$\begin{aligned}
(L_{22} + \Delta)^k - (\Gamma_{11} - \Delta) &= (1 + \tau(\ell_{22} + \delta) + o(\tau))^k - (1 + \tau(\gamma_{11} - \delta) + o(\tau)) \\
&= k\tau(\ell_{22} + \delta) - \tau(\gamma_{11} - \delta) + o(\tau).
\end{aligned}$$

This expression is negative for sufficiently small τ if Condition CDR(k) is satisfied. □

7.2 Attractive negatively invariant manifolds

In Section 7.1 the connection between the ODE (7.1) and its time-τ map (7.2) is established. This allows to transfer the results of Part I on invariant manifolds for maps to ODEs. In this section we prove four theorems on attractive invariant manifolds for ODEs.

Theorem 7.5 (Existence and smoothness). *Let the differential equation* (7.1) *satisfy Hypotheses HD, HDA and assume that the constants* γ_{11}, ℓ_{12}, ℓ_{21}, ℓ_{22} *satisfy Conditions CD, CDA and CDA*(k), $k \geq 1$.

Then there is a function $s_A\colon X \times E \to Y$ *of class* C_b^k *such that the following assertions hold for* $\vartheta \in E$.

i) *The set* $M_\vartheta = \{(x, y) \mid x \in X,\ y = s_A(x, \vartheta)\}$ *is a negatively invariant manifold of* (7.1), *i.e., if* $(x, y) \in M_\vartheta$ *then the solution* $(\varphi(t; x, y, \vartheta), \psi(t; x, y, \vartheta))$ *of* (7.1) *remains in* M_ϑ *for* $t \leq 0$. *The function* s_A *satisfies the invariance equation*

$$\left[\frac{\partial}{\partial x} s_A(x, \vartheta)\right] f(x, s_A(x, \vartheta), \vartheta) = g(x, s_A(x, \vartheta), \vartheta) \tag{7.8}$$

for $x \in X$.

ii) *The function s_A is bounded and uniformly λ_A-Lipschitz continuous with respect to x and uniformly ν_A-Lipschitz continuous with respect to ϑ where*

$$\lambda_A = \frac{2\ell_{21}}{\gamma_{11} - \ell_{22} + \sqrt{(\gamma_{11} - \ell_{22})^2 - 4\ell_{12}\ell_{21}}}, \qquad \nu_A = \frac{\ell_{23} + \ell_{13}\lambda_A}{-(\ell_{22} + \ell_{12}\lambda_A)}.$$

iii) *The manifold M_ϑ is uniformly attractive. More precisely, every solution $(x(t), y(t))$ of* (7.1) *with $(x(0), y(0)) \in X \times Y$ satisfies*

$$|y(t) - s_A(x(t), \vartheta)| \le e^{\alpha_A t} \, |y(0) - s_A(x(0), \vartheta)|$$

with

$$\alpha_A = \ell_{22} + \delta < 0, \quad \delta = \frac{2\ell_{12}\ell_{21}}{\gamma_{11} - \ell_{22} + \sqrt{(\gamma_{11} - \ell_{22})^2 - 4\ell_{12}\ell_{21}}} = \ell_{12}\lambda_A$$

for all $t \ge 0$ as long as $x(t) \in X$.

iv) *Let $(x(t), y(t))$ be a solution of* (7.1) *with $(x(0), y(0)) \in X \times Y$ satisfying $(x(t), y(t)) \in X \times Y_0$, $t \le 0$, for some bounded $Y_0 \subset Y$. Then $(x(t), y(t))$ lies in M_ϑ as long as $x(t) \in X$.*

v) *If there is a map $\kappa \colon X \to X$ such that f and g are κ-invariant then s_A is κ-invariant, i.e., if for all $(x, y) \in X \times Y$*

$$f(\kappa x, y, \vartheta) = f(x, y, \vartheta),$$
$$g(\kappa x, y, \vartheta) = g(x, y, \vartheta)$$

then $s_A(\kappa x, \vartheta) = s_A(x, \vartheta)$ holds.

vi) *If the function g has the form $g(x, y, \vartheta) = B(x, y, \vartheta)y + \hat{g}(x, y, \vartheta)$ and if for all $x \in X$ the estimate $\mu\big(B(x, s_A(x, \vartheta), \vartheta)\big) \le -b < 0$ holds then*

$$|s_A(x, \vartheta)| \le \frac{1}{b} \sup_{x \in X} |\hat{g}(x, s_A(x, \vartheta), \vartheta)|.$$

Remark 7.6. In the case $\gamma_{11} \le 0$, which may occur in applications, Condition CD implies Condition CDA.

Proof. i) Lemma 7.3 implies that for τ sufficiently small the time-τ map (7.2) satisfies Hypotheses HM and HMA. Lemma 7.4 implies that the "upper and lower Lipschitz constants" of F and G of the time-τ map (7.2) satisfy Conditions CM, CMA and CMA(k). It follows from Theorem 1.5 that there is a function s_A such that the set M_ϑ is invariant under the time-τ map (7.2). Since Hypothesis HD implies Hypothesis HD0, we may apply Theorem 6.1 a) implying assertion i) of Theorem 7.5. Moreover, all assertions of Theorem 1.5 hold for the time-τ map. These assertions are used in what follows.

ii) The function s_A defining M_ϑ is $\lambda_A(\tau)$-Lipschitz continuous with respect to x and $\nu_A(\tau)$-Lipschitz continuous with respect to ϑ with

$$\begin{aligned}\lambda_A(\tau) &= \frac{2L_{21}}{\Gamma_{11} - L_{22} + \sqrt{(\Gamma_{11} - L_{22})^2 - 4L_{12}L_{21}}} \\ &= \frac{2\ell_{21}}{\gamma_{11} - \ell_{22} + \sqrt{(\gamma_{11} - \ell_{22})^2 - 4\ell_{12}\,\ell_{21}}} + o(1), \quad \tau \to 0, \qquad (7.9) \\ &= \lambda_A + o(1), \quad \tau \to 0,\end{aligned}$$

and

$$\nu_A(\tau) = \frac{L_{23} + L_{13}\,\lambda_A(\tau)}{1 - L_{22} - L_{12}\,\lambda_A(\tau)} = \frac{\ell_{23} + \ell_{13}\,\lambda_A}{-\ell_{22} - \ell_{12}\,\lambda_A} + o(1) = \nu_A + o(1), \quad \tau \to 0.$$

Taking the limit $\tau \to 0$ proves assertion ii).

iii) We take t such that $x(t) := \varphi(t; x, y, \vartheta) \in X$ and we take N large enough such that $\tau := t/N < \tau_1$ with τ_1 as in Lemma 7.4. Assertion iii) of Theorem 1.5 then implies that

$$|y(t) - s_A(x(t), \vartheta)| \le \chi_A(\tau)^N\, |y(0) - s_A(x(0), \vartheta)|$$

with

$$\chi_A(\tau) := L_{22} + L_{12}\,\lambda_A(\tau) = 1 + \tau(\ell_{22} + \ell_{12}\,\lambda_A(\tau)) + o(\tau) < 1, \quad \tau \to 0. \quad (7.10)$$

Hence,

$$\begin{aligned}\chi_A(\tau)^N &= e^{N \log \chi_A(\tau)} = e^{N\tau(\ell_{22}+\delta) + N\,o(\tau)} \\ &= e^{t(\ell_{22}+\delta)} + o(1), \quad N \to \infty,\end{aligned}$$

proves assertion iii).

iv) Define $\Lambda := \{(x(j\tau_1), y(j\tau_1)) \mid j \in \mathbb{Z},\ j \le 0\}$ with τ_1 given in Lemma 7.4. Note that $\Lambda \subset X \times Y_0$ and $P_\vartheta^{\tau_1}(\Lambda) \supset \Lambda$. Theorem 1.5 iv) implies that $\Lambda \subset M_\vartheta$. Now, the assertion follows from Theorem 6.1 a).

v) The function s_A satisfies the invariance equation (7.8). Using the κ-invariance of f and g we find

$$s_{A,x}(\kappa x, \vartheta)\, f(x, s_A(\kappa x, \vartheta), \vartheta) = g(x, s_A(\kappa x, \vartheta), \vartheta).$$

Hence, the function $\tilde{s}_A(x, \vartheta) := s_A(\kappa x, \vartheta)$ obeys the invariance equation and by Theorem 7.8 proved later in this chapter it holds that $\tilde{s}_A(x, \vartheta) = s_A(x, \vartheta)$ for $x \in X, \vartheta \in E$.

vi) We omit the dependence on the parameter ϑ for simplicity. The function s_A satisfies the invariance equation for the time-τ map (7.2), $\tau \le \tau_1$, with τ_1 as in Lemma 7.4,

$$G(x, s_A(x)) = s_A(F(x, s_A(x))), \; x \in X \text{ with } F(x, s_A(x)) \in X,$$

as given in Theorem 1.5. We rewrite this equations as

$$s_A(x) + \int_0^\tau \big[B(\varphi(t), \psi(t))\, \psi(t) + \hat{g}(\varphi(t), \psi(t))\big]\, dt = s_A\big(x + \int_0^\tau f(\varphi(t), \psi(t))\, dt\big)$$

where $(\varphi(t), \psi(t))$ is the solution of the ODE (7.1) with $(\varphi(0), \psi(0)) = (x, s_A(x))$. Using $\varphi(t) = x + O(t)$, $\psi(t) = s_A(x) + O(t)$ we get

$$\begin{aligned} s_A(x + O(\tau)) &= s_A(x) + \int_0^\tau \big(B(x, s_A(x)) s_A(x) + \hat{g}(x, s_A(x)) + O(t)\big)\, dt \\ &= \big(I + \tau\, B(x, s_A(x))\big)\, s_A(x) + \tau \hat{g}(x, s_A(x)) + O(\tau^2). \end{aligned}$$

Taking norms and then taking the supremum with respect to x first on the right-hand side and then on the left-hand side we obtain

$$\begin{aligned} |s_A| &\le \Big(1 + \tau \sup_x \frac{|I + \tau\, B(x, s_A(x))| - 1}{\tau}\Big) |s_A| + \tau \sup_x |\hat{g}(x, s_A(x))| + O(\tau^2) \\ &= (1 + \tau\, \mu(B) + o(\tau))\, |s_A| + \tau \sup_x |\hat{g}(x, s_A(x))| + O(\tau^2). \end{aligned}$$

Taking into account that $\mu(B) \le -b$ we get for $\tau \to 0$,

$$b|s_A| \le \sup_x |\hat{g}(x, s_A(x))|. \qquad \square$$

Next we formulate a perturbation result. We consider the ODE

$$\begin{aligned} \dot{x} &= \bar{f}(x, y, \bar{\vartheta}), \\ \dot{y} &= \bar{g}(x, y, \bar{\vartheta}) \end{aligned} \tag{7.11}$$

as a perturbation of the differential equation (7.1). For fixed ϑ, $\bar{\vartheta}$ we define

$$\begin{aligned} \delta_1 &:= \sup_{(x,y) \in X \times Y} |f(x, y, \vartheta) - \bar{f}(x, y, \bar{\vartheta})|, \\ \delta_2 &:= \sup_{(x,y) \in X \times Y} |g(x, y, \vartheta) - \bar{g}(x, y, \bar{\vartheta})|. \end{aligned} \tag{7.12}$$

Theorem 7.7 (Perturbation). *Fix $\vartheta, \bar{\vartheta} \in E$ and let the differential equations* (7.1), (7.11) *satisfy Hypotheses HD, HDA and assume that the constants γ_{11}, ℓ_{12}, ℓ_{21}, ℓ_{22} and $\bar{\gamma}_{11}$, $\bar{\ell}_{12}$, $\bar{\ell}_{21}$, $\bar{\ell}_{22}$, respectively, satisfy Conditions CD and CDA. Let δ_1, δ_2 be defined as in* (7.12).

Then for the functions s_A and $\bar{s}_A$ of Theorem 7.5 *defining the negatively invariant manifolds M_ϑ, $\bar{M}_{\bar{\vartheta}}$, respectively, the estimate*

$$|s_A(x, \vartheta) - \bar{s}_A(x, \bar{\vartheta})| \le \frac{1}{-(\ell_{22} + \ell_{12}\lambda_A)} (\lambda_A \delta_1 + \delta_2)$$

holds for λ_A defined as in Theorem 7.5.

Proof. We apply Theorem 2.1 to the time-τ map (7.2) for $\tau \leq \tau_1$, with τ_1 as in Lemma 7.4. We have to estimate $|\bar{F}(x,y,\bar{\vartheta}) - F(x,y,\vartheta)|$ and $|\bar{G}(x,y,\bar{\vartheta}) - G(x,y,\vartheta)|$. We obtain

$$\begin{aligned} |\bar{\varphi}(\tau) - \varphi(\tau)| &\leq \delta_1 \tau + \textstyle\int_0^\tau \big(|f_x|\,|\bar{\varphi}(t) - \varphi(t)| + |f_y|\,|\bar{\psi}(t) - \psi(t)|\big)\,dt, \\ |\bar{\psi}(\tau) - \psi(\tau)| &\leq \delta_2 \tau + \textstyle\int_0^\tau \big(|g_x|\,|\bar{\varphi}(t) - \varphi(t)| + |g_y|\,|\bar{\psi}(t) - \psi(t)|\big)\,dt, \end{aligned} \tag{7.13}$$

where we have used the notation $\bar{\varphi}(t) := \bar{\varphi}(t; x, y, \bar{\vartheta})$ etc. Adding the two equations and applying Gronwall's lemma yields the estimate

$$|\bar{\varphi}(\tau) - \varphi(\tau)| + |\bar{\psi}(\tau) - \psi(\tau)| \leq (\delta_1 + \delta_2)\,\tau + o(\tau), \quad \tau \to 0.$$

Inserting this estimate into (7.13) yields

$$\begin{aligned} |\bar{\varphi}(\tau) - \varphi(\tau)| &= |\bar{F}(x,y,\bar{\vartheta}) - F(x,y,\vartheta)| \leq \delta_1 \tau + o(\tau), \quad \tau \to 0, \\ |\bar{\psi}(\tau) - \psi(\tau)| &= |\bar{G}(x,y,\bar{\vartheta}) - G(x,y,\vartheta)| \leq \delta_2 \tau + o(\tau), \quad \tau \to 0. \end{aligned}$$

From Theorem 2.1 we conclude, taking into account Lemma 7.2, that

$$|\bar{s}_A(x,\bar{\vartheta}) - s_A(x,\vartheta)| \leq \frac{(\lambda_A \delta_1 + \delta_2)\tau + o(\tau)}{-(\ell_{22} + \ell_{12}\lambda_A)\tau + o(\tau)}, \quad \tau \to 0,$$

where we have used the formulas (7.9) and (7.10). Taking the limit $\tau \to 0$ completes the proof. □

We state an approximation result. The function s_A describing the manifold M_ϑ in Theorem 7.5 satisfies the invariance equation (7.8). We show that if a function approximately satisfies this invariance equation then its graph approximates M_ϑ.

Theorem 7.8 (Approximation). *Let the differential equation* (7.1) *satisfy Hypotheses HD, HDA and assume that the constants γ_{11}, ℓ_{12}, ℓ_{21}, ℓ_{22} satisfy Conditions CD and CDA. Moreover, let $\sigma\colon X \times E \to Y$ be a bounded function of class C_b^1 satisfying*

$$\sigma_x(x,\vartheta) f(x,\sigma(x,\vartheta),\vartheta) = g(x,\sigma(x,\vartheta),\vartheta) + \rho(x,\vartheta) \tag{7.14}$$

for some bounded function $\rho\colon X \times E \to Y$.

Then for the function s_A obtained from Theorem 7.5 *the estimate*

$$|\sigma - s_A| \leq \frac{1}{-(\ell_{22} + \ell_{12}\lambda_A)}\,|\rho|$$

holds.

Proof. We consider a solution $x(t)$ of the ODE $\dot{x} = f(x,\sigma(x,\vartheta),\vartheta)$. Then we have

$$\frac{d}{dt}\sigma(x(t),\vartheta) = \sigma_x(x(t),\vartheta)\, f(x(t),\sigma(x(t),\vartheta),\vartheta)$$

and

$$\frac{d}{dt} s_A(x(t),\vartheta) = s_{A,x}(x(t),\vartheta)\, f(x(t),\sigma(x(t),\vartheta),\vartheta).$$

Taking the difference of the two equations and integrating from 0 to $\tau > 0$, τ small enough, and taking into account (7.14) we obtain (suppressing ϑ, for short, and writing x_0 for $x(0)$)

$$\begin{aligned}\big|\sigma(x(\tau)) - s_A(x(\tau))\big| = \Big|\sigma(x_0) - s_A(x_0) + \int_0^\tau \big[g\big(x(t),\sigma(x(t))\big) + \rho(x(t))\\ - s_{A,x}(x(t)) f\big(x(t),\sigma(x(t))\big)\big]dt\Big|.\end{aligned}$$

Using the invariance equation of s_A, cf. (7.8), we get

$$\begin{aligned}&\big|\sigma(x(\tau)) - s_A(x(\tau))\big|\\ &\quad = \Big|\sigma(x_0) - s_A(x_0) + \int_0^\tau \big[g\big(x(t),\sigma(x(t))\big) - g\big(x(t),s_A(x(t))\big)\\ &\qquad\qquad + s_{A,x}(x(t)) f\big(x(t),s_A(x(t))\big)\\ &\qquad\qquad - s_{A,x}(x(t)) f\big(x(t),\sigma(x(t))\big)\\ &\qquad\qquad + \rho(x(t))\big]dt\Big|\\ &\quad \le \Big|\sigma(x_0) - s_A(x_0) + \int_0^\tau\int_0^1 g_2\big(x(t), s_A(x(t)) + r\big(\sigma(x(t)) - s_A(x(t))\big)\big)\\ &\qquad \cdot \big(\sigma(x(t)) - s_A(x(t))\big)dr\,dt\Big| + \tau\lambda_A\ell_{12}|\sigma - s_A| + \tau\,|\rho|.\end{aligned}$$

Since $\sigma(x(t)) - s_A(x(t)) = \sigma(x_0) - s_A(x_0) + O(t)$ it follows that

$$\begin{aligned}&|\sigma(x(\tau)) - s_A(x(\tau))|\\ &\quad \le \Big|I + \int_0^\tau\int_0^1 g_2\Big(x(t), s_A(x(t)) + r\big(\sigma(x(t)) - s_A(x(t))\big)\Big)dr\,dt\Big|\\ &\qquad \cdot \big|\sigma(x_0) - s_A(x_0)\big| + K\tau^2 + \tau\lambda_A\ell_{12}|\sigma - s_A| + \tau|\rho|\end{aligned}$$

for some $K > 0$. The first norm on the right-hand side may be estimated by $1 + \tau\,\ell_{22}$. The estimate claimed follows easily. □

Next we state a result on the existence of a stable foliation with smooth fibers.

Theorem 7.9 (Stable foliation). *Let the differential equation* (7.1) *satisfy Hypotheses HD, HDA and assume that the constants γ_{11}, ℓ_{12}, ℓ_{21}, ℓ_{22} satisfy Conditions CD, CDA and CDA(k), $k \ge 1$. Let $\vartheta \in E$ and let $\Omega_\vartheta \subset X \times Y$ be a positively invariant set of the* *flow* P_ϑ^t *of the differential equation* (7.1)*, i.e., for all* $t \geq 0$ *the inclusion* $P_\vartheta^t(\Omega_\vartheta) \subset \Omega_\vartheta$ *holds.*

Then there is a continuous function $w_\vartheta^s \colon \Omega_\vartheta \times Y \to X$ *such that the following assertions hold.*

i) $w_\vartheta^s(x, y, y) = x$ *for all* $(x, y) \in \Omega_\vartheta$.

ii) *For all* $(x, y) \in \Omega_\vartheta$ *the function* $w_\vartheta^s(x, y, \cdot)$ *is of class* C_b^k *and is uniformly* λ_R*-Lipschitz continuous with*

$$\lambda_R := \frac{2\ell_{12}}{\gamma_{11} - \ell_{22} + \sqrt{(\gamma_{11} - \ell_{22})^2 - 4\ell_{12}\ell_{21}}}.$$

iii) *The stable fibers* $W_\vartheta^s(x, y) := \{(\xi, \eta) \mid \eta \in Y,\ \xi = w_\vartheta^s(x, y, \eta)\}$, $(x, y) \in \Omega_\vartheta$, *form a positively invariant family under the flow of the differential equation* (7.1)*, i.e.,*

$$P_\vartheta^t\,(W_\vartheta^s(x, y)) \subset W_\vartheta^s(P_\vartheta^t(x, y))$$

for all $t \geq 0$.

iv) *The stable fibers are disjoint, i.e., for* $(x_i, y_i) \in \Omega_\vartheta$, $i = 1, 2$, *either*

$$W_\vartheta^s(x_1, y_1) \cap W_\vartheta^s(x_2, y_2) = \emptyset$$

or

$$W_\vartheta^s(x_1, y_1) = W_\vartheta^s(x_2, y_2).$$

v) *The flow is contracting along the stable fibers: For* $(x, y) \in \Omega_\vartheta$ *let* $(u(t), v(t))$ *and* $(\tilde{u}(t), \tilde{v}(t))$*, respectively, be solutions of the differential equation* (7.1) *with initial values* $(u_0, v_0) \in W_\vartheta^s(x, y)$ *and* $(\tilde{u}_0, \tilde{v}_0) \in W_\vartheta^s(x, y)$*, respectively. Then the estimates*

$$|u(t) - \tilde{u}(t)| \leq \lambda_R e^{\alpha_A t} |v_0 - \tilde{v}_0|,$$

$$|v(t) - \tilde{v}(t)| \leq e^{\alpha_A t} |v_0 - \tilde{v}_0|$$

hold with $\alpha_A := \ell_{22} + \delta < 0$, $\delta = \ell_{21}\lambda_R$, *for all* $t \geq 0$.

vi) *The manifold* M_ϑ *of Theorem* 7.5 *has the property of asymptotic phase: For every solution* $(x(t), y(t))$ *of the differential equation* (7.1) *with initial values* $(x_0, y_0) \in W_\vartheta^s(\Omega_\vartheta) := \bigcup_{(x,y)\in\Omega_\vartheta} W_\vartheta^s(x, y)$ *there is a solution* $(\tilde{x}(t), \tilde{y}(t)) = (\tilde{x}(t), s_A(\tilde{x}(t), \vartheta)) \in M_\vartheta$ *such that for all* $t \geq 0$ *the estimates*

$$|x(t) - \tilde{x}(t)| \leq q_A\, e^{\alpha_A t} |y_0 - s_A(x_0, \vartheta)|,$$

$$|y(t) - \tilde{y}(t)| \leq (1 + \lambda_A q_A)\, e^{\alpha_A t} |y_0 - s_A(x_0, \vartheta)|$$

hold with

$$q_A = \frac{\ell_{12}}{\sqrt{(\gamma_{11} - \ell_{22})^2 - 4\ell_{12}\ell_{21}}}.$$

vii) *Let* $(x(t), y(t))$ *be a solution of the differential equation* (7.1) *with initial value* $(x_0, y_0) \in \Omega_\vartheta$. *The stable fiber* $W^s_\vartheta(x_0, y_0)$ *contains all points which under the flow of* (7.1) *exponentially tend to* $(x(t), y(t))$ *with rate* $e^{\alpha_A t}$, $t \geq 0$, *i.e., every solution* $(\tilde{x}(t), \tilde{y}(t))$ *satisfying*

$$|x(t) - \tilde{x}(t)| + |y(t) - \tilde{y}(t)| \leq c\, e^{\alpha_A t}, \quad t \geq 0,$$

for some constant c, *has initial value* $(\tilde{x}(0), \tilde{y}(0)) \in W^s_\vartheta(x_0, y_0)$ *and* $\tilde{x}(t) = w^s_\vartheta(x(t), y(t), \tilde{y}(t))$ *for* $t \geq 0$.

viii) *If there is a map* $\kappa \colon X \to X$ *such that* f *and* g *are* κ*-invariant and* Ω_ϑ *is invariant with respect to* $(x, y) \mapsto (\kappa x, y)$ *then* w^s_ϑ *is* κ*-equivariant, i.e., if for all* $(u, v) \in X \times Y$ *and for all* $(x, y) \in \Omega_\vartheta$

$$\begin{aligned} f(\kappa u, v, \vartheta) &= f(u, v, \vartheta), \\ g(\kappa u, v, \vartheta) &= g(u, v, \vartheta), \\ (\kappa x, y) &\in \Omega_\vartheta, \end{aligned}$$

then $w^s_\vartheta(\kappa x, y, \cdot) = \kappa\, w^s_\vartheta(x, y, \cdot)$ *holds for all* $(x, y) \in \Omega_\vartheta$.

ix) *If* $X = \mathbb{R}^m$ *and* $\Omega_\vartheta = M_\vartheta$ *and if Hypotheses HDB, HDAB and Condition CDAB*$(k-1)$ *are satisfied then the function*

$$w \colon (x, \eta, \vartheta) \in X \times Y \times E \longmapsto w(x, \eta, \vartheta) := w^s_\vartheta(x, s_A(x, \vartheta), \eta) \in X$$

describing the stable fibers $W^s_\vartheta(x, s_A(x, \vartheta)) = \{(\xi, \eta) \mid \eta \in Y,\ \xi = w(x, \eta, \vartheta)\}$ *with base point* $(x, s_A(x, \vartheta)) \in M_\vartheta$ *is of class* C^{k-1}_b *and is of the form*

$$w(x, \eta, \vartheta) = x + R(x, \eta, \vartheta)[\eta - s_A(x, \vartheta)]$$

with $R \in C^{k-1}_b(X \times Y \times E, \mathcal{L}_b(\mathbb{R}^m, \mathbb{R}^n))$.

Proof. As in the previous theorems of this chapter we consider the time-τ map (7.2) for τ sufficiently small. It is easy to verify that this time-τ map satisfies the assumptions of Theorems 4.1 and 5.1. This implies the assertions in Theorem 7.9. Since the details of the proof are similar to the details of the previous proofs we omit them here. □

7.3 Repulsive positively invariant manifolds

In this section we state four theorems analogous to those of Section 7.2 but for repulsive invariant manifolds. We omit the proofs since they are completely analogous to those of Section 7.2.

Theorem 7.10 (Existence and smoothness). *Let the differential equation* (7.1) *satisfy Hypotheses HD, HDR and assume that the constants* γ_{11}, ℓ_{12}, ℓ_{21}, ℓ_{22} *satisfy Conditions CD, CDR and CDR(k),* $k \geq 1$.

Then there is a function $s_R \colon Y \times E \to X$ *of class* C_b^k *such that the following assertions hold for* $\vartheta \in E$.

i) *The set* $N_\vartheta = \{(x, y) \mid y \in Y, x = s_R(y, \vartheta)\}$ *is a positively invariant manifold of* (7.1)*, i.e., if* $(x, y) \in N_\vartheta$ *then the solution* $(\varphi(t; x, y, \vartheta), \psi(t; x, y, \vartheta))$ *of* (7.1) *remains in* N_ϑ *for* $t \geq 0$. *The function* s_R *satisfies the invariance equation*
$$\left[\frac{\partial}{\partial y} s_R(y, \vartheta)\right] g(s_R(y, \vartheta), y, \vartheta) = f(s_R(y, \vartheta), y, \vartheta) \tag{7.15}$$
for $y \in Y$.

ii) *The function* s_R *is bounded and uniformly* λ_R*-Lipschitz continuous with respect to* y *and uniformly* ν_R*-Lipschitz continuous with respect to* ϑ *where*
$$\lambda_R = \frac{2\ell_{12}}{\gamma_{11} - \ell_{22} + \sqrt{(\gamma_{11} - \ell_{22})^2 - 4\ell_{12}\ell_{21}}}, \qquad \nu_R = \frac{\ell_{13} + \ell_{23}\lambda_R}{\gamma_{11} - \ell_{21}\lambda_R}.$$

iii) *The manifold* N_ϑ *is uniformly repulsive. More precisely, every solution* $(x(t), y(t))$ *of* (7.1) *with* $(x(0), y(0)) \in X \times Y$ *satisfies*
$$|x(t) - s_R(y(t), \vartheta)| \geq e^{\alpha_R t}\, |x(0) - s_R(y(0), \vartheta)|$$
with
$$\alpha_R = \gamma_{11} - \delta > 0, \quad \delta = \frac{2\ell_{12}\ell_{21}}{\gamma_{11} - \ell_{22} + \sqrt{(\gamma_{11} - \ell_{22})^2 - 4\ell_{12}\ell_{21}}} = \ell_{21}\lambda_R$$
for all $t \geq 0$ *as long as* $x(t) \in X$.

iv) *Let* $(x(t), y(t))$ *be a solution of* (7.1) *with* $(x(0), y(0)) \in X \times Y$ *satisfying* $(x(t), y(t)) \in X_0 \times Y$, $t \geq 0$, *for some bounded* $X_0 \subset X$. *Then* $(x(t), y(t))$ *lies in* N_ϑ *for* $t \geq 0$.

v) *If there is a map* $\kappa \colon Y \to Y$ *such that* f *and* g *are* κ*-invariant then* s_R *is* κ*-invariant, i.e., if for all* $(x, y) \in X \times Y$,
$$f(x, \kappa y, \vartheta) = f(x, y, \vartheta),$$
$$g(x, \kappa y, \vartheta) = g(x, y, \vartheta)$$
then $s_R(\kappa y, \vartheta) = s_R(y, \vartheta)$ *holds.*

vi) *If the function* f *has the form* $f(x, y, \vartheta) = A(x, y, \vartheta)x + \hat{f}(x, y, \vartheta)$ *and if for all* $y \in Y$ *the estimate* $\mu(-A(s_R(y, \vartheta), y, \vartheta)) \leq -a < 0$ *holds then*
$$|s_R(y, \vartheta)| \leq \frac{1}{a} \sup_{y \in Y} |\hat{f}(s_R(y, \vartheta), y, \vartheta)|.$$

Remark 7.11. In the case $\ell_{22} \geq 0$, which may occur in applications, Condition CD implies Condition CDR.

Next we formulate a perturbation result. We consider the ODE

$$\begin{aligned} \dot{x} &= \bar{f}(x, y, \bar{\vartheta}), \\ \dot{y} &= \bar{g}(x, y, \bar{\vartheta}) \end{aligned} \tag{7.16}$$

as a perturbation of (7.1). For fixed ϑ, $\bar{\vartheta}$ we define

$$\begin{aligned} \delta_1 &:= \sup_{(x,y)\in X\times Y} |f(x, y, \vartheta) - \bar{f}(x, y, \bar{\vartheta})|, \\ \delta_2 &:= \sup_{(x,y)\in X\times Y} |g(x, y, \vartheta) - \bar{g}(x, y, \bar{\vartheta})|. \end{aligned} \tag{7.17}$$

Theorem 7.12 (Perturbation). *Fix $\vartheta, \bar{\vartheta} \in E$ and let the differential equations* (7.1), (7.16) *satisfy Hypotheses HD, HDR and assume that the constants γ_{11}, ℓ_{12}, ℓ_{21}, ℓ_{22} and $\bar{\gamma}_{11}$, $\bar{\ell}_{12}$, $\bar{\ell}_{21}$, $\bar{\ell}_{22}$, respectively, satisfy Conditions CD and CDR. Let δ_1, δ_2 be defined as in* (7.17).

Then for the functions s_R and $\bar{s}_R$ of Theorem 7.10 *defining the positively invariant manifolds N_ϑ, $\bar{N}_{\bar{\vartheta}}$, respectively, the estimate*

$$|s_R(y, \vartheta) - \bar{s}_R(y, \bar{\vartheta})| \leq \frac{1}{\gamma_{11} - \ell_{21}\lambda_R} (\delta_1 + \lambda_R \delta_2)$$

holds for λ_R defined as in Theorem 7.10.

We state an approximation result. The function s_R describing the manifold N_ϑ in Theorem 7.10 satisfies the invariance equation (7.15). We show that if a function approximately satisfies this invariance equation then its graph approximates N_ϑ.

Theorem 7.13 (Approximation). *Let the differential equation* (7.1) *satisfy Hypotheses HD, HDR and assume that the constants γ_{11}, ℓ_{12}, ℓ_{21}, ℓ_{22} satisfy Conditions CD and CDR. Moreover, let $\sigma \colon Y \times E \to X$ be a bounded function of class C_b^1 satisfying*

$$\sigma_y(y, \vartheta) g(\sigma(y, \vartheta), y, \vartheta) = f(\sigma(y, \vartheta), y, \vartheta) + \rho(y, \vartheta) \tag{7.18}$$

for some bounded function $\rho \colon Y \times E \to X$.

Then for the function s_R obtained from Theorem 7.10 *the estimate*

$$|\sigma - s_R| \leq \frac{1}{\gamma_{11} - \ell_{21}\lambda_R} |\rho|$$

holds.

Next we state a result on the existence of an unstable foliation with smooth fibers.

Theorem 7.14 (Unstable foliation). *Let $\vartheta \in E$ and let the differential equation* (7.1) *satisfy Hypotheses HD, HDR, HDRF and assume that the constants $\gamma_{11}, \ell_{12}, \ell_{21}, \ell_{22}$ satisfy Conditions CD, CDR and CDR(k), $k \geq 1$.*

Then there is a continuous function $w_\vartheta^u \colon \Omega_\vartheta \times X \to Y$ such that the following assertions hold.

i) $w_\vartheta^u(x, y, x) = y$ *for all* $(x, y) \in \Omega_\vartheta$.

ii) *For all $(x, y) \in \Omega_\vartheta$ the function $w_\vartheta^u(x, y, \cdot)$ is of class C_b^k and is uniformly λ_A-Lipschitz continuous with*

$$\lambda_A = \frac{2\ell_{21}}{\gamma_{11} - \ell_{22} + \sqrt{(\gamma_{11} - \ell_{22})^2 - 4\ell_{12}\ell_{21}}}.$$

iii) *The unstable fibers $W_\vartheta^u(x, y) := \{(\xi, \eta) \mid \xi \in X,\ \eta = w_\vartheta^u(x, y, \xi)\}$, $(x, y) \in \Omega_\vartheta$, form a negatively invariant family under the flow of the differential equation* (7.1), *i.e.,*

$$P_\vartheta^t\,(W_\vartheta^u(x, y)) \subset W_\vartheta^u\,(P_\vartheta^t(x, y))$$

for all $t \leq 0$.

iv) *The unstable fibers are disjoint, i.e., for $(x_i, y_i) \in \Omega_\vartheta$, $i = 1, 2$, either*

$$W_\vartheta^u(x_1, y_1) \cap W_\vartheta^u(x_2, y_2) = \emptyset$$

or

$$W_\vartheta^u(x_1, y_1) = W_\vartheta^u(x_2, y_2).$$

v) *The inverse flow is contracting along the unstable fibers: For $(x, y) \in \Omega_\vartheta$ let $(u(t), v(t))$ and $(\tilde{u}(t), \tilde{v}(t))$, respectively, be solutions of the differential equation* (7.1) *with initial values $(u_0, v_0) \in W_\vartheta^u(x, y)$ and $(\tilde{u}_0, \tilde{v}_0) \in W_\vartheta^u(x, y)$, respectively. Then the estimates*

$$|u(t) - \tilde{u}(t)| \leq e^{\alpha_R t}|u_0 - \tilde{u}_0|,$$

$$|v(t) - \tilde{v}(t)| \leq \lambda_A e^{\alpha_R t}|u_0 - \tilde{u}_0|$$

hold with $\alpha_R := \gamma_{11} - \delta > 0$, $\delta = \ell_{12}\lambda_A$, for all $t \leq 0$.

vi) *The manifold N_ϑ of Theorem* 7.10 *has the property of asymptotic phase: For every solution $(x(t), y(t))$ of the differential equation* (7.1) *with initial values $(x_0, y_0) \in W_\vartheta^u(\Omega_\vartheta) := \bigcup_{(x,y)\in\Omega_\vartheta} W_\vartheta^u(x, y)$ there is a solution $(\tilde{x}(t), \tilde{y}(t)) = (s_R(\tilde{y}(t), \vartheta), \tilde{y}(t)) \in N_\vartheta$ such that for all $t \leq 0$ the estimates*

$$|x(t) - \tilde{x}(t)| \leq (1 + \lambda_R q_R)\, e^{\alpha_R t}|x_0 - s_R(y_0, \vartheta)|,$$

$$|y(t) - \tilde{y}(t)| \leq q_R\, e^{\alpha_R t}|x_0 - s_R(y_0, \vartheta)|$$

hold with

$$q_R = \frac{\ell_{21}}{\sqrt{(\gamma_{11} - \ell_{22})^2 - 4\ell_{12}\ell_{21}}}.$$

vii) *Let* $(x(t), y(t))$ *be a solution of the differential equation* (7.1) *with initial value* $(x_0, y_0) \in \Omega_\vartheta$. *The unstable fiber* $W^u_\vartheta(x_0, y_0)$ *contains all points which under the flow of* (7.1) *exponentially tend to* $(x(t), y(t))$ *with rate* $e^{\alpha_R t}$, $t \le 0$, *i.e., every solution* $(\tilde{x}(t), \tilde{y}(t))$ *satisfying*

$$|x(t) - \tilde{x}(t)| + |y(t) - \tilde{y}(t)| \le c\, e^{\alpha_R t}, \quad t \le 0,$$

for some constant c, *has initial value* $(\tilde{x}(0), \tilde{y}(0)) \in W^u_\vartheta(x_0, y_0)$ *and* $\tilde{y}(t) = w^u_\vartheta(x(t), y(t), \tilde{x}(t))$ *for* $t \le 0$.

viii) *If there is a map* $\kappa \colon Y \to Y$ *such that* f *and* g *are* κ*-invariant and* Ω_ϑ *is invariant with respect to* $(x, y) \mapsto (x, \kappa y)$ *then* w^u_ϑ *is* κ*-equivariant, i.e., if for all* $(u, v) \in X \times Y$ *and for all* $(x, y) \in \Omega_\vartheta$

$$\begin{aligned} f(u, \kappa v, \vartheta) &= f(u, v, \vartheta), \\ g(u, \kappa v, \vartheta) &= g(u, v, \vartheta), \\ (x, \kappa y) &\in \Omega_\vartheta, \end{aligned}$$

then $w^u_\vartheta(x, \kappa y, \cdot) = \kappa\, w^u_\vartheta(x, y, \cdot)$ *holds for all* $(x, y) \in \Omega_\vartheta$.

Part III

Applications

In Part III we present applications of the theory developed in Part I and Part II. We consider two main application areas, problems of regular and singular perturbation type and discrete dynamical systems stemming from applying a numerical integration method to an ordinary differential equation (ODE). In all applications we show that the dynamical system admits an attractive (negatively) invariant manifold. This means that the dynamics of the system essentially takes place on the manifold. Thus, the dynamical system is essentially reduced to lower dimension.

In Chapter 8 we prove the existence of local stable and local unstable manifolds of fixed points and equilibria. In Chapter 9 we show that to every strictly stable linear multistep method there is an associated one-step method having the same long-time behaviour as the linear multistep method. This result is used to estimate the global error of a linear multistep method. Chapter 10 deals with invariant manifolds for singularly perturbed ODEs. In Chapter 11 we apply a one-step method of order p to a singularly perturbed ODE with perturbation parameter ε. In a first section of Chapter 11 we prove that if the ODE is expressed in the fast time $\tau = t/\varepsilon$ and the step size is taken as h, h small, then the invariant manifold of the discrete dynamical system is $O(\varepsilon^2 h^p)$-close to the invariant manifold of the continuous dynamical system. In a second section we apply appropriate implicit Runge–Kutta methods to singularly perturbed ODEs (in the original slow time t) with step size h independent of ε and large compared to ε. For this case we again prove the existence of an attractive invariant manifold of the discrete dynamical system close to the invariant manifold of the ODE and we estimate the global error. Chapter 12 deals with invariant manifolds for perturbed harmonic oscillators. We investigate the ODE and apply on the one hand a numerical integration method that is area preserving for the harmonic oscillator and on the other hand a method that is not area preserving. For these three dynamical systems we show the existence of an attractive invariant curve. As examples we take the van der Pol equation and we apply the symplectic Euler method and the Euler method to this equation. In Chapter 13 we consider a singularly perturbed system admitting a so-called fold point. As a model example we take the stiff van der Pol equation with perturbation parameter ε and its discretisation by a one-step method. We use the so-called blow-up approach in the vicinity of the fold point to prove the existence of a (negatively) invariant manifold both of the ODE and of the one-step method. This is done by using several charts independent of the perturbation parameter ε for ε in a whole interval. In Chapter 14 we apply an appropriate Runge–Kutta method to a differential-algebraic equation (DAE) of index 2 and show that the generated discrete dynamical system admits an invariant manifold close to the invariant manifold of the DAE.

Chapter 8
Fixed points and equilibria

8.1 The local stable and unstable manifold of a hyperbolic fixed point

Consider a map $P\colon \mathbb{R}^\ell \to \mathbb{R}^\ell$ of class C^2 and assume that P has a strictly hyperbolic fixed point (meaning that at the fixed point the Jacobian has eigenvalues of modulus larger than 1 and of modulus smaller than 1 and there is none of modulus equal to 1). Without loss of generality we suppose that this fixed point is in the origin. Introducing appropriate coordinates in $\mathbb{R}^\ell$ it may be achieved that for $m+n=\ell$ and $m,n \geq 1$,

$$DP(0) = \begin{pmatrix} A & 0 \\ 0 & B \end{pmatrix},$$

where A is an $m \times m$-matrix satisfying $|A^{-1}| \leq 1/a < 1$ and where B is an $n \times n$-matrix satisfying $|B| \leq b < 1$. Hence, the map P is of the form

$$P\colon \begin{pmatrix} x \\ y \end{pmatrix} \longmapsto \begin{pmatrix} \bar{x} \\ \bar{y} \end{pmatrix} = \begin{pmatrix} F(x,y) \\ G(x,y) \end{pmatrix} = \begin{pmatrix} Ax + \hat{F}(x,y) \\ By + \hat{G}(x,y) \end{pmatrix} \tag{8.1}$$

with

$$\hat{F}(0,0) = 0, \quad \hat{G}(0,0) = 0,$$
$$\hat{F}_x(0,0) = 0, \quad \hat{F}_y(0,0) = 0, \quad \hat{G}_x(0,0) = 0, \quad \hat{G}_y(0,0) = 0.$$

For $d > 0$ we introduce the sets $X_d = \{x \in \mathbb{R}^m \mid |x| < d\}$, $Y_d = \{y \in \mathbb{R}^n \mid |y| < d\}$. Since the map P is of class C^2 we conclude from the Taylor formula that there are constants c_1, c_2 such that for all $(x,y) \in X_d \times Y_d$ the estimates

$$|\hat{F}(x,y)|,\ |\hat{G}(x,y)| \leq c_1\, d^2,$$
$$|\hat{F}_x(x,y)|,\ |\hat{F}_y(x,y)|,\ |\hat{G}_x(x,y)|,\ |\hat{G}_y(x,y)| \leq c_2\, d$$

hold.

We aim at applying Theorem 1.7. Since there is no z-coordinate the maps P_A and P_R have the same splitting and are of the form (8.1).

We verify Hypothesis HM. Hypothesis HM a): Under the condition $d < (1-b)/c_1$ it follows that $G(x,y) \in Y_d$ for all $(x,y) \in X_d \times Y_d$ since

$$|G(x,y)| \leq bd + c_1\, d^2 < d.$$

Hypothesis HM b): Under the condition $d < \min\{(a-1)/c_1,\ a/c_2\}$ we show that for every $\bar{x} \in X_d$, $y \in Y_d$ there is $x \in X_d$ such that $F(x, y) = \bar{x}$. By means of the contraction principle we verify that the fixed point equation

$$x = A^{-1}(\bar{x} - \widehat{F}(x, y)) =: R(x)$$

has a unique solution $x \in X_d$. We have $R\colon X_d \to X_d$ since for $x \in X_d$,

$$|R(x)| \le \frac{1}{a}\,(d + c_1\, d^2) < d.$$

The map R is a contraction since $|R'(x)| \le c_2 d/a < 1$. Hypothesis HM c): We find the following constants for the functions F and G:

$$\Gamma_{11} = a - c_2 d, \quad L_{12} = L_{21} = c_2 d, \quad L_{22} = b + c_2 d.$$

We verify Hypothesis HMA. We choose $y^* = 0$. Under the condition $d < 1/c_1$ it follows that $G(\cdot, 0) \in Y_d$ since for $x \in X_d$,

$$|G(x,0)| = |\widehat{G}(x,0)| \le c_1\, d^2 < d.$$

We verify Hypothesis HMR. For $x^* = 0$ Hypothesis HMR holds under the same condition on d as in HMA.

We verify Condition CM. Under the condition $d < (a-b)/4c_2$ we find

$$2\sqrt{L_{12}L_{21}} = 2c_2\, d < a - b - 2c_2\, d = \Gamma_{11} - L_{22}.$$

We verify Condition CMA. If Condition CM holds we may estimate

$$\Delta < \frac{2L_{12}L_{21}}{\Gamma_{11} - L_{22}} = \frac{2(c_2\, d)^2}{a - b - 2c_2\, d} < c_2\, d.$$

Under the additional condition $d < (1-b)/2c_2$ we have

$$L_{22} + \Delta < b + 2c_2\, d < 1.$$

We verify Condition CMR. Under the condition $d < (a-1)/2c_2$ we have

$$\Gamma_{11} - \Delta > a - 2c_2\, d > 1.$$

We verify the assumption $\lambda_A \lambda_R < 1$. For

$$\lambda_A = \frac{2L_{21}}{\Gamma_{11} - L_{22} + \sqrt{(\Gamma_{11} - L_{22})^2 - 4L_{12}L_{21}}},$$

$$\lambda_R = \frac{2L_{12}}{\Gamma_{11} - L_{22} + \sqrt{(\Gamma_{11} - L_{22})^2 - 4L_{12}L_{21}}}$$

the inequality

$$\lambda_A \lambda_R < \frac{4L_{12}L_{21}}{(\Gamma_{11} - L_{22})^2} < 1$$

holds if Condition CM is satisfied.

If d is chosen such that all conditions above are satisfied then the map P admits the negatively invariant manifold

$$M = \{(x, y) \mid |x| < d,\ y = s_A(x)\}$$

and the positively invariant manifold

$$N = \{(x, y) \mid |y| < d,\ x = s_R(y)\}.$$

The hyperbolic invariant manifold of the map P reduces to $K = M \cap N = \{0\} \subset \mathbb{R}^\ell$. The functions s_A and s_R have Lipschitz constant $\lambda_A = \lambda_R < 2c_2\, d/(a - b - 2c_2\, d)$. The manifold M is attractive with attractivity constant $\chi_A = L_{22} + \Delta < b + 2c_2 d < 1$. The manifold N is repulsive with repulsivity constant $\chi_R = \Gamma_{11} - \Delta > a - 2c_2 d > 1$. The manifold M is the local unstable manifold of the fixed point 0 and the manifold N is the local stable manifold of 0, cf. Figure 8.1.

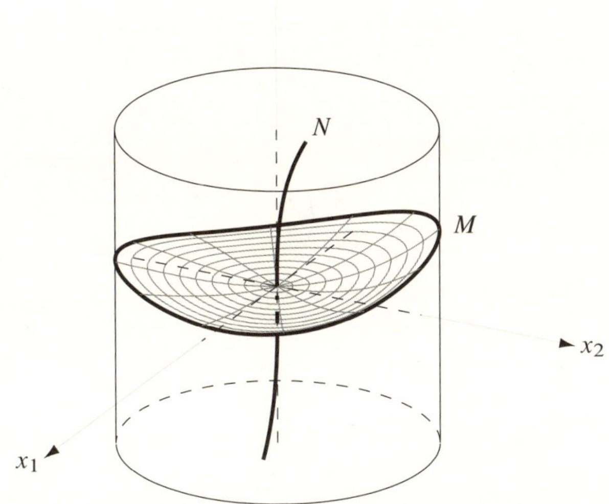

Figure 8.1. The local stable manifold N and the local unstable manifold M of the fixed point 0.

Example. We apply the theory derived above to the Hénon map

$$\begin{aligned} \bar{x} &= 1 + \frac{3}{4}x^2 + y, \\ \bar{y} &= -x \end{aligned} \tag{8.2}$$

introduced in the introduction of Part I. With the coordinate change

$$\begin{pmatrix} x \\ y \end{pmatrix} = \begin{pmatrix} 2 \\ -2 \end{pmatrix} + \begin{pmatrix} 1 & -\delta \\ -\delta & 1 \end{pmatrix} \begin{pmatrix} u \\ v \end{pmatrix}, \quad \delta = \frac{3-\sqrt{5}}{2} = 0.381\ldots, \tag{8.3}$$

we obtain the map

$$\begin{pmatrix} \bar{u} \\ \bar{v} \end{pmatrix} = \frac{1}{1-\delta^2} \begin{pmatrix} (3-2\delta)u + \frac{3}{4}(u-\delta v)^2 \\ \delta(2-3\delta)v + \frac{3}{4}\delta(u-\delta v)^2 \end{pmatrix} \tag{8.4}$$

which is of the form (8.1) with $A = 1/\delta$, $B = \delta$. We find $a = A$, $b = B$, $c_1 = 3(1+\delta)^2/4$, $c_2 = 3(1+\delta)/2$. The constant d defining the sets $U_d = \{u \mid |u| < d\} = V_d$ has to satisfy the six conditions

$$\begin{aligned} &d < (1-b)/c_1, && d < (a-1)/c_1, && d < a/c_2, \\ &d < (a-b)/(4c_2), && d < (1-b)/(2c_2), && d < (a-1)/(2c_2). \end{aligned}$$

The fifth condition is the most restrictive one and yields $d < \sqrt{5}/15 = 0.149\ldots$. It follows that in $U_d \times V_d$ the map (8.4) admits a local unstable (attractive) and a local stable (repulsive) manifold, respectively, given as the graph of a function $v = \tilde{s}_A(u)$ and $u = \tilde{s}_R(v)$, respectively. These functions have Lipschitz constant $\tilde{\lambda}_A = \tilde{\lambda}_R = 0.198\ldots$.

Transforming the quadratic domain $U_d \times V_d$ back to the original variables by

$$\begin{pmatrix} u \\ v \end{pmatrix} = \frac{1}{1-\delta^2} \begin{pmatrix} 1 & \delta \\ \delta & 1 \end{pmatrix} \begin{pmatrix} x-2 \\ y+2 \end{pmatrix} \tag{8.5}$$

yields a domain containing the quadratic domain $(x-2, y+2) \in X_{d(1-\delta)} \times Y_{d(1-\delta)}$ with $d(1-\delta) = 0.092\ldots$. Setting $v = \tilde{s}_A(u)$ and expressing u, v in x, y by (8.5) yields

$$y = -2 - \delta(x-2) + (1-\delta^2)\tilde{s}_A\Big(\frac{1}{1-\delta^2}(x - 2 + \delta(y+2))\Big).$$

By a contraction argument this equation has a unique smooth solution $y = s_A(x)$, $x - 2 \in X_{d(1-\delta)}$. The function s_A describes the local unstable manifold of the fixed point $(2, -2)$ of (8.2). Taking the derivative leads to the estimate

$$|s_A' + \delta| \le \frac{1-\delta^2}{1-\delta\tilde{\lambda}_A}\tilde{\lambda}_A = 0.183\ldots.$$

In the same way one obtains the function $x = s_R(y)$ describing the local stable manifold of $(2, -2)$. For the derivative s_R' the same estimate holds as for s_A'. This agrees with the picture shown in Figure 1 in the introduction of Part I.

8.2 The strongly stable manifold of an equilibrium

We consider an ordinary differential equation

$$\begin{aligned} \dot{x} &= f(x, y, \varepsilon), \quad x \in \mathbb{R}^m, \\ \dot{y} &= g(x, y, \varepsilon), \quad y \in \mathbb{R}^n, \end{aligned} \qquad \varepsilon \in (0, \varepsilon_0],$$

admitting $p = 0$ as an equilibrium where p is strongly attracting in y-direction and weakly attracting in x-direction. Weakly means that the attractivity is of order $O(\varepsilon)$ and strongly means that it is of order $O(1)$ as $\varepsilon \to 0$. In this situation, Theorem 7.5 may be applied and yields the existence of a strongly attractive invariant manifold M_ε containing the equilibrium p. In this setting, p has a strongly stable manifold. This manifold cannot be obtained by applying Theorem 7.10 since Condition CDR is violated. However, this manifold may be obtained as the stable fiber of the equilibrium p.

As an example we consider a simplified version of a model for a miniature synchronous motor, cf. Tognola [128]:

$$\begin{aligned} \dot{x}_1 &= x_2, \\ \dot{x}_2 &= -x_1 + \varepsilon f(x_1, x_2, y, \varepsilon), \\ \dot{y} &= -qy + \varepsilon g(x_1, x_2, y, \varepsilon) \end{aligned} \tag{8.6}$$

with $x_1, x_2, y \in \mathbb{R}$, $q > 0$, f and g bounded and of class C_b^k, $k \geq 1$, $f(0,0,0,0) = 0$, $g(0,0,0,0) = 0$, $f_{x_2}(0,0,0,0) < 0$ and $\varepsilon \in (0, \varepsilon_0]$ with ε_0 small enough. The assumption $f_{x_2}(0,0,0,0) < 0$ implies that for small ε the equilibrium $p = 0$ is weakly attractive in (x_1, x_2)-direction.

We apply Theorems 7.5 and 7.9. Hypothesis HD is satisfied with

$$\begin{aligned} \ell_{12} &= O(\varepsilon), \quad \ell_{13} = O(1), \\ \ell_{21} &= O(\varepsilon), \quad \ell_{23} = O(1), \end{aligned}$$

and

$$\gamma_{11} = O(\varepsilon) < 0, \quad \ell_{22} = -q + O(\varepsilon).$$

Hypothesis HDA is satisfied with $y^* = 0$. Conditions CD, CDA and CDA(k) are satisfied for ε small enough. Theorem 7.5 implies that there is a smooth attractive invariant manifold $M_\varepsilon = \{(x, y) \mid x \in \mathbb{R}^2,\ y = s_A(x, \varepsilon)\}$ containing the equilibrium $p = 0$. The attractivity constant is $\alpha_A = -q + O(\varepsilon)$. In contrast to Section 8.1 where Theorem 1.3 may be applied, the strongly stable manifold of p cannot be obtained by applying Theorem 7.10 since Condition CDR is violated. We apply Theorem 7.9 with $\Omega_\vartheta := M_\varepsilon$, instead. It follows that the whole space $\mathbb{R}^3$ is foliated by stable fibers with base points on M_ε. Since $p \in M_\varepsilon$ is an equilibrium, the fiber through p is invariant and hence is the strongly stable manifold $W_\varepsilon^s(p)$ of p. This situation is sketched in Figure 8.2.

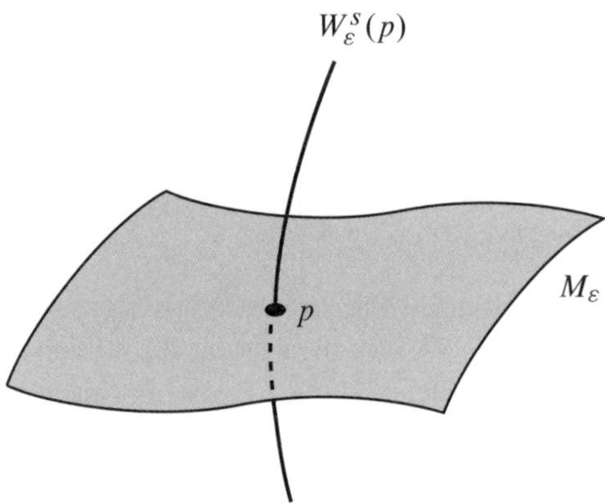

Figure 8.2. The equilibrium p on M_ε and the strongly stable manifold $W^s_\varepsilon(p)$.

Remark 8.1. For ε sufficiently small equation (8.6) admits a strongly attractive invariant manifold M_ε containing the equilibrium p and p has a strongly stable manifold transversal to M_ε. This geometric property is preserved under discretisation by a numerical integration method for small ε and small step size h, i.e., the discrete dynamical system admits an attractive invariant manifold $M_{h,\varepsilon}$ close to M_ε and a strongly stable manifold of the fixed point p. However, the dynamics on M_ε might not be correctly described by the discretisation. E.g., the Euler method applied to (8.6) reproduces the weak attractivity of p only if $h = O(\varepsilon)$. If ε is small compared to h then p is repelling on $M_{h,\varepsilon}$. In contrast, the symplectic Euler method reproduces the correct dynamics near the fixed point p for small step sizes h independent of ε. Related results (in polar coordinates) are shown in Chapter 12, see also Stoffer [120].

Chapter 9

The one-step method associated to a linear multistep method

Consider a differential equation $\dot{x} = f(x)$, $x \in \mathbb{R}^\ell$, and let $x(t)$ be a solution. A linear multistep method is a map from $\mathbb{R}^{k\ell}$ to $\mathbb{R}^\ell$. Given k approximations $x_n, \dots, x_{n+k-1}$ of $x(nh), \dots, x((n+k-1)h)$ an approximation x_{n+k} of $x((n+k)h)$ is computed. In the subsequent step the approximations $x_{n+1}, \dots, x_{n+k}$ are used to compute x_{n+k+1}, etc. Hence, to start a linear multistep method, starting values $x_0, \dots, x_{k-1}$ have to be provided. According to Kirchgraber [64] a one-step method can be associated to any strictly stable linear multistep method. The connection of one-step methods to linear multistep methods is also treated in Stuart, Humphries [125]. In our presentation we follow Stoffer [119] where stronger results are derived. A one-step method of order p and step size h applied to $\dot{x} = f(x)$ is a map $\Phi(h, x)$ approximating the time-h map of the ordinary differential equation up to an error of order $O(h^{p+1})$. For the one-step method Φ associated to the linear multistep method the following holds.

- If the linear multistep method is started with the one-step method Φ, i.e., $x_i = \Phi(h, x_{i-1})$, $i = 1, \dots, k-1$, then the linear multistep method and the one-step method generate the same orbit $(x_i)_{i \geq 0}$.

- Let $(x_i)_{i \geq 0}$ be the orbit generated by the linear multistep method with arbitrary starting values $x_0, \dots, x_{k-1}$. Then there is $\tilde{x}_0$ such that for $\tilde{x}_n := \Phi(h, \tilde{x}_{n-1})$, $n \geq 1$, the estimate

$$|x_n - \tilde{x}_n| \leq K \, {\chi_A}^n \sum_{i=1}^{k-1} |x_i - \Phi(h, x_{i-1})|, \quad \chi_A < 1,$$

  holds.

This means that the dynamics of the linear multistep method is essentially determined by the one-step method Φ. In this chapter we further investigate the relation between the linear multistep method and its associated one-step method Φ. In Section 9.1 we give the basic facts on linear multistep methods. In Section 9.2 we show that the one-step method has the same order as the linear multistep method and we also determine the leading term of the local error of Φ. In Section 9.3 we give a detailed estimate of the global error of a linear multistep method valid for arbitrary starting values.

9.1 Basic facts on linear multistep methods

Consider the autonomous differential equation

$$\dot{x} = f(x), \quad x \in \mathbb{R}^\ell, \tag{9.1}$$

where f is bounded and of class C_b^r. We denote the solution of equation (9.1) with initial value x_0 by $\varphi(t, x_0)$. If the choice of the initial value is obvious from the context we often write $x(t)$ instead of $\varphi(t, x_0)$. We apply a linear multistep method (LMM) of k steps to equation (9.1). For given starting values $x_0, \dots, x_{k-1}$ approximating the solution $x(t)$ at $t = 0, h, \dots, (k-1)h$ an LMM provides an approximation x_k of $x(kh)$. The approximation x_k is obtained by solving the equation

$$\sum_{j=0}^{k} \alpha_j x_j = h \sum_{j=0}^{k} \beta_j f(x_j), \quad \alpha_k = 1, \tag{9.2}$$

for x_k. If $\beta_k = 0$ the LMM is explicit, otherwise it is an implicit method. We show that for given starting values $x_0, \dots, x_{k-1}$ equation (9.2) has a unique solution x_k for sufficiently small h. In order to apply the contraction principle we consider the map

$$G\colon x_k \longmapsto \bar{x}_k = h \sum_{j=0}^{k} \beta_j f(x_j) - \sum_{j=0}^{k-1} \alpha_j x_j .$$

For $h = 0$ equation (9.2) has a unique solution x_k^0. We consider a ball $B_d(x_k^0)$. Since f is bounded by some constant m_f we have

$$|G(x_k) - x_k^0| = |h \sum_{j=0}^{k} \beta_j f(x_j)| \le h\, m_f \sum_{j=0}^{k} |\beta_j|.$$

Hence, for any $h \le d/\big(m_f \sum_{j=0}^{k} |\beta_j|\big)$ the ball $B_d(x_k^0)$ is mapped into itself by the map G. Since $f(x)$ is uniformly Lipschitz continuous we have for $x_k^1, x_k^2 \in \mathbb{R}^\ell$

$$|G(x_k^1) - G(x_k^2)| = h\, \beta_k\, |f(x_k^1) - f(x_k^2)| \le h\, \beta_k \operatorname{Lip}_f |x_k^1 - x_k^2|.$$

Hence, there is $h_0 > 0$ such that G is a contraction in $B_d(x_k^0)$ for $h \le h_0$. It follows from the contraction principle that for $h \le h_0$ equation (9.2) has a unique solution x_k which is $O(h)$-close to x_k^0. From the implicit function theorem it follows that this solution is a smooth function of $h, x_0, \dots, x_{k-1}$ with bounded derivatives.

For a general overview on LMMs see Hairer, Nørsett, Wanner [52]. We give the basic definitions and statements used in this paper.

An LMM is said do be of *order* p if

$$\sum_{j=0}^{k} \alpha_j = 0 \quad \text{and} \quad \sum_{j=0}^{k} \alpha_j\, j^q = q \sum_{j=0}^{k} \beta_j\, j^{q-1} \quad \text{for } q = 1, \dots, p. \tag{9.3}$$

The *local error* of an LMM is defined by $\ell(h, x_0) := x_k - x(kh)$ where $x(t) = \varphi(t, x_0)$ is a solution of equation (9.1) and where x_k is obtained by using the exact starting values $x_j = x(jh)$, $j = 0, \ldots, k-1$. The local error satisfies

$$\ell(h, x_0) = (I + O(h)) \sum_{j=0}^{k} \big[h\,\beta_j\, f(x(jh)) - \alpha_j\, x(jh)\big].$$

Moreover, it holds that $\ell(h, x_0) = C_{p+1}\, h^{p+1}\, x^{(p+1)}(0) + O(h^{p+2})$ where the constant C_{p+1} is given by

$$C_{p+1} = \frac{1}{(p+1)!} \sum_{j=0}^{k} \big[(p+1)\,\beta_j\, j^p - \alpha_j\, j^{p+1}\big]. \tag{9.4}$$

An LMM is called *convergent* if for fixed T,

$$|x_n - x(nh)| \longrightarrow 0 \quad \text{for } h \to 0,\ nh \le T,$$

holds for all starting values satisfying $x_i - x(ih) \to 0$ for $h \to 0$, $i = 0, \ldots, k-1$.

An LMM is called *convergent of order* p if for $h \to 0$, $nh \le T$, the estimate $|x_n - x(nh)| = O(h^p)$ holds for all starting values satisfying $x_i - x(ih) = O(h^p)$, $i = 0, \ldots, k-1$.

We make the following assumptions on the LMM.

1) The LMM is of order $p \ge 1$.

2) The LMM is ϱ_1-strictly stable, i.e., the polynomial $\varrho(z) := \sum_{j=0}^{k} \alpha_j\, z^j$ has 1 as simple zero and all other zeros have modulus smaller than $\varrho_1 < 1$.

Note that these assumptions imply that the LMM is convergent of order p.

9.2 The associated one-step method

It is helpful to introduce the vectors

$$X := \begin{pmatrix} x_0 \\ \vdots \\ x_{k-1} \end{pmatrix}, \quad \bar{X} := \begin{pmatrix} x_1 \\ \vdots \\ x_k \end{pmatrix}, \quad f(X) := \begin{pmatrix} f(x_0) \\ \vdots \\ f(x_{k-1}) \end{pmatrix}$$

in $\mathbb{R}^{k\ell}$ and the $k \times k$-matrices

$$L_\alpha := \begin{pmatrix} & 0 & \\ \alpha_0 & \ldots & \alpha_{k-1} \end{pmatrix}, \quad L_\beta := \begin{pmatrix} & 0 & \\ \beta_0 & \ldots & \beta_{k-1} \end{pmatrix}, \quad R := \begin{pmatrix} 0 & 1 & & 0 \\ & \ddots & \ddots & \\ & & \ddots & 1 \\ 0 & & & 0 \end{pmatrix}$$

Using these notations and the vector $e_k := (0,\dots,0,1)^T \in \mathbb{R}^k$ the LMM (9.2) may be written as

$$P_h\colon \mathbb{R}^{k\ell} \longrightarrow \mathbb{R}^{k\ell}, \quad X \longmapsto \bar{X} \tag{9.5}$$

with $\bar{X} = ((R - L_\alpha) \otimes I_\ell)\, X + h(L_\beta \otimes I_\ell)\, f(X) + h\, \beta_k (e_k \otimes f(x_k))$.

We show that the map P_h admits an ℓ-dimensional attractive invariant manifold.

Theorem 9.1. *Let f in the differential equation (9.1) be bounded and of class C_b^r and let $x(t) := \varphi(t, \xi_0)$, $\xi_0 \in \mathbb{R}^\ell$, be the solution with initial condition $x(0) = \xi_0$. Let the LMM (9.2) be ρ_1-strictly stable, of order p, and assume $1 \le p < r$. Let $x_j \in \mathbb{R}^\ell$, $j = 0,\dots,k-1$, and let $(x_n,\dots,x_{n+k-1}) = P_h^n(x_0,\dots,x_{k-1})$ where P_h^n denotes the n-th iterate of P_h.*

Then there are positive constants h_0, c, K and a map

$$\Phi\colon (0, h_0] \times \mathbb{R}^\ell \longrightarrow \mathbb{R}^\ell, \quad (h, x) \longmapsto \Phi(h, x),$$

of class C_b^r such that with $\Delta(x_0,\dots,x_{k-1},h) := \sum_{j=1}^{k-1} |x_j - \Phi^j(h, x_0)|$, where Φ^j denotes the j-th iterate of Φ, the following assertions hold for all $h \le h_0$.

i) *The map P_h of equation (9.5) admits an invariant manifold*

$$\begin{aligned} M_h &= \left\{ X = \begin{pmatrix} x_0 \\ \vdots \\ x_{k-1} \end{pmatrix} \;\middle|\; \Delta(x_0,\dots,x_{k-1},h) = 0 \right\} \\ &= \left\{ X = \begin{pmatrix} x_0 \\ \vdots \\ x_{k-1} \end{pmatrix} \;\middle|\; x_0 \in \mathbb{R}^\ell, x_j := \Phi^j(h, x_0),\ j = 1,\dots,k-1 \right\}, \end{aligned} \tag{9.6}$$

i.e., $\Delta(x_0,\dots,x_{k-1},h) = 0$ implies $\Delta(x_n,\dots,x_{n+k-1},h) = 0$ for all $n \in \mathbb{N}$, or, equivalently, $x_j = \Phi^j(h, x_0)$ for $j = 1,\dots,k-1$ implies $x_n = \Phi^n(h, x_0)$ for $n \in \mathbb{N}$.

ii) *For $n \in \mathbb{N}$ the estimate*

$$\Delta(x_n,\dots,x_{n+k-1},h) \le K\, {\chi_A}^n\, \Delta(x_0,\dots,x_{k-1},h)$$

holds with $\chi_A(h) = \rho_1 + ch < 1$.

iii) *The manifold M_h has the property of asymptotic phase, i.e., there is $\tilde{x}_0$ such that for $\tilde{x}_n := \Phi^n(h, \tilde{x}_0)$, $n \in \mathbb{N}$,*

$$|x_n - \tilde{x}_n| \le K\, {\chi_A}^n\, \Delta(x_0,\dots,x_{k-1},h).$$

iv) *The map Φ is a one-step method of order p, i.e., the local error of Φ is of order $O(h^{p+1})$.*

v) *The leading term* $\ell_{p+1}(\xi_0)$ *of the local error of* Φ *is*

$$\ell_{p+1}(\xi_0) = \frac{C_{p+1}}{\sum_{j=0}^{k} \beta_j} \, x^{(p+1)}(0),$$

where C_{p+1} *is defined in* (9.4).

Remark 9.2. (1) Assertion i) states: If the LMM is started using the one-step method Φ then the LMM and the one-step method Φ generate the same orbit.

(2) Assertion ii) states: For arbitrary starting values the generated LMM orbit approaches the invariant manifold M_h with attractivity constant χ_A.

Proof. i) Since the LMM is ρ_1-strictly stable, the matrix $R - L_\alpha$ has $\lambda_0 = 1$ as a simple eigenvalue with eigenvector $v_0 = (1, \dots, 1)^T$ and all other eigenvalues λ_i satisfy $|\lambda_i| < \varrho_1, i = 1, \dots, k-1$. We choose eigenvectors corresponding to λ_i (or generalized eigenvectors, if necessary) and denote them by v_i. For the $k \times k$-matrix $V := (v_0, \dots, v_{k-1})$ we have

$$V^{-1}(R - L_\alpha)\, V = \begin{pmatrix} 1 & 0 \\ 0 & P_s \end{pmatrix} \quad \text{with } |P_s| < \varrho_1.$$

We introduce new coordinates (x^*, X_s^*) with $x^* \in \mathbb{R}^\ell$, $X_s^* \in \mathbb{R}^{\ell(k-1)}$ by

$$X = (V \otimes I_\ell) \begin{pmatrix} x^* \\ X_s^* \end{pmatrix}. \tag{9.7}$$

In these coordinates the map (9.5) may be written as implicit equation

$$\begin{aligned} \bar{x}^* &= x^* + h\, f^*(x^*, X_s^*, \bar{x}^*, \bar{X}_s^*, h), \\ \bar{X}_s^* &= (P_s \otimes I_\ell)\, X_s^* + h\, f_s^*(x^*, X_s^*, \bar{x}^*, \bar{X}_s^*, h). \end{aligned}$$

Hence, for h small enough, solving for $\bar{x}^*$, $\bar{X}_s^*$ the map P_h in the new coordinates has the form

$$P_h^* \colon \begin{pmatrix} x^* \\ X_s^* \end{pmatrix} \longmapsto \begin{pmatrix} \bar{x}^* \\ \bar{X}_s^* \end{pmatrix} = \begin{pmatrix} F(x^*, X_s^*, h) \\ G(x^*, X_s^*, h) \end{pmatrix}$$

with F and G being bounded and of class C_b^r.

We apply Theorems 3.6 and 4.1. We verify Hypothesis HM. For given d the set $|X_s^*| \le d$ is mapped into itself for h small enough verifying Hypothesis HM a). Hypothesis HM b) is also satisfied for h small enough. Hypothesis HM c) is satisfied with

$$\begin{aligned} \Gamma_{11} &= 1 + O(h), & L_{12} &= O(h), \\ L_{21} &= O(h), & L_{22} &= \rho_1 + O(h). \end{aligned}$$

Hypothesis HMA is satisfied for $X_s^* = 0$. Conditions CM, CMA and CMA(r) hold for h sufficiently small. Theorems 3.6 and 4.1 imply the following.

There is a λ_A-Lipschitz continuous function σ_A^* of class C_b^r, $\lambda_A = O(h)$, such that

- $M_h^* = \{(x^*, X_s^*) \mid x^* \in \mathbb{R}^\ell,\ X_s^* = \sigma_A^*(x^*, h)\}$ *is an invariant manifold of the map P_h^*, i.e., $P_h^*(M_h^*) = M_h^*$.*
- M_h^* *is uniformly attractive for P_h^* with attractivity constant $\chi_A = L_{22} + L_{12}\lambda_A \le \varrho_1 + O(h) < 1$.*
- *The property of asymptotic phase holds.*
- $\sigma_A^*(x^*, h) = O(h)$.

We express the invariant manifold in the original variables X:

$$M_h = \Big\{X \mid X = (V \otimes I_\ell)\begin{pmatrix} x^* \\ X_s^* \end{pmatrix},\ x^* \in \mathbb{R}^\ell,\ X_s^* = \sigma_A^*(x^*, h)\Big\}.$$

The first three properties of M_h^* also hold for M_h. Since the entries in the first column of the matrix V are all 1 we have for $X \in M_h$

$$x_i = x^* + \sum_{j=1}^{k-1} v_{ij}\,\sigma_{A,j}^*(x^*, h), \quad i = 0, \dots, k-1. \tag{9.8}$$

Since $\sigma_A^*(x^*, h) = O(h)$ equation (9.8) may be solved for x^* for h sufficiently small. Hence, there are functions w_i, $i = 0, \dots, k-1$, such that $x^* = w_i(x_i, h)$ holds with $w_i(x_i, 0) = x_i$. By inserting $x^* = w_{i-1}(x_{i-1}, h)$ into equation (9.8) we obtain functions Φ_i of class C_b^r such that

$$x_i = \Phi_i(h, x_{i-1}), \quad i = 1, \dots, k-1,$$

with $\Phi_i(0, x_{i-1}) = x_{i-1}$. Hence, the manifold M_h may be described as

$$M_h = \left\{X = \begin{pmatrix} x_0 \\ \vdots \\ x_{k-1} \end{pmatrix} \,\middle|\, x_0 \in \mathbb{R}^\ell,\ x_i = \Phi_i(h, x_{i-1}),\ i = 1, \dots, k-1\right\}.$$

A point $X \in M_h$ is characterised by $(X)_{i+1} = \Phi_i(h, (X)_i)$, $i = 1, \dots, k-1$. Since M_h is invariant under P_h any $X \in M_h$ is mapped to $P_h(X) = \bar{X} \in M_h$. Therefore, $(\bar{X})_{i+1} = (X)_{i+2} = \Phi_i(h, (\bar{X})_i) = \Phi_i(h, (X)_{i+1})$ holds for $i = 1, \dots, k-2$. On the other hand, we know from $X \in M_h$ that $(X)_{i+2} = \Phi_{i+1}(h, (X)_{i+1})$, $i = 1, \dots, k-2$. This implies $\Phi_1 = \Phi_2 = \cdots = \Phi_{k-1} =: \Phi$. We conclude $\Phi_i(h, \cdot) = \Phi^i(h, \cdot)$, $i = 1, \dots, k-1$, where Φ^i denotes the i-th iterate of Φ, and thus the manifold M_h may be written as

$$M_h = \left\{X = \begin{pmatrix} x_0 \\ \vdots \\ x_{k-1} \end{pmatrix} \,\middle|\, x_0 \in \mathbb{R}^\ell,\ x_i := \Phi^i(h, x_0),\ i = 1, \dots, k-1\right\}.$$

Since $\Delta(x_0, \dots, x_{k-1}, h) = \sum_{i=1}^{k-1} |x_i - \Phi^i(h, x_0)|$ we have

$$M_h = \left\{ X = \begin{pmatrix} x_0 \\ \vdots \\ x_{k-1} \end{pmatrix} \;\middle|\; \Delta(x_0, \dots, x_{k-1}, h) = 0 \right\}.$$

This concludes the proof of assertion i).

ii), iii) The assertions follow from Theorems 1.5 and 4.1 as $\Delta(x_n, \dots, x_{n+k-1}, h)$ measures the distance of the point $(x_n, \dots, x_{n+k-1}) \in \mathbb{R}^{k\ell}$ to M_h.

iv), v) Since $\Phi(0, x) = x$ holds for $x \in \mathbb{R}^\ell$, the local error ℓ_Φ of Φ may be expanded with respect to h as follows:

$$\ell_\Phi(h, x) := \Phi(h, x) - \varphi(h, x) = \sum_{j=q}^{r-1} h^j \, \ell_j(x) + h^r \, \hat{\ell}(h, x) \tag{9.9}$$

for some $q \geq 1$. We show that $q = p + 1$ and determine the leading term ℓ_{p+1}. Define $x_j := \Phi^j(h, \xi_0)$, $j = 0, \dots, k$. For the LMM the local error $\ell(h, \xi_0) = x_k - x(kh)$ where $x(t) := \varphi(t, \xi_0)$ satisfies

$$\begin{aligned} \ell(h, \xi_0) &= C_{p+1} \, h^{p+1} \, x^{(p+1)}(0) + O(h^{p+2}) \\ &= \sum_{j=0}^{k} [h \, \beta_j \, f(x(jh)) - \alpha_j \, x(jh)] + O(h^{p+2}) \\ &= \sum_{j=0}^{k} [h \, \beta_j \, (f(x(jh)) - f(x_j)) - \alpha_j (x(jh) - x_j)] + O(h^{p+2}) \end{aligned} \tag{9.10}$$

where we have used that $\sum_{j=0}^{k} [h \, \beta_j \, f(x_j) - \alpha_j \, x_j] = 0$, cf. equation (9.2). We compute the terms $x_j - x(jh)$, $j = 1, \dots, k$:

$$\begin{aligned} x_j - x(jh) &= [\Phi(h, x_{j-1}) - \Phi(h, x((j-1)h))] + [\Phi(h, x((j-1)h)) - x(jh)] \\ &= (I + O(h)) \, [x_{j-1} - x((j-1)h))] + \ell_\Phi(h, x((j-1)h)). \end{aligned}$$

By induction it follows that

$$x_j - x(jh) = \sum_{i=0}^{j-1} (I + O(h)) \, \ell_\Phi(h, x(ih)).$$

Since $x(ih) = \xi_0 + O(h)$, $i \leq k$, we get from (9.9)

$$\begin{aligned} \ell_\Phi(h, x(ih)) &= h^q \, \ell_q(x(ih)) + O(h^{q+1}) \\ &= h^q \, \ell_q(\xi_0) + O(h^{q+1}) \end{aligned}$$

and therefore

$$x_j - x(jh) = j\, h^q\, \ell_q(\xi_0) + O(h^{q+1}),\ j \le k.$$

Inserting these expressions into equation (9.10) we obtain

$$\begin{aligned}\ell(h,\xi_0) &= \sum_{j=0}^{k} j\alpha_j h^q \ell_q(\xi_0) + O(h^{q+1}) + O(h^{p+2}) \\ &= h^q \ell_q(\xi_0) \sum_{j=0}^{k} j\alpha_j + O(h^{q+1}) + O(h^{p+2}).\end{aligned}$$

Since the LMM is a stable method of order p we know from (9.3) that

$$\sum_{j=0}^{k} j\,\alpha_j = \sum_{j=0}^{k} \beta_j \neq 0.$$

We have used that 1 is a simple zero of the polynomial ρ, implying $\sum_{j=0}^{k} j\alpha_j \neq 0$. Hence, using the first equation of (9.10) we conclude that $q = p+1$ and

$$\ell_{p+1}(\xi_0) = \frac{C_{p+1}}{\sum\limits_{j=0}^{k} \beta_j}\, x^{(p+1)}(0). \qquad \square$$

9.3 The global error of linear multistep methods

It follows from Theorem 9.1 that the dynamics of the LMM is essentially described by the method restricted to the manifold M_h or, equivalently, by the one-step method Φ. Due to this equivalence the global error of an LMM may be estimated by the global error of the associated one-step method.

Theorem 9.3. *Let f in equation (9.1) be bounded and of class C_b^r, and let $\varphi(t,x_0)$ be the solution with initial condition $\varphi(0,x_0) = x_0$. Let the LMM (9.2) be ρ_1-strictly stable, of order p, and assume $1 \le p < r$. Let $(x_n)_{n\in\mathbb{N}_0}$ be the orbit generated by the LMM with given starting values $x_0,\dots,x_{k-1}$ and let $T > 0$. For $n \in \mathbb{N}$ let $e_n(x_0,\dots,x_{k-1},h) := x_n - \varphi(nh,x_0)$ be the global error of the LMM and let $\Lambda := \sum_{i=1}^{k-1} |e_i|$ measure the total starting error.*

Then for h small enough and for all $n \ge k$ with $nh \le T$ the global error e_n of the LMM satisfies

$$e_n(x_0,\dots,x_{k-1},h) = E(nh)\, h^p + O(\Lambda) + O(h^{p+1}),$$

where the function $E(t)$ is the solution of the initial value problem

$$\frac{dE}{dt} = Df(\varphi(t,x_0))\, E + \frac{C_{p+1}}{\sum_{j=0}^{k} \beta_j}\, \varphi^{(p+1)}(t,x_0), \quad E(0) = 0.$$

Proof. By means of Theorem 9.1 iii) there is $\tilde{x}_0$ such that

$$|x_n - \Phi^n(h, \tilde{x}_0)| \le K {\chi_A}^n \Delta(x_0, \dots, x_{k-1}, h).$$

We consider

$$e_n = [x_n - \Phi^n(h, \tilde{x}_0)] + [\Phi^n(h, \tilde{x}_0) - \Phi^n(h, x_0)] + [\Phi^n(h, x_0) - \varphi(nh, x_0)]. \quad (9.11)$$

The third bracket is the global error of the one-step method and may be estimated by $\Phi^n(h, x_0) - \varphi(nh, x_0) = E(nh)\, h^p + O(h^{p+1})$ (cf., e.g., Hairer, Nørsett, Wanner [52]). We estimate the second bracket

$$R := \Phi^n(h, \tilde{x}_0) - \Phi^n(h, x_0) = \int_0^1 D_x \Phi^n(h, x_0 + \tau(\tilde{x}_0 - x_0)) d\tau\, (\tilde{x}_0 - x_0).$$

We show that the integral is bounded. Since the method Φ is of order p we know that $D_x \Phi(h, x) = D_x \varphi(h, x) + O(h^{p+1})$. It follows by an induction argument that

$$D_x \Phi^n(h, x) = D_x \varphi(nh, x) + O(h^p)$$

holds for $nh \le T$. Note that $D_x \varphi(t, x)$ is the solution of the variational equation

$$\frac{dW}{dt} = Df(\varphi(t, x_0))W, \quad W(0) = I, \quad t \in [0, T],$$

and therefore is bounded. It follows that the integral is bounded and that $R = O(\Delta(x_0, \dots, x_{k-1}, h)$. Assertion iii) of Theorem 9.1 implies that the first bracket in (9.11) is estimated as

$$|x_n - \Phi^n(h, \tilde{x}_0)| \le K {\chi_A}^n \Delta(x_0, \dots, x_{k-1}, h).$$

Since by Theorem 9.1 iv) the one-step method Φ is of order p one has

$$\Delta(x_0, \dots, x_{k-1}, h) - \Lambda = O(h^{p+1}).$$

This concludes the proof of Theorem 9.3. □

Remark 9.4. The estimate of Theorem 9.3 may be refined as follows:

$$e_n(x_0, \dots, x_{k-1}, h) = E(nh)\, h^p + \delta(nh) + O(({\chi_A}^n + h + \Lambda)\Lambda) + O(h^{p+1})$$

with $\chi_A = \rho_1 + O(h) < 1$. The function $\delta(t)$ is the solution of the initial value problem

$$\frac{d\delta}{dt} = Df(\varphi(t, x_0))\, \delta, \quad \delta(0) = \sum_{i=0}^{k-1} \gamma_i\, e_i$$

for some constants γ_i only depending on the method, cf. Stoffer [119].

Chapter 10
Invariant manifolds for singularly perturbed ODEs

In this chapter we consider singularly perturbed systems of ordinary differential equations (ODEs) of the form $\dot{x} = f(x, y, \varepsilon), \quad \varepsilon \dot{y} = g(x, y, \varepsilon)$ where ε is a small perturbation parameter. Commonly, x is called a "slow variable" while y is called a "fast variable". Such systems arise, e.g., in mechanical multi-body systems or in chemical reaction systems, cf. Lubich [83], Stumpp [126], Gorban, Karlin [46]. A system of singular perturbation type exhibits regions of fast and of slow motion. For $\varepsilon = 0$ the algebraic equation $g(x, y, 0) = 0$ defines the so-called reduced manifold. In an ε-neighbourhood of the reduced manifold the motion is slow. Away from the reduced manifold there is fast motion either approaching or leaving the region of slow motion. ε-close to the reduced manifold there exists an invariant manifold of the ODE. If the manifold is attractive the essential motion of the dynamical system takes place on the manifold. There is a vast literature on invariant manifolds for singularly perturbed ODEs. We mention Fenichel [41], Purfürst [109], Knobloch, Aulbach [70], Nipp [92], [96], Sakamoto [113], Krupa, Smolyan [73], Anosova [1], Battelli, Palmer [14], Liu [80], Tin, Kopell, Jones [127].

In Section 10.1 we investigate attractive manifolds. As an example we prove that the stiff van der Pol equation admits an attractive negatively invariant manifold close to the upper branch of the reduced manifold. In Section 10.2 we deal with hyperbolic manifolds.

10.1 Attractive manifolds

We apply the results of Part II to the singularly perturbed system

$$\begin{aligned} \frac{dx}{dt} &= f(x, y, \varepsilon), \\ \varepsilon \frac{dy}{dt} &= g(x, y, \varepsilon), \end{aligned} \tag{10.1}$$

where ε is a small parameter. We assume that the system is contracting in y-direction. More precisely, we make the following

Assumption ASA$(k + 1)$

Let $X \subset \mathbb{R}^m, Y \subset \mathbb{R}^n$ be nonempty open convex sets, let $\varepsilon_0 > 0$ and let $(f, g)\colon X \times Y \times (-\varepsilon_0, \varepsilon_0) \to \mathbb{R}^m \times \mathbb{R}^n$.

a) *The flow of the differential equation* (10.1) *is outflowing with respect to* X, *i.e., if* X *has a boundary* ∂X, *then it is piecewise of class* C^1 *and* $n_X(x) \cdot f(x, y, \varepsilon) > 0$ *for all* $(x, y, \varepsilon) \in \partial X \times Y \times (-\varepsilon_0, \varepsilon_0)$, n_X *being an outer normal with respect to* X.

b) *There is a function* $s_A^0 \in C_b^{k+1}(X, Y)$, $k \geq 1$, *and positive constants* b_A, C *such that for* $x \in X$,

   i) $g(x, s_A^0(x), 0) = 0$,

   ii) $|f(x, s_A^0(x), 0)| \leq C$,

   iii) *the logarithmic norm of* g_y *satisfies*

$$\mu\big(g_y(x, s_A^0(x), 0)\big) \leq -b_A.$$

c) *There is a constant* d_0 *such that*

$$U^0 := \{(x, y) \mid x \in X,\ |y - s_A^0(x)| < d_0\} \subset X \times Y$$

and such that the functions f, g *are of class* C_b^{k+1} *in* $U^0 \times (-\varepsilon_0, \varepsilon_0)$.

Theorem 10.1. *Let the differential equation* (10.1) *satisfy Assumption ASA*$(k+1)$ *and let* $\beta_A \in (0, b_A)$.

Then there are constants ε_1, d, K *and a function* $s_A \in C_b^k(X, Y)$ *such that the following assertions hold for* $\varepsilon \in (0, \varepsilon_1)$.

i) *The set* $M_\varepsilon = \{(x, y) \mid x \in X,\ y = s_A(x, \varepsilon)\}$ *is a strongly attractive negatively invariant manifold of the differential equation* (10.1) *with attractivity rate* $-\beta_A/\varepsilon$. *More precisely, every solution* $(x(t), y(t))$ *of* (10.1) *with* $(x(0), y(0)) \in M_\varepsilon$ *stays in* M_ε *for* $t < 0$, *and every solution* $(x(t), y(t))$ *of* (10.1) *with* $|y(0) - s_A^0(x(0))| < d$ *satisfies*

$$|y(t) - s_A(x(t), \varepsilon)| \leq K\, e^{-\beta_A t/\varepsilon} |y(0) - s_A(x(0), \varepsilon)|$$

for $t \geq 0$ *as long as* $x(t) \in X$. *The function* s_A *satisfies the invariance equation*

$$\varepsilon \left[\frac{\partial}{\partial x} s_A(x, \varepsilon)\right] f(x, s_A(x, \varepsilon), \varepsilon) = g(x, s_A(x, \varepsilon), \varepsilon).$$

ii) *The invariant manifold* M_ε *is* ε*-close to the so-called reduced manifold* $M_0 = \{(x, y) \mid x \in X,\ y = s_A^0(x)\}$, *i.e.,* $s_A(x, \varepsilon) = s_A^0(x) + O(\varepsilon)$ *as* $\varepsilon \to 0$, *uniformly for* $x \in X$.

iii) *The flow of* (10.1) *on the manifold* M_ε *is given by the differential equation*

$$\dot{x} = f(x, s_A(x, \varepsilon), \varepsilon). \tag{10.2}$$

iv) *Every solution* $(x(t), y(t))$ *of* (10.1) *satisfying* $|y(t) - s_A^0(x(t))| < d$ *for all* $t < 0$ *lies in* M_ε.

v) *If* $X = \mathbb{R}^m$ *then there is a function* $w^s : (x, \eta, \varepsilon) \mapsto w^s(x, \eta, \varepsilon) \in \mathbb{R}^m$ *of class* C_b^{k-1} *defined for* $x \in \mathbb{R}^m, |\eta - s_A^0(x)| < d, \varepsilon \in (0, \varepsilon_1)$ *with the following properties.*

- *The stable fibers* $W^s(x, \varepsilon) = \{(\xi, \eta) \mid \eta \in Y,\ \xi = w^s(x, \eta, \varepsilon)\}$, $x \in \mathbb{R}^m$, *have base points in* M_ε, *i.e.,* $w^s(x, s_A(x, \varepsilon), \varepsilon) = x$.
- *The function* $w^s(x, \cdot, \varepsilon)$ *is* λ_R*-Lipschitz continuous with* $\lambda_R = O(\varepsilon)$.
- *The stable fibers* $W^s(x, \varepsilon)$, $x \in \mathbb{R}^m$, *form a positively invariant family.*
- *The property of asymptotic phase holds: For every solution* $(x(t), y(t))$ *of the differential equation* (10.1) *with* $|y(0) - s_A(x(0), \varepsilon)| < d$ *there is a solution* $\tilde{x}(t)$ *of the differential equation* (10.2) *such that* $w^s(\tilde{x}(t), y(t), \varepsilon) = x(t)$ *and*

$$
\begin{aligned}
|x(t) - \tilde{x}(t)| &\le \varepsilon K\, e^{-\beta_A t/\varepsilon}\, |y(0) - s_A(x(0), \varepsilon)|, \\
|y(t) - s_A(\tilde{x}(t), \varepsilon)| &\le (1 + \varepsilon K)\, e^{-\beta_A t/\varepsilon}\, |y(0) - s_A(x(0), \varepsilon)|
\end{aligned}
$$

holds for $t \ge 0$.

Remark 10.2. (1) We prove Theorem 10.1 under Assumption ASA($k+1$). A refined argument shows that the same results follow from the slightly weaker Assumption ASA(k).

(2) Theorem 10.1 also holds if Assumption ASA($k+1$) b) iii) is replaced by

$$
\mathcal{Re}\big\{\sigma\big(g_y(x, s_A^0(x), 0)\big)\big\} \le -b_A, \tag{10.3}
$$

where σ denotes the spectrum. This substantially relies on the fact that the ODE (10.1) is singularly perturbed. Note that Assumption ASA($k+1$) b) iii) implies assumption (10.3). The latter is often found in the literature.

Proof. We introduce the new variables τ, z by

$$
t = \varepsilon\tau, \quad y = s_A^0(x) + z.
$$

In the so-called fast time τ the singularly perturbed system (10.1) is still a differential equation in the limit $\varepsilon = 0$. In the new variables, (10.1) takes the form

$$
\begin{aligned}
\frac{dx}{d\tau} &= \varepsilon \bar{f}(x, z, \varepsilon), \\
\frac{dz}{d\tau} &= \bar{g}(x, z, \varepsilon),
\end{aligned} \tag{10.4}
$$

where

$$
\begin{aligned}
\bar{f}(x, z, \varepsilon) &:= f(x, s_A^0(x) + z, \varepsilon), \\
\bar{g}(x, z, \varepsilon) &:= g(x, s_A^0(x) + z, \varepsilon) - \varepsilon s_{A,x}^0(x)\, f(x, s_A^0(x) + z, \varepsilon).
\end{aligned}
$$

The functions $\bar{f}, \bar{g}$ are defined for $x \in X, z \in \{z \in \mathbb{R}^n \mid |z| < d_0\}$, $\varepsilon \in (-\varepsilon_0, \varepsilon_0)$ and are bounded and of class C_b^k.

We show that for ε small enough the assumptions of Theorems 7.5 and 7.9 are satisfied for the differential equation (10.4).

We verify Hypothesis HD for $x \in X$ and $z \in \{z \in \mathbb{R}^n \mid |z| < d\}$ with $d \leq d_0$. Hypothesis HD a) is a consequence of Assumption ASA($k+1$) b), c) for ε, d small enough. Hypothesis HD b) is a consequence of ASA($k+1$) a). Hypothesis HD c) is satisfied with

$$\begin{aligned} \ell_{12} &= O(\varepsilon), & \ell_{13} &= O(1), \\ \ell_{21} &= O(1), & \ell_{23} &= O(1), \\ \gamma_{11} &= O(\varepsilon), & \ell_{22} &= -b_A + O(d+\varepsilon). \end{aligned}$$

Hypothesis HDA for $z^* = 0$ is a consequence of Assumption ASA($k+1$) b) and c).

Conditions CD, CDA and CDA(k) are satisfied for d and ε small enough. Hence, all assumptions of Theorems 7.5 and 7.9 are satisfied for (10.4). In order to apply assertion ix) of Theorem 7.9 we verify Hypotheses HDB, HDAB and Condition CDAB($k-1$). Hypothesis HDB is satisfied for some constant γ_{22}. Hypothesis HDAB is satisfied with $\ell_{11} = O(\varepsilon)$ and Condition CDAB($k-1$) has the form

$$\gamma_{11} - \ell_{22} - 2\delta = b_A + O(d+\varepsilon) > (k-1)(\ell_{11} + \delta) = O(\varepsilon)$$

and is satisfied for d and ε small enough.

Going back to the original variables of (10.1) completes the proof of Theorem 10.1. □

Example. We apply Theorem 10.1 to the stiff van der Pol differential equation, cf. Section 13.1,

$$\begin{aligned} \frac{dx}{dt} &= 1 + y =: f(x,y), \\ \varepsilon\frac{dy}{dt} &= -x - y^2 - y^3/3 =: g(x,y). \end{aligned} \tag{10.5}$$

The algebraic equation $g(x,y) = 0$ has a solution $y = s_+^0(x)$ defined for $x \in (-\infty, 0]$. We show that ε-close to this upper branch $\mathcal{G}_+$ of the so-called reduced manifold, cf. Figure 13.1, the differential equation (10.5) admits a strongly attractive negatively invariant manifold.

Theorem 10.3. *Let $\xi > 0$ and let the function $s_+^0 \colon X := (-\infty, -\xi] \to \mathbb{R}^+$ satisfy $g(x, s_+^0(x)) = 0$. Let $k \in \mathbb{N}$.*

Then there are positive constants ε_1, β_A, d, K such that there exists a function $s_+^\infty \colon (-\infty, -\xi] \times (0, \varepsilon_1] \to \mathbb{R}^+$ such that the following assertions hold.

i) *The set* $M^{\infty}_{+,\varepsilon} = \{(x, y) \mid x \in (-\infty, -\xi],\ y = s^{\infty}_{+}(x, \varepsilon)\}$ *is a strongly attractive negatively invariant manifold of the differential equation* (10.5)*. More precisely, every solution* $(x(t), y(t))$ *of* (10.5) *with* $|y(0) - s^0_+(x(0))| < d$ *satisfies*

$$|y(t) - s^{\infty}_{+}(x(t), \varepsilon)| \leq K\, e^{-\beta_A t/\varepsilon} |y(0) - s^{\infty}_{+}(x(0), \varepsilon)|$$

for $t \geq 0$ *as long as* $x(t) < -\xi$.

ii) *The function* s^{∞}_{+} *satisfies the invariance equation*

$$\varepsilon\Big[\frac{\partial}{\partial x} s^{\infty}_{+}(x, \varepsilon)\Big] f(x, s^{\infty}_{+}(x, \varepsilon)) = g(x, s^{\infty}_{+}(x, \varepsilon)).$$

iii) *The function* s^{∞}_{+} *is of class* C^k_b.

iv) *The function* s^{∞}_{+} *has the following expansion with respect to* ε*:*

$$s^{\infty}_{+}(x, \varepsilon) = s^0_+(x) + \varepsilon \frac{(1 + s^0_+(x))}{[s^0_+(x)(2 + s^0_+(x))]^2} + O(\varepsilon^2).$$

Proof. The function s^0_+ satisfies $x = -s^0_+(x)^2 - s^0_+(x)^3/3$. Introducing the new time $\bar{t} = s^0_+(x)^2 t$ we obtain the ODE

$$\frac{dx}{d\bar{t}} = \frac{1 + y}{s^0_+(x)^2} =: \tilde{f}(x, y),$$

$$\varepsilon\frac{dy}{d\bar{t}} = \frac{-x - y^2 - y^3/3}{s^0_+(x)^2} =: \tilde{g}(x, y).$$

We apply Theorem 10.1 to this ODE. For any $k \in \mathbb{N}$, Assumption ASA($k + 1$) is satisfied for $s^0_A := s^0_+$, for $\varepsilon_0, d_0 > 0$ small enough and $b_A = 1$. □

10.2 Hyperbolic manifolds

In this section we state a theorem for hyperbolic invariant manifolds of singularly perturbed ordinary differential equations and give a sketch of the proof.

We consider the ODE

$$\begin{aligned}\frac{dx}{dt} &= f(x, y, z, \varepsilon),\\ \varepsilon\frac{dy}{dt} &= g(x, y, z, \varepsilon),\\ \varepsilon\frac{dz}{dt} &= h(x, y, z, \varepsilon),\end{aligned} \tag{10.6}$$

where ε is a small parameter. We assume that the system is contracting in y-direction and expanding in z-direction. More precisely, we make the following

Assumption ASH($k+1$)

Let $X = \mathbb{R}^m$ and let $Y \subset \mathbb{R}^n$, $Z \subset \mathbb{R}^\ell$ be nonempty open convex sets, let $\varepsilon_0 > 0$, $d_0 > 0$ and let $(f, g, h)\colon X \times Y \times Z \times (-\varepsilon_0, \varepsilon_0) \to \mathbb{R}^m \times \mathbb{R}^n \times \mathbb{R}^\ell$ be of class C_b^{k+1}, $k \geq 1$.

a) *There are functions $r_A^0 \in C_b^{k+1}(\mathbb{R}^m, Y)$ and $r_R^0 \in C_b^{k+1}(\mathbb{R}^m, Z)$ and positive constants b_A, b_R, C such that for $x \in \mathbb{R}^m$*

   i) $|y - r_A^0(x)| < d_0$ *implies* $y \in Y$,
   $|z - r_R^0(x)| < d_0$ *implies* $z \in Z$,
   $g(x, r_A^0(x), r_R^0(x), 0) = 0$ *and* $h(x, r_A^0(x), r_R^0(x), 0) = 0$.

   ii) $|f(x, r_A^0(x), r_R^0(x), 0)| \leq C$.

   iii) *The logarithmic norms of*

$$g_y^0(x) := g_y(x, r_A^0(x), r_R^0(x), 0), \quad -h_z^0(x) := -h_z(x, r_A^0(x), r_R^0(x), 0)$$

   satisfy

$$\begin{aligned} \mu\big(g_y^0(x)\big) &\leq -b_A, \quad x \in \mathbb{R}^m, \\ \mu\big(-h_z^0(x)\big) &\leq -b_R, \quad x \in \mathbb{R}^m, \end{aligned}$$

   and the norms of

$$g_z^0(x) := g_z(x, r_A^0(x), r_R^0(x), 0), \quad h_y^0(x) := h_y(x, r_A^0(x), r_R^0(x), 0)$$

   obey the estimate

$$\sqrt{\sup_{x\in\mathbb{R}^m} \{|g_z^0(x)|\} \sup_{x\in\mathbb{R}^m} \{|h_y^0(x)|\}} < \frac{1}{2}(b_A + b_R) \min\{b_A, b_R\}. \tag{10.7}$$

b) *The functions f, g and h are of class C_b^{k+1} in $Q \times (-\varepsilon_0, \varepsilon_0)$ with*

$$Q := \{(x, y, z) \mid x \in \mathbb{R}^m,\ |y - r_A^0(x)| < d_0,\ |z - r_R^0(x)| < d_0\}.$$

Our main goal is to show that for small $\varepsilon > 0$ the differential equation (10.6) admits an attractive negatively invariant manifold M_ε and a repulsive positively invariant manifold N_ε intersecting transversally. The intersection is a hyperbolic invariant manifold K_ε. This situation is sketched in Figure 10.1.

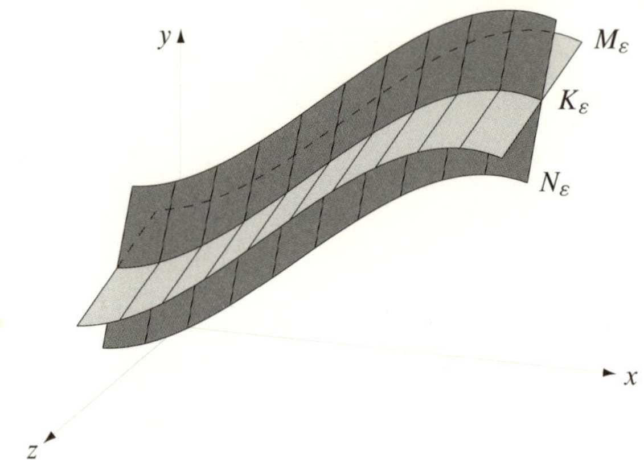

Figure 10.1. The geometric picture for the differential equation (10.6): $K_\varepsilon = M_\varepsilon \cap N_\varepsilon$.

Theorem 10.4. *Let the differential equation* (10.6) *satisfy Assumption ASH*$(k+1)$.
Then for $\beta_A \in (0, b_A)$, $\beta_R \in (0, b_R)$ *there are constants* ε_1, c, d, d_1 *and functions*

$$\begin{aligned} s_A &: (x, z, \varepsilon) \longmapsto s_A(x, z, \varepsilon) \in Y, \\ s_R &: (x, y, \varepsilon) \longmapsto s_R(x, y, \varepsilon) \in Z, \\ r_A &: (x, \varepsilon) \longmapsto r_A(x, \varepsilon) \in Y, \\ r_R &: (x, \varepsilon) \longmapsto r_R(x, \varepsilon) \in Z \end{aligned}$$

of class C_b^k *defined for* x, y, z, ε *with*

$$x \in \mathbb{R}^m, \quad |y - r_A^0(x)| < d, \quad |z - r_R^0(x)| < d, \quad \varepsilon \in (0, \varepsilon_1)$$

and functions

$$\begin{aligned} w^s &: (x, \eta, \varepsilon) \longmapsto w^s(x, \eta, \varepsilon) \in \mathbb{R}^m, \\ w^u &: (x, \zeta, \varepsilon) \longmapsto w^u(x, \zeta, \varepsilon) \in \mathbb{R}^m \end{aligned}$$

of class C_b^{k-1} *defined for* x, η, ζ, ε *with*

$$x \in \mathbb{R}^m, \quad |\eta - r_A^0(x)| < d, \quad |\zeta - r_R^0(x)| < d, \quad \varepsilon \in (0, \varepsilon_1)$$

such that the following assertions hold for $\varepsilon \in (0, \varepsilon_1)$.

i) *The set* $M_\varepsilon = \{(x, y, z) \mid x \in \mathbb{R}^m, |z - r_R^0(x)| < d, y = s_A(x, z, \varepsilon)\}$ *is an attractive negatively invariant manifold of the differential equation* (10.6) *with attractivity rate* $-\beta_A/\varepsilon$. *More precisely, every solution* $(x(t), y(t), z(t))$ *of*

(10.6) *with* $(x(0), y(0), z(0)) \in M_\varepsilon$ *stays in* M_ε *for all* $t < 0$, *and every solution* $(x(t), y(t), z(t))$ *of* (10.6) *with* $|y(0) - r_A^0(x(0))| < d$ *satisfies*

$$|y(t) - s_A(x(t), z(t), \varepsilon)| \le K\, e^{-\beta_A t/\varepsilon} |y(0) - s_A(x(0), z(0), \varepsilon)|$$

for $t \ge 0$ *as long as* $|z(t) - r_R^0(x(t))| < d$.

ii) *The set* $N_\varepsilon = \{(x, y, z) \mid x \in \mathbb{R}^m,\ |y - r_A^0(x)| < d,\ z = s_R(x, y, \varepsilon)\}$ *is a repulsive positively invariant manifold of the differential equation* (10.6) *with repulsivity rate* β_R/ε. *More precisely, every solution* $(x(t), y(t), z(t))$ *of* (10.6) *with* $(x(0), y(0), z(0)) \in N_\varepsilon$ *stays in* N_ε *for all* $t > 0$, *and every solution* $(x(t), y(t), z(t))$ *of* (10.6) *with* $|z(0) - r_R^0(x(0))| < d$ *satisfies*

$$|z(t) - s_R(x(t), y(t), \varepsilon)| \ge K e^{\beta_R t/\varepsilon} |z(0) - s_R(x(0), y(0), \varepsilon)|$$

for $t \ge 0$ *as long as* $|z(t) - r_R^0(x(t))| < d$.

iii) *The set* $K_\varepsilon = \{(x, y, z) \mid x \in \mathbb{R}^m,\ y = r_A(x, \varepsilon),\ z = r_R(x, \varepsilon)\}$ *is a hyperbolic invariant manifold of the differential equation* (10.6). *The manifold* K_ε *is the intersection of the manifolds* M_ε *and* N_ε. N_ε *is a local stable manifold of the hyperbolic invariant manifold* K_ε *and* M_ε *is a local unstable manifold of* K_ε.

Moreover, the identities

$$r_A(x, \varepsilon) = s_A(x, r_R(x, \varepsilon), \varepsilon),$$
$$r_R(x, \varepsilon) = s_R(x, r_A(x, \varepsilon), \varepsilon)$$

hold for all $x \in \mathbb{R}^m$.

The invariant manifold K_ε *is* ε*-close to the reduced manifold* $K_0 = \{(x, y, z) \mid x \in \mathbb{R}^m,\ y = r_A^0(x),\ z = r_R^0(x)\}$, *i.e.,*

$$r_A(x, \varepsilon) = r_A^0(x) + O(\varepsilon),$$
$$r_R(x, \varepsilon) = r_R^0(x) + O(\varepsilon)$$

for $\varepsilon \to 0$.

The differential equation

$$\dot{x} = f(x, r_A(x, \varepsilon), r_R(x, \varepsilon), \varepsilon) \tag{10.8}$$

describes the flow on the manifold K_ε.

iv) *There is a positively invariant family of stable fibers*

$$W^s(x, \varepsilon) = \big\{(\xi, \eta, \zeta) \mid |\eta - r_A^0| < d_1,\ \xi = w^s(x, \eta, \varepsilon),\ \zeta = s_R(\xi, \eta, \varepsilon)\big\} \subset N_\varepsilon,$$

$x \in \mathbb{R}^m$, *and a negatively invariant family of unstable fibers*

$$W^u(x, \varepsilon) = \big\{(\xi, \eta, \zeta) \mid |\zeta - r_R^0| < d_1,\ \xi = w^u(x, \zeta, \varepsilon),\ \eta = s_A(\xi, \zeta, \varepsilon)\big\} \subset M_\varepsilon,$$

$x \in \mathbb{R}^m$. The functions $w^s(x,\cdot,\varepsilon)$ and $w^u(x,\cdot,\varepsilon)$ are λ_R-Lipschitz continuous and λ_A-Lipschitz continuous, respectively, with $\lambda_R = O(\varepsilon)$, $\lambda_A = O(\varepsilon)$. Moreover,

$$\bigcup_{x\in\mathbb{R}^m} W^s(x,\varepsilon) \subset N_\varepsilon \quad \text{and} \quad \bigcup_{x\in\mathbb{R}^m} W^u(x,\varepsilon) \subset M_\varepsilon.$$

v) *The manifold K_ε has the property of asymptotic phase: For every solution $(x(t), y(t), z(t))$ of the differential equation* (10.6) *with $(x(0), y(0), z(0)) \in N_\varepsilon$ there is a solution $\tilde{x}(t)$ of the differential equation* (10.8) *such that*

$$|x(t) - \tilde{x}(t)| \le \varepsilon c\, e^{-\beta_A t/\varepsilon}\, |y(0) - r_A(x(0),\varepsilon)|$$

$$|y(t) - r_A(\tilde{x}(t),\varepsilon)| \le (1+\varepsilon c)\, e^{-\beta_A t/\varepsilon}\, |y(0) - r_A(x(0),\varepsilon)|$$

$$|z(t) - r_R(\tilde{x}(t),\varepsilon)| \le c e^{-\beta_A t/\varepsilon}\, |y(0) - r_A(x(0),\varepsilon)|$$

holds for $t \ge 0$. For every solution $(x(t), y(t), z(t))$ of the differential equation (10.6) *with $(x(0), y(0), z(0)) \in M_\varepsilon$ there is a solution $\tilde{x}(t)$ of the differential equation* (10.8) *such that*

$$|x(t) - \tilde{x}(t)| \le \varepsilon c\, e^{\beta_R t/\varepsilon}\, |z(0) - r_R(x(0),\varepsilon)|,$$

$$|y(t) - r_A(\tilde{x}(t),\varepsilon)| \le c e^{\beta_R t/\varepsilon}\, |z(0) - r_R(x(0),\varepsilon)|,$$

$$|z(t) - r_R(\tilde{x}(t),\varepsilon)| \le (1+\varepsilon c)\, e^{\beta_R t/\varepsilon}\, |z(0) - r_R(x(0),\varepsilon)|$$

holds for $t \le 0$.

vi) *Every solution $(x(t), y(t), z(t))$ of the differential equation* (10.6) *satisfying $|y(t) - r_A^0(x(t))| < d$ and $|z(t) - r_R^0(x(t))| < d$ for $t \in \mathbb{R}$, lies in K_ε.*

Proof (*sketch*). i) We transform the time by $t = \varepsilon\tau$ and introduce new coordinates (x_A, y_A) by

$$x_A = \begin{pmatrix} x \\ \zeta_A \end{pmatrix} \quad \text{with } z = r_R^0(x) + \zeta_A,$$

$$y = r_A^0(x) + y_A.$$

With the notation

$$f_A(x_A, y_A, \varepsilon)$$
$$:= \begin{pmatrix} \varepsilon f(x, r_A^0(x) + y_A, r_R^0(x) + \zeta_A, \varepsilon) \\ h(x, r_A^0(x) + y_A, r_R^0(x) + \zeta_A, \varepsilon) - \varepsilon r_{R,x}^0(x) f(x, r_A^0(x) + y_A, r_R^0(x) + \zeta_A, \varepsilon) \end{pmatrix},$$

$$g_A(x_A, y_A, \varepsilon)$$
$$:= g(x, r_A^0(x) + y_A, r_R^0(x) + \zeta_A, \varepsilon) - \varepsilon r_{A,x}^0(x) f(x, r_A^0(x) + y_A, r_R^0(x) + \zeta_A, \varepsilon)$$

the ODE (10.6) is taken to

$$\begin{aligned}\frac{dx_A}{d\tau} &= f_A(x_A, y_A, \varepsilon),\\ \frac{dy_A}{d\tau} &= g_A(x_A, y_A, \varepsilon).\end{aligned} \tag{10.9}$$

We apply Theorem 7.5 to the differential equation (10.9). Hypothesis HD is satisfied with

$$\begin{aligned}\ell_{12} &= \sup_{x\in\mathbb{R}^m} |h_y^0(x)| + O(d) + O(\varepsilon), & \ell_{21} &= \sup_{x\in\mathbb{R}^m} |g_z^0(x)| + O(d) + O(\varepsilon),\\ \gamma_{11} &= b_R + O(d) + O(\varepsilon), & \ell_{22} &= -b_A + O(d) + O(\varepsilon).\end{aligned} \tag{10.10}$$

Hypothesis HDA is satisfied with $y_A^* := 0$. Under the condition (10.7), Conditions CD, CDA and CDA(k) are satisfied for d and ε small enough.

ii) We transform the time by $t = \varepsilon\tau$ and introduce new coordinates (x_R, y_R) by

$$z = r_R^0(x) + x_R,$$

$$y_R = \begin{pmatrix} x \\ \eta_R \end{pmatrix} \quad \text{with } y = r_A^0(x) + \eta_R.$$

With the notation

$$\begin{aligned}&f_R(x_R, y_R, \varepsilon)\\ &:= h(x, r_A^0(x) + \eta_R, r_R^0(x) + x_R, \varepsilon) - \varepsilon r_{R,x}^0(x) f(x, r_A^0(x) + \eta_R, r_R^0(x) + x_R, \varepsilon),\end{aligned}$$

$$\begin{aligned}&g_R(x_R, y_R, \varepsilon)\\ &:= \begin{pmatrix} \varepsilon f(x, r_A^0(x) + \eta_R, r_R^0(x) + x_R, \varepsilon) \\ g(x, r_A^0(x) + \eta_R, r_R^0(x) + x_R, \varepsilon) - \varepsilon r_{A,x}^0(x) f(x, r_A^0(x) + \eta_R, r_R^0(x) + x_R, \varepsilon) \end{pmatrix}\end{aligned}$$

the ODE (10.6) is taken to

$$\begin{aligned}\frac{dx_R}{d\tau} &= f_R(x_R, y_R, \varepsilon),\\ \frac{dy_R}{d\tau} &= g_R(x_R, y_R, \varepsilon).\end{aligned} \tag{10.11}$$

We apply Theorem 7.10 to the differential equation (10.11). Hypothesis HD is again satisfied with the constants (10.10). Hypothesis HDR is satisfied with $x_R^* := 0$. Under the condition (10.7), Conditions CD, CDR and CDR(k) are satisfied for d and ε small enough.

iii) The hyperbolic invariant manifold K_ε is obtained as the intersection $K_\varepsilon = M_\varepsilon \cap N_\varepsilon$.

iv), v) We consider The ODE (10.6) reduced to N_ε and get

$$\begin{aligned} \dot{x} &= f(x, y, s_R(x, y, \varepsilon), \varepsilon), \\ \varepsilon\dot{y} &= g(x, y, s_R(x, y, \varepsilon), \varepsilon). \end{aligned}$$

The stable fibers of K_ε and the asymptotic phase property are obtained from Theorem 10.1. The unstable fibers are obtained analogously.

vi) Cf. Theorem 10.1 iv). □

Chapter 11
Runge–Kutta methods applied to singularly perturbed ODEs

We consider singularly perturbed autonomous systems of the form

$$\begin{aligned}\frac{dx}{dt} &= f(x, y, \varepsilon),\\ \varepsilon\frac{dy}{dt} &= g(x, y, \varepsilon).\end{aligned} \tag{11.1}$$

As in Section 10.1 we assume

Assumption ASA$(k + 1)$

Let $X \subset \mathbb{R}^m, Y \subset \mathbb{R}^n$ be nonempty open convex sets, let $\varepsilon_0 > 0$ and let $(f, g)\colon X \times Y \times (-\varepsilon_0, \varepsilon_0) \to \mathbb{R}^m \times \mathbb{R}^n$.

a) *The flow of the differential equation* (10.1) *is outflowing with respect to X, i.e., if X has a boundary ∂X, then it is piecewise of class C^1 and $n_X(x) \cdot f(x, y, \varepsilon) > 0$ for all $(x, y, \varepsilon) \in \partial X \times Y \times (-\varepsilon_0, \varepsilon_0)$, n_X being an outer normal with respect to X.*

b) *There is a function $s_A^0 \in C_b^{k+1}(X, Y)$, $k \geq 1$, and positive constants b_A, C such that for $x \in X$*

   i) $g(x, s_A^0(x), 0) = 0,$

   ii) $|f(x, s_A^0(x), 0)| \leq C,$

   iii) *the logarithmic norm of g_y satisfies*

$$\mu\big(g_y(x, s_A^0(x), 0)\big) \leq -b_A.$$

c) *There is a constant d_0 such that*

$$U^0 := \{(x, y) \mid x \in X,\ |y - s_A^0(x)| < d_0\} \subset X \times Y$$

   and such that the functions f, g are of class C_b^{k+1} in $U^0 \times (-\varepsilon_0, \varepsilon_0)$.

For simplicity we take $X = \mathbb{R}^m$. It follows that the differential equation (11.1) admits a highly attractive invariant manifold $M_\varepsilon = \{(x, y) \mid x \in \mathbb{R}^m,\ y = s_A(x, \varepsilon)\}$ with properties stated in Theorem 10.1.

We apply a Runge–Kutta method (RKM) to the differential equation (11.1) and show that the generated discrete dynamical system admits an attractive invariant manifold close to the invariant manifold of the differential equation. Singularly perturbed systems of the type (11.1) are a model class for stiff systems. The smaller the perturbation parameter ε the stiffer the system. Stiff systems are considerably more difficult to solve numerically than nonstiff ones.

In Section 11.1 we apply nonstiff RKMs of order p with constant step size to (11.1). For stability reasons the step size has to be taken as εh with sufficiently small h. We prove that the RK-map admits an invariant manifold $O(\varepsilon^2 h^p)$-close to the invariant manifold of the differential equation. In Section 11.2 we apply a stiff RKM of order p and stage order q with constant step size h independent of ε and large compared to ε. In this case the RK-manifold is $O(h^{q+1})$-close to the manifold of the differential equation in general and $O(\varepsilon h^q)$-close for stiffly accurate methods. On the RK-manifold the x-component of the RKM is a one-step method of the differential equation $\dot{x} = f(x, s_A(x,\varepsilon), \varepsilon)$ with local error $O(h^{p+1} + \varepsilon h^{q+1})$. We show that this one-step method is conjugate to a one-step method with local error $O(h^{p+1} + \varepsilon h^{q+2})$. These results are used to estimate the global error of the RKM.

For our approach it is essential that an RKM is considered as a map in phase space. An RKM with s stages applied to the differential equation $\dot{w} = F(w)$, $w \in \mathbb{R}^\ell$, is the map which takes $w \in \mathbb{R}^\ell$ to

$$\bar{w} = w + h \sum_{i=1}^{s} b_i \, F(W_i),$$

where the stages W_i are defined by

$$W_i = w + h \sum_{j=1}^{s} a_{ij} \, F(W_j), \quad i = 1, \ldots, s.$$

The RKM is characterised by the $s \times s$-matrix $A = (a_{ij})$ and the s-vector $b = (b_i)$. If $a_{ij} = 0$ for all $i \le j$ the method is explicit, otherwise implicit. It is convenient to introduce the following vectors in $\mathbb{R}^{s\ell}$:

$$W := \begin{pmatrix} W_1 \\ \vdots \\ W_s \end{pmatrix}, \quad F(W) := \begin{pmatrix} F(W_1) \\ \vdots \\ F(W_s) \end{pmatrix}, \quad \mathbb{1} \otimes w = \begin{pmatrix} w \\ \vdots \\ w \end{pmatrix}.$$

In this notation the RK-map may be written as

$$\begin{aligned} \bar{w} &= w + h\,(b^T \otimes I_\ell)\, F(W), \\ W &= \mathbb{1} \otimes w + h\,(A \otimes I_\ell)\, F(W). \end{aligned}$$

11.1 Nonstiff methods

Applying an RKM of order p with step size εh to (11.1) we obtain the RK-map

$$\begin{aligned}\bar{x} &= x + \varepsilon h(b^T \otimes I_m) f(X, Y, \varepsilon),\\ \bar{y} &= y + h(b^T \otimes I_n) g(X, Y, \varepsilon),\end{aligned} \tag{11.2}$$

where X and Y are given by

$$\begin{aligned}X &= \mathbb{1} \otimes x + \varepsilon h(A \otimes I_m) f(X, Y, \varepsilon),\\ Y &= \mathbb{1} \otimes y + h(A \otimes I_n) g(X, Y, \varepsilon).\end{aligned}$$

This map is of the form

$$P_{\mathrm{RK}}\colon \begin{pmatrix} x \\ y \end{pmatrix} \longmapsto \begin{pmatrix} \bar{x} \\ \bar{y} \end{pmatrix} = \begin{pmatrix} x + \varepsilon h F(x, y, \varepsilon, h) \\ y + h G(x, y, \varepsilon, h) \end{pmatrix} \tag{11.3}$$

and it admits an invariant manifold as stated in the following theorem.

Theorem 11.1. *Let the differential eqution* (11.1) *satisfy Assumption ASA*$(k + 1)$*. Apply an RKM of order p with step size* εh *to* (11.1)*, let* $\beta_A \in (0, b_A)$ *and assume* $k > p$*.*

Then there are constants d, ν, δ, K *and functions* s_A *and* σ_A *of class* C_b^k *such that for* $h \in (0, \nu]$, $\varepsilon \in (0, \delta]$ *the following assertions hold.*

i) *The set* $M_\varepsilon = \{(x, y) \mid x \in \mathbb{R}^m,\ y = s_A(x, \varepsilon)\}$ *is an attractive invariant manifold of the differential equation* (11.1) *with properties given in Theorem* 10.1.

ii) *The set* $M_{h,\varepsilon} = \{(x, y) \mid x \in \mathbb{R}^m,\ y = \sigma_A(x, \varepsilon, h)\}$ *is an invariant manifold of the map* P_{RK} *given in* (11.3)*, i.e.,* $P_{\mathrm{RK}}(M_{h,\varepsilon}) = M_{h,\varepsilon}$.

iii) *The manifold* $M_{h,\varepsilon}$ *is uniformly attractive for* P_{RK} *with attractivity constant* $\chi_A(h) = 1 - h\beta_A < 1$*, i.e., for all* (x, y) *with* $|y - s_A(x, \varepsilon)| < d$ *and for* $(\bar{x}, \bar{y}) = P_{\mathrm{RK}}(x, y)$ *the inequality*

$$|\bar{y} - \sigma_A(\bar{x}, \varepsilon, h)| \le \chi_A(h)\,|y - \sigma_A(x, \varepsilon, h)|$$

holds.

iv) *The manifold* $M_{h,\varepsilon}$ *is* $O(\varepsilon^2 h^p)$*-close to the manifold* M_ε*, i.e., for all* $x \in \mathbb{R}^m$ *the estimate*

$$|\sigma_A(x, \varepsilon, h) - s_A(x, \varepsilon)| \le K\varepsilon^2 h^p$$

holds.

Remark 11.2. (1) The distance of the manifolds $M_{h,\varepsilon}$ and M_ε is smaller than one would expect at first glance. Since one RK-step on $M_{h,\varepsilon}$ leads to an increment of x of order $O(\varepsilon h)$ also the increment of y has to be of order $O(\varepsilon h)$. Hence, one would expect that σ_A is $O(\varepsilon)$-close to s_A^0. Since the RKM is of order p one would expect that $\sigma_A - s_A = O(\varepsilon h^p)$. It is remarkable that the distance of the two manifolds is of order $O(\varepsilon^2 h^p)$.

(2) The invariant manifold M_ε of the differential equation (11.1) has all properties stated in Theorems 7.5 and 7.9. and the invariant manifold $M_{h,\varepsilon}$ of the map (11.3) has all properties stated in Theorems 3.6 and 4.1.

Proof. i) In the proof of Theorem 10.1 we have shown that Assumption ASA$(k+1)$ implies Hypotheses HD, HDA and Conditions CD, CDA, CDA(k) for d and ε small enough. Hence, the assertions of Theorem 10.1 hold for the differential equation (11.1).

ii), iii), iv) The time-εh map of (11.1) with starting value on the invariant manifold M_ε has the form

$$p^h\colon \begin{pmatrix} x \\ s_A(x,\varepsilon) \end{pmatrix} \longmapsto \begin{pmatrix} \tilde{x} \\ s_A(\tilde{x},\varepsilon) \end{pmatrix} = \begin{pmatrix} x + \varepsilon h \tilde{F}(x, s_A(x,\varepsilon), \varepsilon, h) \\ s_A(x,\varepsilon) + h\tilde{G}(x, s_A(x,\varepsilon), \varepsilon, h) \end{pmatrix}. \tag{11.4}$$

Since the RKM has order p it follows that

$$\begin{aligned} \varepsilon F(x, s_A(x,\varepsilon), \varepsilon, h) - \varepsilon \tilde{F}(x, s_A(x,\varepsilon), \varepsilon, h) &= O(h^p), \\ G(x, s_A(x,\varepsilon), \varepsilon, h) - \tilde{G}(x, s_A(x,\varepsilon), \varepsilon, h) &= O(h^p). \end{aligned} \tag{11.5}$$

For simplicity of notation we omit in all functions the dependence on ε and h. In the new variables (x,z), where $y = s_A(x) + z$, the map P_{RK} defines a map $\check{P}_{\mathrm{RK}}\colon (x,z) \in \mathbb{R}^m \times Z_d \mapsto (\bar{x},\bar{z})$ with $Z_d = \{z \in \mathbb{R}^n \mid |z| < d\}$, where d will be chosen later. This map is of the form

$$\check{P}_{\mathrm{RK}}\colon \quad \begin{aligned} \bar{x} &= x + \varepsilon h F(x, s_A(x) + z), \\ \bar{z} &= \big[I_n + h\big(G_y(x, s_A(x)) + O(d) + O(\varepsilon)\big)\big]z - Q(x), \end{aligned} \tag{11.6}$$

where $Q(x) = s_A\big(x + \varepsilon h F(x, s_A(x))\big) - s_A(x) - hG(x, s_A(x))$.

We estimate $Q(x)$.

Assertion 1 $Q(x) = Q_1(x)h^{p+1}$ *with* $|Q_1| \le K_1$, K_1 *independent of* ε *and* h.

Using (11.4) we get

$$\begin{aligned} Q(x) = s_A\big(x + \varepsilon h F(x, s_A(x))\big) - s_A\big(x + \varepsilon h \tilde{F}(x, s_A(x))\big) \\ - h\big[G(x, s_A(x)) - \tilde{G}(x, s_A(x))\big] \end{aligned}$$

implying

$$Q(x) = O(h)\big(\varepsilon F(x, s_A(x)) - \varepsilon \tilde{F}(x, s_A(x))\big) + O(h)\big(G(x, s_A(x)) - \tilde{G}(x, s_A(x))\big).$$

Now, Assertion 1 follows from (11.5).

Assertion 2 $Q(x) = Q_2(x)\varepsilon^2$ *with* $|Q_2| \le K_2$, K_2 *independent of* ε *and* h.

Setting $y = s_A(x)$ in (11.2) yields $(\hat{x}, \hat{y}) = P_{\mathrm{RK}}(x, s_A(x))$ with

$$\hat{x} = x + \varepsilon h(b^T \otimes I_m)\, f(\widehat{X}, \widehat{Y}),$$
$$\hat{y} = s_A(x) + h(b^T \otimes I_n)\, g(\widehat{X}, \widehat{Y}),$$

where

$$\widehat{X} = \mathbb{1} \otimes x + \varepsilon h(A \otimes I_m)\, f(\widehat{X}, \widehat{Y}),$$
$$\widehat{Y} = \mathbb{1} \otimes s_A(x) + h(A \otimes I_n)\, g(\widehat{X}, \widehat{Y}),$$

and Q may be written as

$$Q(x) = s_A(\hat{x}) - s_A(x) - h(b^T \otimes I_n)\, g(\widehat{X}, \widehat{Y}). \tag{11.7}$$

We first estimate $s_A(\widehat{X}) - \widehat{Y}$. We have

$$\begin{aligned} s_A(\widehat{X}) - \widehat{Y} &= s_A\big(\mathbb{1} \otimes x + \varepsilon h(A \otimes I_m)\, f(\widehat{X}, s_A(\widehat{X}))\big) - \mathbb{1} \otimes s_A(x) \\ &\quad - h(A \otimes I_n)\, g(\widehat{X}, s_A(\widehat{X})) + O(h)\big(s_A(\widehat{X}) - \widehat{Y}\big). \end{aligned} \tag{11.8}$$

By means of the invariance equation for s_A, cf. Theorem 7.5 i), we have

$$\begin{aligned} g(\widehat{X}, s_A(\widehat{X})) &= \varepsilon \operatorname{diag}\big[s_{A,x}(\widehat{X})\big] f(\widehat{X}, s_A(\widehat{X})) \\ &= \varepsilon\big(I_s \otimes s_{A,x}(x) + O(\varepsilon h)\big) f(\widehat{X}, s_A(\widehat{X})), \end{aligned} \tag{11.9}$$

where $\operatorname{diag}[s_{A,x}(\widehat{X})]$ denotes the $sn \times sn$-block-diagonal matrix with $n \times n$-blocks $s_{A,x}(\widehat{X}_1), \ldots, s_{A,x}(\widehat{X}_s)$. Expanding the first expression on the right-hand side of (11.8) about $\mathbb{1} \otimes x$ and using (11.9) we find $s_A(\widehat{X}) - \widehat{Y} = O(\varepsilon^2 h^2)$. Using this estimate in (11.7) we get

$$\begin{aligned} Q(x) &= s_A\big(x + \varepsilon h(b^T \otimes I_m)\, f(\widehat{X}, s_A(\widehat{X}))\big) \\ &\quad - s_A(x) - h(b^T \otimes I_n)\, g(\widehat{X}, s_A(\widehat{X})) + O(\varepsilon^2 h^2). \end{aligned}$$

Using (11.9) we find

$$\begin{aligned} Q(x) &= \varepsilon h(b^T \otimes s_{A,x}(x))\, f(\widehat{X}, s_A(\widehat{X})) \\ &\quad - \varepsilon h(b^T \otimes I_n)(I_s \otimes s_{A,x}(x))\, f(\widehat{X}, s_A(\widehat{X})) + O(\varepsilon^2 h^2) \\ &= O(\varepsilon^2 h^2). \end{aligned}$$

This proves Assertion 2.

Assertions 1 and 2 imply the estimate

$$Q(x) = O(\varepsilon^2 h^{p+1}). \tag{11.10}$$

In order to apply Theorems 3.6 and 4.1 to the map (11.6) we verify Hypotheses HM, HMA and Conditions CM, CMA and CMA(k). Note that the map P_{RK} of (11.3) is of class C_b^{k+1} since the vector field (f, g) is of class C_b^{k+1} by Assumption ASA($k+1$). Since $G_y(x, s_A(x)) = g_y(x, s_A(x)) + O(h)$ the map $\check{P}_{\mathrm{RK}}$ of (11.6) is of the form

$$\begin{aligned} \bar{x} &= x + O(\varepsilon h), \\ \bar{z} &= B(x,z)z + O(\varepsilon^2 h^{p+1}) \end{aligned}$$

with $B(x,z) = I_n + h\big(g_y(x, s_A(x)) + O(d + h + \varepsilon)\big)$. Since $|B(x,z)| \le 1 - h\beta_A$ for d, ε, h sufficiently small, Hypothesis HM is satisfied with

$$\begin{aligned} \Gamma_{11} &= 1 + O(\varepsilon h), & L_{12} &= O(\varepsilon h), \\ L_{21} &= O(hd), & L_{22} &= 1 - h\beta_A. \end{aligned}$$

Hypothesis HMA is satisfied with $z^* = 0$. Conditions CM, CMA and CMA(k) are satisfied for ε small enough. It follows that Theorems 3.6 and 4.1 apply to the map (11.6) which proves the assertions ii), iii) and iv). □

11.2 Stiff methods

11.2.1 The invariant manifold

The differential equation (11.1) under Assumption ASA($k+1$) admits a highly attractive invariant manifold with properties stated in Theorem 10.1. We apply a stiff RKM to (11.1) with step size h independent of ε and large compared to ε. We show that the RKM admits an attractive invariant manifold close to the manifold of the differential equation. For simplicity we take $X = \mathbb{R}^m$.

We make the following assumptions on the RKM which are appropriate to integrate stiff systems.

Assumption ASARK

a) *The RKM has order p and stage order q with $1 \le q \le p$.*

b) *The RK-matrix A is invertible.*

c) *The stability function $R(z) := 1 + zb^T(I_s - zA)^{-1}\mathbb{1}$, $z \in \mathbb{C}$, where $\mathbb{1} = (1, \dots, 1)^T \in \mathbb{R}^s$, satisfies $|R(\infty)| < 1$.*

Remark 11.3. Since the RK-matrix A is invertible, $R(\infty)$ may be written as $R(\infty) = 1 - b^T A^{-1}\mathbb{1}$.

An RKM applied to the differential equation (11.1) yields the equations

$$\begin{aligned}\bar{x} &= x + h(b^T \otimes I_m)\, f(X,Y,\varepsilon),\\ \bar{y} &= y + \frac{h}{\varepsilon}(b^T \otimes I_n)\, g(X,Y,\varepsilon),\\ X &= \mathbb{1} \otimes x + h(A \otimes I_m)\, f(X,Y,\varepsilon),\\ Y &= \mathbb{1} \otimes y + \frac{h}{\varepsilon}(A \otimes I_n)\, g(X,Y,\varepsilon).\end{aligned} \tag{11.11}$$

In the next theorem we show that these equations define a smooth map and we state invariant manifold results of this map, cf. Figure 11.1.

Theorem 11.4. *Let the differential equation* (11.1) *satisfy Assumption ASA*$(k+1)$. *Apply an RKM with step size h and satisfying Assumption ASARK to* (11.1) *and assume* $k > p$.

Then there are constants ν, δ, d, c, K *and functions* s_A *and* σ_A *of class* C_b^k *such that for all* $h \in (0,\nu]$ *and all* ε *with* $\varepsilon/h \in (0,\delta]$ *the following assertions hold.*

i) *The set* $M_\varepsilon = \{(x,y) \mid x \in X,\ y = s_A(x,\varepsilon)\}$ *is an attractive invariant manifold of the differential equation* (11.1) *with properties given in Theorem* 10.1.

ii) *The equations* (11.11) *define a smooth map*

$$P_{\mathrm{RK}}\colon U_d \ni \begin{pmatrix} x \\ y \end{pmatrix} \longmapsto \begin{pmatrix} \bar{x} \\ \bar{y} \end{pmatrix} = \begin{pmatrix} F(x,y,\varepsilon,h) \\ G(x,y,\varepsilon,h) \end{pmatrix} \in \mathbb{R}^m \times \mathbb{R}^n$$

of class C_b^k *where* $U_d = \{(x,y) \mid x \in \mathbb{R}^m,\ |y - s_A(x,\varepsilon)| < d\}$.

iii) *The set* $M_{h,\varepsilon} = \{(x,y) \mid x \in \mathbb{R}^m,\ y = \sigma_A(x,\varepsilon,h)\}$ *is an invariant manifold of the map* P_{RK}, *i.e.,* $P_{\mathrm{RK}}(M_{h,\varepsilon}) = M_{h,\varepsilon}$.

iv) *The manifold* $M_{h,\varepsilon}$ *is uniformly attractive for* P_{RK} *with attractivity constant* $\chi_A(\varepsilon,h) = |R(\infty)| + c\,\varepsilon/h < 1$, *i.e., for all* $(x,y) \in U_d$ *and for* $(\bar{x},\bar{y}) := P_{\mathrm{RK}}(x,y)$ *the estimate*

$$|\bar{y} - \sigma_A(\bar{x},\varepsilon,h)| \le \chi_A(\varepsilon,h)\, |y - \sigma_A(x,\varepsilon,h)|$$

holds.

v) *There is a function* $w_{h,\varepsilon}^s\colon U_d \to \mathbb{R}^m$ *such that the stable fibers* $W_{h,\varepsilon}^s(x) := \{(\xi,\eta) \mid (x,\eta) \in U_d,\ \xi = w_{h,\varepsilon}^s(x,\eta)\}$, $x \in \mathbb{R}^m$, *form a positively invariant family, i.e.,*

$$P_{\mathrm{RK}}(W_{h,\varepsilon}^s(x)) \subset W_{h,\varepsilon}^s(P_{\mathrm{RK}}(x,\sigma_A(x,\varepsilon,h))).$$

The function $w_{h,\varepsilon}^s(x,\cdot)$ *is uniformly* λ_R*-Lipschitz with* $\lambda_R = O(\varepsilon)$, *of class* C_b^k *and satisfies the invariance equation*

$$F(w_{h,\varepsilon}^s(x,\eta),\eta,\varepsilon,h) = w_{h,\varepsilon}^s(F(x,\sigma_A(x,\varepsilon,h),\varepsilon,h), G(w_{h,\varepsilon}^s(x,\eta),\eta,\varepsilon,h)).$$

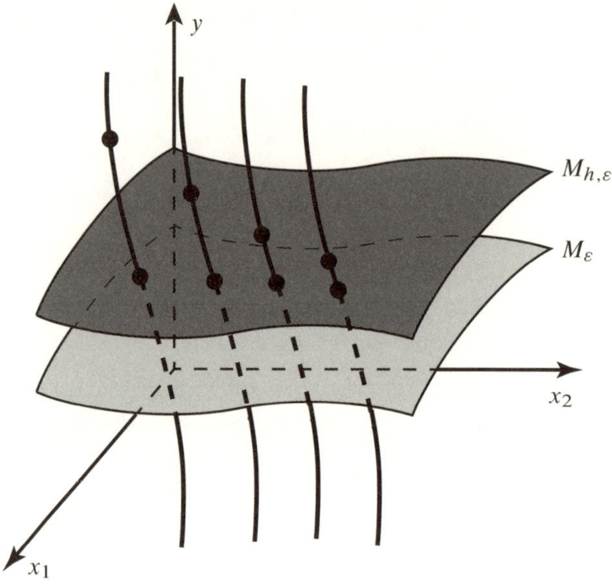

Figure 11.1. The invariant manifold M_ε of the differential equation, the invariant manifold $M_{h,\varepsilon}$ of the RKM and an RKM-orbit with its asymptotic phase orbit and with its corresponding fibers.

vi) *The property of asymptotic phase holds: For every orbit* (x_j, y_j) *of the map* P_{RK} *with* $(x_0, y_0) \in U_d$ *there is an orbit* $(\tilde{x}_j, \sigma_A(\tilde{x}_j, \varepsilon, h))$ *on* $M_{h,\varepsilon}$ *such that*

$$|x_j - \tilde{x}_j| \le \varepsilon K \chi_A(\varepsilon, h)^j |y_0 - \sigma_A(x_0, \varepsilon, h)|,$$
$$|y_j - \sigma_A(\tilde{x}_j, \varepsilon, h)| \le (1 + \varepsilon K)\chi_A(\varepsilon, h)^j |y_0 - \sigma_A(x_0, \varepsilon, h)|$$

holds for $j \ge 0$.

vii) *The manifold* $M_{h,\varepsilon}$ *is close to the manifold* M_ε*, more precisely,*

$$|\sigma_A(x, \varepsilon, h) - s_A(x, \varepsilon)| \le \begin{cases} K\,\varepsilon\, h^q & \text{if } b_i = a_{si},\ i = 1, \dots, s, \\ K\, h^{q+1} & \text{else,} \end{cases}$$

holds for all $x \in \mathbb{R}^m$. *Moreover, the stages of the RKM* (11.11) *satisfy*

$$|Y - s_A(X, \varepsilon)| \le K\, \frac{\varepsilon}{h} \left(|y - s_A(x, \varepsilon)| + h^{q+1}\right)$$

for all $(x, y) \in U_d$.

Proof. i) As in the proof of Theorem 11.1 we conclude that Hypotheses HD, HDA and Conditions CD, CDA, CDA(k) are satisfied for d and ε small enough. Hence the assertions of Theorem 10.1 hold for the differential equation (11.1).

ii) In what follows we mostly suppress the dependence of the functions on ε and h. We verify that the equations (11.11) have a unique solution in U_d, d sufficiently small. We introduce the new variables z, Z measuring the difference to the manifold M_ε of the differential equation (11.1) by

$$y = s_A(x) + z, \quad Y = s_A(X) + Z.$$

We expand the function g about $z = 0$ and get

$$g(x, s_A(x) + z) = g(x, s_A(x)) + (B(x) + \widehat{B}(x, z))\, z$$

with $B(x) := g_y(x, s_A(x)) = g_y(x, s_A^0(x)) + O(\varepsilon)$ and $\widehat{B}(x, z) = O(d)$, d small. In the new variables the equations (11.11) take the form

$$\begin{aligned} \bar{x} &= x + h(b^T \otimes I_m) f(X, s_A(X) + Z), \\ s_A(\bar{x}) + \bar{z} &= s_A(x) + z \\ &\quad + \frac{h}{\varepsilon}(b^T \otimes I_n)\{g(X, s_A(X)) + \operatorname{diag}\big[B(X) + \widehat{B}(X, Z)\big]\, Z\}, \end{aligned} \tag{11.12}$$

$$\begin{aligned} X &= \mathbb{1} \otimes x + h(A \otimes I_m) f(X, s_A(X) + Z), \\ s_A(X) + Z &= \mathbb{1} \otimes s_A(x) + \mathbb{1} \otimes z \\ &\quad + \frac{h}{\varepsilon}(A \otimes I_n)\{g(X, s_A(X)) + \operatorname{diag}\big[B(X) + \widehat{B}(X, Z)\big]\, Z\}, \end{aligned} \tag{11.13}$$

where, e.g., $\operatorname{diag}[B(X)]$ denotes the $sn \times sn$-block-diagonal matrix with $n \times n$-blocks $B(X_1), \dots, B(X_s)$. We consider these equations for $(x, z) \in \mathbb{R}^m \times Z_d$, $Z_d := \{z \mid z \in \mathbb{R}^n,\ |z| < d\}$, where d will be determined later. We suppose that $Z \in [Z_d]^s$. Equation (11.13) yields

$$Z = \frac{\varepsilon}{h}\, C(X, Z)^{-1}(\mathbb{1} \otimes z - E(x, X)) \tag{11.14}$$

with

$$C(X, Z) := -(A \otimes I_n)\operatorname{diag}\big[B(X) + \widehat{B}(X, Z)\big] + \frac{\varepsilon}{h}(I_s \otimes I_n), \tag{11.15}$$

$$E(x, X) := s_A(X) - \mathbb{1} \otimes s_A(x) - \frac{h}{\varepsilon}(A \otimes I_n)\, g(X, s_A(X)). \tag{11.16}$$

Note that due to Assumption ASA($k + 1$) b) iii) and Assumption ASARK b) the matrix C is invertible for d and ε/h small enough. Inserting the expression obtained for Z into (11.12) we find

$$\begin{aligned} \bar{z} &= z - e(x, \bar{x}, X) \\ &\quad + (b^T \otimes I_n)\operatorname{diag}\big[B(X) + \widehat{B}(X, Z)\big]\, C(X, Z)^{-1}\,(\mathbb{1} \otimes z - E(x, X)), \end{aligned} \tag{11.17}$$

where

$$e(x, \bar{x}, X) := s_A(\bar{x}) - s_A(x) - \frac{h}{\varepsilon}\,(b^T \otimes I_n)\, g(X, s_A(X)). \tag{11.18}$$

In what follows we mostly suppress the arguments of E and e for simplicity of notation. Note that E and e are of order $O(h)$. This is due to the fact that $g(X, s_A^0(X)) = 0$ and $s_A(X) - s_A^0(X) = O(\varepsilon)$ and that $X - \mathbb{1} \otimes x$ and $\bar{x} - x$ are $O(h)$. By means of (11.15) we may replace $\operatorname{diag}\big[B(X) + \widehat{B}(X, Z)\big]$ by $(A^{-1} \otimes I_n)\big(\frac{\varepsilon}{h}\,(I_s \otimes I_n) - C(X, Z)\big)$. Hence, using equations (11.14), (11.17) we find that the map P_{RK} may be written in the form

$$\check{P}_{\mathrm{RK}}: \quad \begin{aligned} \bar{x} &= x + h(b^T \otimes I_m) f(X, s_A(X) + Z), \\ \bar{z} &= \Big(R(\infty)\, I_n + \frac{\varepsilon}{h}\, Q(\,\mathbb{1} \otimes I_n)\Big)\, z + \Big((b^T\, A^{-1} \otimes I_n) - \frac{\varepsilon}{h}\, Q\Big)\, E - e, \end{aligned} \tag{11.19}$$

where $Q := (b^T\, A^{-1} \otimes I_n)\, C(X, Z)^{-1}$ and where X, Z are defined by

$$\begin{aligned} X &= \mathbb{1} \otimes x + h(A \otimes I_m) f(X, s_A(X) + Z), \\ Z &= \frac{\varepsilon}{h}\, C(X, Z)^{-1}\, (\mathbb{1} \otimes z - E). \end{aligned} \tag{11.20}$$

By means of the contraction principle it can be shown that for h and ε/h sufficiently small equation (11.20) has a unique solution $(X(x, z, \varepsilon, h), Z(x, z, \varepsilon, h))$ which is $O(h + \varepsilon/h)$-close to $(\mathbb{1} \otimes x, 0)$. From the implicit function theorem it follows that for small h and ε/h this solution is smooth with bounded derivatives. Therefore, the map $\check{P}_{\mathrm{RK}}$ given in (11.19) is well defined and is of the form

$$\check{P}_{\mathrm{RK}}: \begin{pmatrix} x \\ z \end{pmatrix} \longmapsto \begin{pmatrix} \bar{x} \\ \bar{z} \end{pmatrix} = \begin{pmatrix} \check{F}(x, z, \varepsilon, h) \\ \check{G}(x, z, \varepsilon, h) \end{pmatrix},$$

where $\check{F}$ and $\check{G}$ are of class C_b^k.

iii), iv), v), vi) We apply the invariant manifold Theorems 3.6 and 4.1. Hypothesis HM is satisfied for h and ε/h small enough with

$$\begin{aligned} \Gamma_{11} &= 1 + O(h), & L_{12} &= O(\varepsilon), \\ L_{21} &= O\Big(\frac{\varepsilon}{h} d + h\Big), & L_{22} &= |R(\infty)| + O\Big(\frac{\varepsilon}{h}\Big). \end{aligned}$$

The estimates for L_{12} and L_{22} might not be obvious. They may be obtained as follows. Differentiating (11.20) with respect to z one gets $dX/dz = O(h)\, dZ/dz$ and $dZ/dz = O(\varepsilon/h)$. Differentiating (11.19) with respect to z and using these estimates yields the above estimates for L_{12} and L_{22}. Hypothesis HMA holds for $z^* = 0$. Conditions CM, CMA and CMA(k) are satisfied for h and ε/h small enough. Theorem 3.6 implies the existence of a smooth attractive invariant manifold

$$\check{M}_{h,\varepsilon} = \{(x, z) \mid x \in \mathbb{R}^m,\ z = \check{\sigma}_A(x, \varepsilon, h)\}$$

of the map $\check{P}_{\mathrm{RK}}$. In the original variables this gives rise to an invariant manifold $M_{h,\varepsilon} = \{(x, y) \mid x \in \mathbb{R}^m,\ y = \sigma_A(x, \varepsilon, h) := s_A(x, \varepsilon) + \check{\sigma}_A(x, \varepsilon, h)\}$ proving assertions iii), iv). Assertions v) and vi) follow from Theorem 4.1 where we write $w^s_{h,\varepsilon}(x, \eta)$ instead of $w^s_{h,\varepsilon}(x, \sigma_A(x, \varepsilon, h), \eta)$, for short.

vii) In order to estimate the distance of the manifolds $M_{h,\varepsilon}$ and M_ε we estimate $\check{\sigma}_A$.

Assertion. *The following estimates hold:*

$$Z = O\Big(\frac{\varepsilon}{h}\Big)\ (\mathbb{1} \otimes z) + O(\varepsilon\, h^q), \tag{11.21}$$

$$e(x, \bar{x}, X) = O(\varepsilon)\ (\mathbb{1} \otimes z) + O(\varepsilon\, h^{q+1}) + O(h^{p+1}), \tag{11.22}$$

$$E(x, X) = O(\varepsilon)\ (\mathbb{1} \otimes z) + O(h^{q+1}). \tag{11.23}$$

We consider solutions $(u(t), v(t))$ of the differential equation (11.1) on the manifold M_ε. These solutions satisfy the differential equation

$$\begin{aligned} \dot{u} &= f(u, s_A(u)), \\ \dot{v} &= \frac{1}{\varepsilon}\, g(u, s_A(u)) = s_{A,x}(u) f(u, s_A(u)). \end{aligned} \tag{11.24}$$

Applying the RKM to the differential equation (11.24) we obtain

$$\begin{aligned} \bar{u} &= u + h(b^T \otimes I_m)\ f(U, s_A(U)), \\ \bar{v} &= v + h(b^T \otimes I_n)\ \mathrm{diag}\,[s_{A,x}(U)]\ f(U, s_A(U)) \end{aligned} \tag{11.25}$$

with

$$\begin{aligned} U &= \mathbb{1} \otimes u + h(A \otimes I_m)\ f(U, s_A(U)), \\ V &= \mathbb{1} \otimes v + h(A \otimes I_n)\ \mathrm{diag}\,[s_{A,x}(U)]\ f(U, s_A(U)). \end{aligned} \tag{11.26}$$

For $u = x$ and $v = s_A(x)$ we have for $E(x, X)$ of (11.16) and $e(x, \bar{x}, X)$ of (11.18)

$$\begin{aligned} e(x, \bar{x}, X) &= e(u, \bar{u}, U) + O(1)(\bar{x} - \bar{u}) + O(h)(X - U), \\ E(x, X) &= E(u, U) + O(1)(X - U). \end{aligned} \tag{11.27}$$

Since the method is of order p and has stage order q and since f, g are of class C^k_b with $k > p$, we conclude that

$$O(h^{p+1}) = \bar{v} - v(h) = \bar{v} - s_A(u(h)) = \bar{v} - s_A(\bar{u}) + O(h^{p+1}),$$

$$O(h^{q+1}) = V - V(ch) = V - s_A(U(ch)) = V - s_A(U) + O(h^{q+1}),$$

where $V(ch) = \big(v(c_1h)^T, \dots, v(c_sh)^T\big)^T$ and $U(ch) = \big(u(c_1h)^T, \dots, u(c_sh)^T\big)^T$ with $c_i = \sum_{j=1}^{s} a_{ij}$. Using equations (11.16), (11.18) and (11.25), (11.26) it follows that

$$\begin{aligned} e(u, \bar{u}, U) &= s_A(\bar{u}) - \bar{v} = O(h^{p+1}), \\ E(u, U) &= s_A(U) - V = O(h^{q+1}). \end{aligned} \tag{11.28}$$

In order to estimate $e(x, \bar{x}, X)$ and $E(x, X)$ it remains to estimate $\bar{x} - \bar{u}$ and $X - U$ where $\bar{x}$ and X are defined by equations (11.19), (11.20) and $\bar{u}$, U by equations (11.25), (11.26). For $u = x$ we have

$$\begin{aligned} \bar{x} - \bar{u} &= h(b^T \otimes I_m)\big(f(X, s_A(X) + Z) - (f(U, s_A(U))\big), \\ X - U &= h(A \otimes I_m)\big(f(X, s_A(X) + Z) - (f(U, s_A(U))\big) \end{aligned}$$

implying

$$\begin{aligned} \bar{x} - \bar{u} &= O(h)\, Z + O(h)\,(X - U), \\ \big((I_s \otimes I_m) + O(h)\big)(X - U) &= O(h)\, Z \end{aligned}$$

and

$$\begin{aligned} \bar{x} - \bar{u} &= O(h)\, Z, \\ X - U &= O(h)\, Z. \end{aligned} \tag{11.29}$$

Using equations (11.27), (11.28) and (11.29) we find

$$\begin{aligned} e(x, \bar{x}, X) &= O(h^{p+1}) + O(h)\, Z, \\ E(x, X) &= O(h^{q+1}) + O(h)\, Z. \end{aligned}$$

Inserting the second estimate into (11.20) gives the estimate (11.21) and hence the estimates (11.22), (11.23) hold proving the assertion.

Inserting the estimates (11.21), (11.22), (11.23) into the z-equation of the map $\check{P}_{\mathrm{RK}}$, cf. (11.19), yields

$$\bar{z} = \widetilde{B}(x, z, \varepsilon, h)z + O(h^{q+1})$$

with $|\widetilde{B}| = |R(\infty)| + O(\varepsilon/h) \leq \beta$, for some $\beta \in (0, 1)$ and ε/h sufficiently small. Theorem 1.5 vi) implies that $\check{\sigma}_A = O(h^{q+1})$. In the case $b_i = a_{si}$, $i = 1, \dots, s$, one has $\bar{z} = Z_s$. Therefore, for $(x, z) \in \check{M}_{h,\varepsilon}$ equation (11.21) implies $Z = O(\varepsilon\, h^q)$ and hence $\check{\sigma}_A = O(\varepsilon\, h^q)$.

Expressing the results above in the original variables x, y and defining $\sigma_A(x, \varepsilon, h) := s_A(x, \varepsilon) + \check{\sigma}_A(x, \varepsilon, h)$ completes the proof of Theorem 11.4. □

11.2.2 The global error

The results of Theorem 11.4 are used to estimate the global error of the RKM (11.11). Assertion v) implies that through every point $(x, \sigma_A(x, \varepsilon, h)) \in M_{h,\varepsilon}$ there is a stable fiber $W^s_{h,\varepsilon}(x) = \{(\xi, \eta) \mid \xi = w^s_{h,\varepsilon}(x, \eta)\}$. Any point (u, v) near $M_{h,\varepsilon}$ lies on a fiber $W^s_{h,\varepsilon}(x)$ where x is given by the implicit equation

$$u = w^s_{h,\varepsilon}(x, v). \tag{11.30}$$

For short, we write w instead of $w^s_{h,\varepsilon}$ and we drop the dependence of functions on ε and h. With the coordinate transformation

$$\begin{pmatrix} u \\ v \end{pmatrix} \longmapsto \begin{pmatrix} x \\ v \end{pmatrix}, \qquad \text{where } u = w(x, v), \tag{11.31}$$

the fibers $W^s_{h,\varepsilon}$ become vertical. Note that $w(x, \sigma_A(x)) = x$. We expand $w(x, v)$ with respect to $v = \sigma_A(x)$ and get from (11.30)

$$u = w(x, \sigma_A(x) + z) = x + D_2 w(x, \sigma_A(x))z + O(\varepsilon|z|^2) \tag{11.32}$$

since $w(x, \cdot)$ is λ_R-Lipschitz continuous with $\lambda_R = O(\varepsilon)$. Differentiating the invariance equation

$$F(w(x, v), v) = w\big(F(x, \sigma_A(x)), G(w(x, v), v)\big)$$

with respect to v one finds

$$D_2 w(x, \sigma_A(x)) = \frac{d}{dv} w(x, v)\big|_{v=\sigma_A(x)} = \varepsilon f_y(x, \sigma_A(x)) g_y(x, \sigma_A(x))^{-1} + o(\varepsilon)$$

for $h \to 0, \varepsilon/h \to 0$. We apply the transformation (11.31) to $(u, v) = (u, s_A(u)) \in M_\varepsilon$. From (11.32) we obtain $x = u - \varepsilon f_y(u, \sigma_A(u)) g_y(u, \sigma_A(u))^{-1}\big(s_A(u) - \sigma_A(u)\big) + o(\varepsilon)$ where we have used that $\sigma_A = s_A + O(h^{q+1})$, cf. Theorem 11.4 vii). For simplicity we omit the $o(\varepsilon)$-term. We state the precise results in the following theorem.

Theorem 11.5. *Let the differential equation* (11.1) *satisfy Assumption ASA*$(k+1)$. *Apply a RKM satisfying Assumption ASARK to* (11.1) *and assume* $k > p$.

Then there are constants ν, δ, d, c *and functions* s_A *and* σ_A *of class* C^k_b *and a coordinate transformation*

$$\rho\colon u \longmapsto x = u + \varepsilon f_y(u, s_A(u, \varepsilon), \varepsilon) g_y(u, s_A(u, \varepsilon), \varepsilon)^{-1}\big(\sigma_A(u, \varepsilon, h) - s_A(u, \varepsilon)\big)$$

such that for all $h \in (0, \nu]$ *and all* ε *with* $\varepsilon/h \in (0, \delta]$ *the following assertions hold.*

i) *All assertions of Theorem* 11.4 *hold.*

ii) *The map* $P_\rho := \rho^{-1} \circ P_{\mathrm{RK}}\big|_{M_{h,\varepsilon}} \circ \rho$ *conjugate to the map* P_{RK} *of* (11.11) *is a one-step method for the differential equation* $\dot{u} = f(u, s_A(u, \varepsilon))$ *describing the flow of* (11.1) *on the invariant manifold* M_ε. *The map* P_ρ *has local error* $O(h^{p+1} + \varepsilon h^{q+2})$.

iii) *Let* $(x(t), y(t))$ *be the solution of the differential equation* (11.1) *with initial condition* $(x_0, y_0) \in U_d = \{(x, y) \mid x \in \mathbb{R}^m,\ |y - s_A(x, \varepsilon)| < d\}$, *and let* (x_j, y_j) *be its RK-approximation, i.e.,* $(x_j, y_j) = P_{\mathrm{RK}}^j(x_0, y_0)$. *Then the following global error bounds hold for* $jh \le T$, T *fixed:*

$$x_j - x(jh) = O(h^p) + O(\varepsilon\, h^{q+1}) + O\big(\varepsilon\, |y_0 - s_A(x_0, \varepsilon)|\big),$$

$$y_j - y(jh) = O(h^{q+1}) + O(h^p) + O\big((\varepsilon + \chi_A(\varepsilon, h)^j)\, |y_0 - s_A(x_0, \varepsilon)|\big),$$

where $\chi_A(\varepsilon, h) = |R(\infty)| + c\,\varepsilon/h < 1$. *Moreover, if* $b_i = a_{si}$, $i = 1, \dots, s$, *the estimate*

$$y_j - y(jh) = O(h^p) + O(\varepsilon\, h^q) + O\big((\varepsilon + \chi_A(\varepsilon, h)^j)\, |y_0 - s_A(x_0, \varepsilon)|\big)$$

holds.

Remark 11.6. The attractivity of $M_{h,\varepsilon}$ is weak compared to the attractivity of M_ε unless $R(\infty) = 0$. For the global error, however, the strong attractivity of $M_{h,\varepsilon}$ is not crucial. In any case, after a few steps, $\chi_A(\varepsilon, h)^j\, |y_0 - s_A(x_0, \varepsilon)|$ becomes small compared to the other terms in the global error estimate.

Proof. i) All assumptions of Theorem 11.4 are satisfied.

ii) We drop the arguments ε and h in all functions. We consider one step of the map $P_\rho\colon u \mapsto \bar{u}$. The corresponding RK-step on $M_{h,\varepsilon}$ is $P_{\mathrm{RK}}\colon (x, \sigma_A(x)) \mapsto (\bar{x}, \sigma_A(\bar{x}))$ with $x = \rho(u)$ and $\bar{u} = \rho^{-1}(\bar{x})$. As in the proof of Theorem 11.4 we first transform the y-variable by $y = s_A(x) + z$. For $z = \sigma_A(x) - s_A(x)$ and $Y = s_A(X) + Z$ we get from (11.11)

$$\begin{aligned}
\bar{x} &= x + h(b^T \otimes I_m) f(X, s_A(X) + Z),\\
\bar{z} &= z + \frac{h}{\varepsilon}\,(b^T \otimes I_n)\big[g(X, s_A(X) + Z) - g(X, s_A(X))\big] - e(x, \bar{x}, X),\\
X &= \mathbb{1} \otimes x + h(A \otimes I_m) f(X, s_A(X) + Z),\\
Z &= \mathbb{1} \otimes z + \frac{h}{\varepsilon}\,(A \otimes I_n)\big[g(X, s_A(X) + Z) - g(X, s_A(X))\big] - E(x, X),
\end{aligned}$$

where e and E are defined in (11.18) and (11.16) and are estimated in (11.22) and (11.23). From Theorem 11.4 vii) we have $z = O(h^{q+1})$ and hence from (11.21) $Z = O(\varepsilon h^q)$. Transforming the x-equation to the u-variable we get with $D(u) := f_y(u, s_A(u)) g_y(u, s_A(u))^{-1}$

$$\begin{aligned}
\bar{u} + \varepsilon D(\bar{u})\bar{z} = u &+ \varepsilon D(u) z + h(b^T \otimes I_m) f(X, s_A(X))\\
&+ h(b^T \otimes I_m)\,\mathrm{diag}\big[f_y(X, s_A(X))\big] Z + O(|Z|^2).
\end{aligned}$$

From $x - u = \varepsilon D(u) z = O(\varepsilon h^{q+1})$ we have $X - U = O(\varepsilon h^{q+1})$ where U is defined as in (11.26) and where we have used (11.29). Hence,

$$\begin{aligned}
\bar{u} + \varepsilon\big[D(\bar{u})\bar{z} - \varepsilon D(u) z\big] = u &+ h(b^T \otimes I_m) f(U, s_A(U))\\
&+ h(b^T \otimes f_y(u, s_A(u))) Z + O(\varepsilon h^{q+2}).
\end{aligned}$$

We estimate

$$\begin{aligned}\varepsilon\big[D(\bar{u})\bar{z} - D(u)z\big] &= \varepsilon\big[D(\bar{u}) - D(u)\big]\bar{z} + \varepsilon D(u)[\bar{z} - z] \\ &= \varepsilon D(u)[\bar{z} - z] + O(\varepsilon h^{q+2}) \\ &= hD(u)(b^T \otimes g_y(u, s_A(u)))Z + O(\varepsilon h^{q+2}) \\ &= h(b^T \otimes f_y(u, s_A(u)))Z + O(\varepsilon h^{q+2})\end{aligned}$$

implying

$$\bar{u} = u + h(b^T \otimes I_m) f(U, s_A(U)) + O(\varepsilon h^{q+2}). \tag{11.33}$$

Let $u(t)$ be the solution of $\dot{u} = f(u, s_A(u))$. The first two terms of (11.33) are the RK-approximation of $u(h)$ with error $O(h^{p+1})$. Hence, $\bar{u}$ is an approximation of $u(h)$ with error $O(h^{p+1} + \varepsilon h^{q+2})$.

iii) We estimate the global error of the RKM. We consider the orbit (x_j, y_j) of the map P_{RK} and its asymptotic phase orbit $(\tilde{x}_j, \sigma_A(\tilde{x}_j))$ on $M_{h,\varepsilon}$, cf. Theorem 11.4 vi). Moreover, we consider the asymptotic phase solution $\big(u(t), s_A(u(t))\big)$ of the solution $\big(x(t), y(t)\big)$ of the differential equation (11.1), cf. Theorem 10.1 v). In addition, we define $\tilde{u}_0$ by $\rho(\tilde{u}_0) = \tilde{x}_0$ and $\tilde{u}_j = P_\rho^j(\tilde{u}_0) = \rho^{-1}(x_j)$. Moreover, we consider the solution $\tilde{u}(t)$ of the differential equation $\dot{u} = f(u, s_A(u))$ with $\tilde{u}(0) = \tilde{u}_0$. The situation is sketched in Figure 11.2.

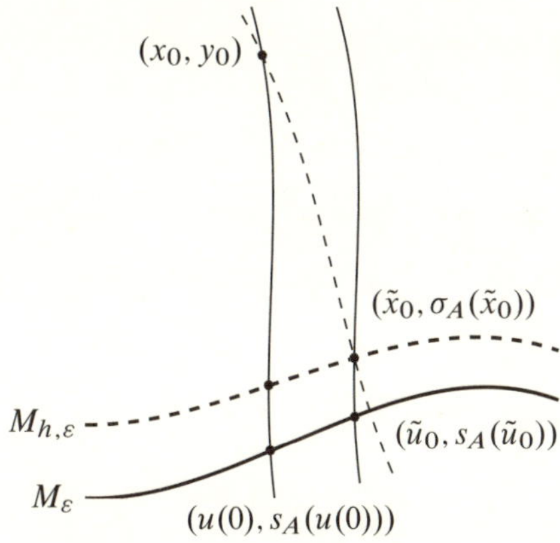

Figure 11.2. The initial conditions of the RK-orbits (x_j, y_j), $(\tilde{x}_j, \tilde{y}_j) = (\tilde{x}_j, \sigma_A(\tilde{x}_j))$ and of the solutions $(x(t), y(t))$, $(u(t), s_A(u(t)))$ and $(\tilde{u}(t), s_A(\tilde{u}(t)))$ of the differential equation (11.1). These orbits are used to estimate the global error of the RKM.

We estimate

$$x_j - x(jh) = [x_j - \tilde{x}_j] + [\tilde{x}_j - \tilde{u}_j] + [\tilde{u}_j - \tilde{u}(jh)] + [\tilde{u}(jh) - u(jh)] + [u(jh) - x(jh)].$$

From Theorem 11.4 vi) and vii) we have

$$x_j - \tilde{x}_j = O(\varepsilon \chi_A(\varepsilon, h)^j |y_0 - \sigma_A(x_0)|) = O\big(\varepsilon \chi_A(\varepsilon, h)^j \big[|y_0 - s_A(x_0)| + h^{q+1}\big]\big).$$

From Theorem 11.4 vii) we have by the definition of ρ

$$\tilde{x}_j - \tilde{u}_j = O(\varepsilon h^{q+1}).$$

Using the estimate of the local error of the one-step method P_ρ in assertion ii) a Gronwall type argument yields

$$\tilde{u}_j - \tilde{u}(jh) = O(h^p + \varepsilon h^{q+1}).$$

The Gronwall Lemma yields $\tilde{u}(jh) - u(jh) = O(|\tilde{u}_0 - u(0)|)$ leading to

$$\tilde{u}(jh) - u(jh) = O\big(\varepsilon\big[|y_0 - s_A(x_0)| + h^{q+1}\big]\big).$$

The asymptotic phase property v) of Theorem 10.1 gives

$$u(jh) - x(jh) = O(\varepsilon e^{-\beta_A t/\varepsilon} |y_0 - s_A(x_0)|).$$

Gathering all estimates we obtain

$$x_j - x(jh) = O(h^p) + O(\varepsilon h^{q+1}) + O(\varepsilon |y_0 - s_A(x_0)|).$$

It remains to estimate $y_j - y(jh)$, $jh \le T$. Using the attractivity of the manifolds $M_{h,\varepsilon}$ and M_ε, their closeness and the above estimate of the x-component we obtain

$$\begin{aligned} y_j - y(jh) &= [y_j - \sigma_A(x_j)] + [\sigma_A(x_j) - s_A(x_j)] \\ &\qquad + [s_A(x_j) - s_A(x(jh))] + [s_A(x(jh)) - y(jh)] \\ &= [O(\chi_A(\varepsilon, h)^j \, |y_0 - \sigma_A(x_0)|)] + [O(h^{q+1})] \\ &\qquad + [O(h^p) + O(\varepsilon\, h^{q+1}) + O(\varepsilon\, |y_0 - s_A(x_0)|)] \\ &\qquad + [O(e^{-\beta_A jh/\varepsilon} |y_0 - s_A(x_0)|)] \\ &= O(h^{q+1}) + O(h^p) + O\big((\varepsilon + \chi_A(\varepsilon, h)^j)\, |y_0 - s_A(x_0)|\big). \end{aligned}$$

If $b_i = a_{si}$ the $O(h^{q+1})$-term is replaced by $O(\varepsilon h^q)$. □

Chapter 12

Invariant curves of perturbed harmonic oscillators

12.1 The van der Pol equation

Cole [31] considers the ordinary differential equation (ODE)

$$\begin{aligned} \frac{dX}{dt} &= Y, \\ \varepsilon \frac{dY}{dt} &= -X + Y(1 - X^2), \end{aligned} \tag{12.1}$$

where $X, Y \in \mathbb{R}$ and where $\varepsilon > 0$ is a real parameter. This so-called relaxation oscillator of van der Pol describes a linear oscillating circuit with resistance which is coupled inductively to a triode. It effectively provides a negative resistance for small currents. This negative resistance causes small currents to grow, but the eventual amplitude is limited due to saturations of the triode. An oscillation of definite amplitude and period is produced.

The ODE (12.1) is equivalent to the ODE

$$\begin{aligned} \frac{dx}{dt} &= y, \\ \varepsilon \frac{dy}{dt} &= -x + y - y^3/3 \end{aligned} \tag{12.2}$$

in the following sense. If $\big(x(t), y(t)\big)$ is a solution of (12.2) then $\big(X(t), Y(t)\big) := \frac{d}{dt}\big(x(t), y(t)\big)$ satisfies equation (12.1). Conversely, every solution $\big(X(t), Y(t)\big)$ of (12.1) with initial condition $\big(X(0), Y(0)\big)$ is the derivative of the solution $\big(x(t), y(t)\big)$ of (12.2) with initial condition

$$x(0) = -\varepsilon\, Y(0) + X(0) - X^3(0)/3, \quad y(0) = X(0).$$

We investigate equation (12.1) for the parameter regimes $\varepsilon \to 0$ and $\varepsilon \to \infty$. The case $\varepsilon \to 0$ is a singularly perturbed differential equation and is treated in Chapter 13 in the form (12.2). For the case $\varepsilon \to \infty$ we use the scaled variables

$$\tau = \frac{1}{\sqrt{\varepsilon}}\, t, \quad x = X, \quad y = \sqrt{\varepsilon}\, Y$$

and obtain, putting $\mu := 1/\sqrt{\varepsilon}$,

$$\begin{aligned} \frac{dx}{d\tau} &= y, \\ \frac{dy}{d\tau} &= -x + \mu\, y(1 - x^2). \end{aligned} \tag{12.3}$$

For small μ this is a regularly perturbed harmonic oscillator.

The system (12.3) admits a stable limit cycle for μ sufficiently small as shown in this section below. We ask whether this geometric property is preserved under discretisation by a numerical integration method. In Section 12.2 we consider methods providing a linear, area preserving map if applied to the harmonic oscillator ($\mu = 0$). For the sake of a concise presentation we consider the symplectic Euler method with step size h. This is a partitioned Runge–Kutta method. For such methods, in the case $\mu = 0$, all orbits lie on concentric ellipses. Using the method of averaging we will prove the existence of a stable invariant curve of the symplectic Euler method close to the limit cycle of (12.3) for μ and h sufficiently small, independent of each other. Moreover, as μ tends to 0, this invariant curve approaches a specific ellipse out of the family of ellipses in the case $\mu = 0$. In Section 12.3 we consider integration methods applied to (12.3) which are not area preserving for $\mu = 0$. For simplicity we take the Euler method. We will prove that the Euler method with step size h admits an attractive invariant manifold close to the limit cycle of (12.3) if h is small compared to μ and that there is no such manifold close to the limit cycle for μ small compared to h.

In Hairer, Lubich, Wanner [51], Chapter XII, numerical experiments are performed concerning the limit cycle of the van der Pol equation for the implicit, the symplectic and the explicit Euler method, respectively. In a setting different to ours it is proved that the continuous and the discrete dynamical system admit attractive invariant tori close to each other.

In oder to be able to apply the invariant manifold results of Chapter 7 to the ODE (12.3) appropriate coordinates have to be chosen. In a first step we introduce polar coordinates

$$\begin{aligned} x &= r\cos\varphi, \\ y &= -r\sin\varphi \end{aligned} \tag{12.4}$$

and obtain the system

$$\begin{aligned} \dot\varphi &= 1 + \mu\sin\varphi\cos\varphi(1 - r^2\cos^2\varphi) = 1 + \mu\big[2(2 - r^2)\sin 2\varphi - r^2\sin 4\varphi\big]/8, \\ \dot r &= \mu r\sin^2\varphi(1 - r^2\cos^2\varphi) = \mu r\big[(4 - r^2) - 4\cos 2\varphi + r^2\cos 4\varphi\big]/8, \end{aligned} \tag{12.5}$$

where $\dot{}$ denotes the derivative $d/d\tau$. Instead of the system (12.5) we treat the more general system

$$\dot z = f(z,\mu) = \begin{pmatrix} 1 \\ 0 \end{pmatrix} + \mu f^1(z) + \mu^2 f^2(z) + O(\mu^3), \tag{12.6}$$

where $z = (\varphi, r)^T$ and where $f \in C^k, k \in \mathbb{N}$, is 2π-periodic in φ with

$$f^1(\varphi, r) = \begin{pmatrix} a_0(r)/2 + \sum\limits_{j=1}^{k} [a_j(r)\cos j\varphi + b_j(r)\sin j\varphi] \\ c_0(r)/2 + \sum\limits_{j=1}^{k} [c_j(r)\cos j\varphi + d_j(r)\sin j\varphi] \end{pmatrix} \tag{12.7}$$

and with analogous f^2. Such systems may be transformed to a normal form by the so-called method of averaging.

In Subsection 12.1.1 we apply the method of averaging to (12.6). In Subsection 12.1.2 we show that under a certain assumption Theorems 7.5 and 7.9 may be applied to the averaged equation. In Subsection 12.1.3 we apply these results to equation (12.5) and draw conclusions for the van der Pol equation (12.3).

12.1.1 The method of averaging for perturbed harmonic oscillators

We apply a transformation of the form

$$z = \zeta + \mu\, g^1(\zeta) + \mu^2\, g^2(\zeta) \tag{12.8}$$

with $\zeta = (\alpha, a)^T$ to the ODE (12.6). Suppressing the argument ζ in the functions f^1, etc., we get

$$(I + \mu\, Dg^1 + \mu^2\, Dg^2)\,\dot{\zeta} = \begin{pmatrix}1\\0\end{pmatrix} + \mu\, f^1 + \mu^2[Df^1 \cdot g^1 + f^2] + O(\mu^3)\,\cdot$$

Using

$$(I + \mu\, Dg^1 + \mu^2\, Dg^2)^{-1} = I - \mu\, Dg^1 + \mu^2[Dg^1 \cdot Dg^1 - Dg^2] + O(\mu^3)$$

yields

$$\begin{aligned}\dot{\zeta} = \begin{pmatrix}1\\0\end{pmatrix} &+ \mu\Big[f^1 - Dg^1 \cdot \begin{pmatrix}1\\0\end{pmatrix}\Big] \\ &+ \mu^2\Big[Df^1 \cdot g^1 - Dg^1 \cdot \Big(f^1 - Dg^1 \cdot \begin{pmatrix}1\\0\end{pmatrix}\Big) + f^2 - Dg^2 \cdot \begin{pmatrix}1\\0\end{pmatrix}\Big] \\ &+ O(\mu^3).\end{aligned} \tag{12.9}$$

The functions g^1 and g^2 are determined such that the μ-term and the μ^2-term are independent of α. We first seek a function $g^1(\alpha, a)$ of the form

$$g^1(\alpha, a) = \begin{pmatrix} \displaystyle\sum_{j=1}^{k} [A_j(a)\cos j\alpha + B_j(a)\sin j\alpha] \\ \displaystyle\sum_{j=1}^{k} [C_j(a)\cos j\alpha + D_j(a)\sin j\alpha] \end{pmatrix} \tag{12.10}$$

such that for f^1 as in (12.7)

$$f^1(\alpha, a) - Dg^1(\alpha, a) \cdot \begin{pmatrix}1\\0\end{pmatrix} = \hat{f}^1(a)$$

holds where $\hat{f}^1(\cdot)$ denotes the average of $f^1(\alpha,\cdot)$ for $\alpha \in [0,2\pi]$. A brief calculation shows that

$$\begin{pmatrix} A_j & B_j \\ C_j & D_j \end{pmatrix} = \frac{1}{j}\begin{pmatrix} -b_j & a_j \\ -d_j & c_j \end{pmatrix}, \tag{12.11}$$

where we have suppressed the dependence of the coefficients on a. It follows that the average of $Dg^1 \cdot (f^1 - Dg^1 \cdot \binom{1}{0})$ in (12.9) is zero. Second, we similarly choose g^2 such that the μ^2-term in (12.9) is $\widehat{Df^1 \cdot g^1} + \hat{f}^2$. By this choice of g^1 and g^2 the transformation (12.8) takes the ODE (12.6) to the ODE

$$\dot{\zeta} = \begin{pmatrix} 1 \\ 0 \end{pmatrix} + \mu \hat{f}^1 + \mu^2(\widehat{Df^1 \cdot g^1} + \hat{f}^2) + O(\mu^3). \tag{12.12}$$

Note that the μ- and the μ^2-term are independent of α and that the μ^3-term is 2π-periodic with respect to α.

12.1.2 The invariant manifold for perturbed harmonic oscillators

The following assumption on $\hat{f}^1$ allows to apply the invariant manifold Theorems 7.5 and 7.9 to (12.12).

Assumption ARA

There is a closed interval $[a^-, a^+]$ and real numbers $\tilde{a} \in (a^-, a^+)$ and $\ell > 0$ such that the second component $\hat{f}_2^1$ of $\hat{f}^1$ satisfies

$$\hat{f}_2^1(\tilde{a}) = 0 \quad \textit{and} \quad \frac{d}{da}\hat{f}_2^1(a) \le -\ell \quad \textit{for } a \in [a^-, a^+].$$

We verify the assumptions of Theorem 7.5 for the ODE (12.12).

Hypothesis HD a) is satisfied for μ sufficiently small since $\hat{f}_2^1(a^-) > 0$ and $\hat{f}_2^1(a^+) < 0$.

Hypothesis HD b) holds since $X = \mathbb{R}$.

Hypothesis HD c) holds with

$$\ell_{12} = O(\mu), \quad \ell_{21} = O(\mu^3),$$
$$\gamma_{11} = O(\mu^3), \quad \ell_{22} = -\mu\ell/2$$

for μ small enough.

Hypothesis HDA is satisfied for $y^* = \tilde{a}$.

It follows that for $k \in \mathbb{N}$, Conditions CD, CDA and CDA(k) hold for μ sufficiently small. Theorem 7.5 implies that there is a C_b^k-function $(\alpha,\mu) \mapsto s_a(\alpha,\mu)$ such that the set $M_a = \{(\alpha,a) \mid \alpha \in \mathbb{R},\ a = s_a(\alpha,\mu)\}$ is invariant under the flow of (12.12). Setting $\kappa(\alpha) = \alpha + 2\pi$ in Theorem 7.5 v) implies that the function s_a is 2π-periodic with respect to α. Every solution $\big(\alpha(\tau), a(\tau)\big)$ of (12.12) with initial conditions $(\alpha(0), a(0))$ adjacent to M_a, i.e., $a(0) \in [a^-, a^+]$, tends exponentially to M_a, i.e.,

$$\big|a(\tau) - s_a\big(\alpha(\tau),\mu\big)\,\big| \le e^{-\mu\ell\tau/4}\big|a(0) - s_a\big(\alpha(0),\mu\big)\big|.$$

Theorem 7.9 implies that the space adjacent to M_a is foliated by stable fibers.

We take a closer look at s_a. By means of the contraction principle Assumption ARA implies that for μ sufficiently small the equation, cf. (12.12),

$$\hat{f}_2^1(a) + \mu\big(\widehat{Df^1(\alpha,a)\cdot g^1(\alpha,a)} + \hat{f}^2(a)\big)_2 = 0 \tag{12.13}$$

(the index 2 denotes the second component) has a unique solution $a^*(\mu) \in (a^-, a^+)$ with $a^*(\mu) = \tilde{a} + O(\mu)$. Applying Theorem 7.8 with $\sigma = a^*$ leads to $\rho = O(\mu^3)$ and implies

$$s_a(\alpha,\mu) = a^*(\mu) + O(\mu^2).$$

We formulate the above results in the variables (φ, r) of equation (12.6).

Theorem 12.1. *Let the vector field of the differential equation* (12.6) *be of class* C^k, $k \in \mathbb{N}$, *and let the average* $\hat{f}^1$ *of* f^1 *satisfy Assumption ARA.*

Then there are $\mu_0 > 0$, $c > 0$ *and a* C_b^k*-function* $s_r \colon \mathbb{R} \times (0,\mu_0] \to \mathbb{R}$, 2π-*periodic with respect to the first argument, such that the following assertions hold for* $\mu \in (0,\mu_0]$.

i) *The set* $M_r := \{(\varphi,r) \mid \varphi \in \mathbb{R},\ r = s_r(\varphi,\mu)\}$ *is an invariant manifold of* (12.6) *where*

$$s_r(\varphi,\mu) = a^*(\mu) + \mu\, g_2^1(\varphi,\tilde{a}) + O(\mu^2)$$

with a^* *being the unique solution of* (12.13) *and* g^1 *being defined by* (12.10), (12.11).

ii) *The manifold* M_r *is attractive in the sense that the estimate*

$$|r(\tau) - s_r(\varphi(\tau),\mu)| \le \big(1 + O(\mu)\big)\, e^{-\mu\ell\tau/4}|r(0) - s_r(\varphi(0),\mu)|$$

holds for every solution $\big(\varphi(\tau), r(\tau)\big)$ *with initial condition* $\big(\varphi(0), r(0)\big) \in \mathbb{R} \times [a^- + c\mu_0,\ a^+ - c\mu_0]$.

iii) *The space* $\mathbb{R} \times [a^- + c\mu_0,\ a^+ - c\mu_0]$ *is foliated by stable fibers* $W^s(\varphi,r) = \{(\xi,\eta) \mid \eta \in [a^- + c\mu_0,\ a^+ - c\mu_0],\ \xi = w^s(\varphi,r,\eta)\}$. *The function* $w^s(\varphi,r,\cdot)$ *is of class* C_b^k *and* λ_R*-Lipschitz continuous with* $\lambda_R = 2\ell_{12}/(\mu\ell) + O(\mu)$.

12.1.3 Application to the van der Pol equation

We apply Theorem 12.1 to the system (12.5) which is of the form (12.6) with $f(\varphi, r) = \binom{1}{0} + \mu f^1(\varphi, r)$ and where

$$f^1(\varphi, r) = \begin{pmatrix} \frac{1}{4}(2-r^2)\sin 2\varphi - \frac{r^2}{8}\sin 4\varphi \\ \frac{r}{8}(4-r^2) - \frac{r}{2}\cos 2\varphi + \frac{r^3}{8}\cos 4\varphi \end{pmatrix}$$

is π-periodic with respect to φ.

Using (12.10) and (12.11) we find

$$g^1(\alpha, a) = \begin{pmatrix} -\frac{1}{8}(2-a^2)\cos 2\alpha + \frac{a^2}{32}\cos 4\alpha \\ -\frac{a}{4}\sin 2\alpha + \frac{a^3}{32}\sin 4\alpha \end{pmatrix} \tag{12.14}$$

and

$$\overline{Df^1(\alpha, a)\cdot g^1(\alpha, a)} = \begin{pmatrix} -\frac{1}{8}\left(1 - \frac{3}{2}a^2 + \frac{11}{32}a^4\right) \\ 0 \end{pmatrix}. \tag{12.15}$$

We obtain as averaged equation of (12.5) the ODE

$$\begin{aligned} \dot{\alpha} &= -\frac{\mu^2}{8}\left(1 - \frac{3}{2}a^2 + \frac{11}{32}a^4\right) + O(\mu^3), \\ \dot{a} &= \mu\frac{a}{8}(4-a^2) + O(\mu^3), \end{aligned} \tag{12.16}$$

where the right-hand side is π-periodic with respect to α. Setting $\kappa(\alpha) = \alpha + \pi$ in Theorem 7.5 v) implies that the function s_a describing the invariant manifold of (12.16) is π-periodic with respect to α. Assumption ARA is satisfied with $\tilde{a} = 2$, and with the choice $a^- = 3/2$, $a^+ = 5/2$ and $\ell = 1/3$. We compute $g_2^1(\alpha, \tilde{a})$, $a^*(\mu)$ and ℓ_{12} for our example and find $g_2^1(\alpha, \tilde{a}) = -1/2 \cdot \sin 2\alpha + 1/4 \cdot \sin 4\alpha$, $a^*(\mu) \equiv 2$, $\ell_{12} = O(\mu^2)$ and therefore, in the variables (φ, r) we get

$$s_r(\varphi, \mu) = 2 + \mu\left(-\frac{1}{2}\sin 2\varphi + \frac{1}{4}\sin 4\varphi\right) + O(\mu^2), \tag{12.17}$$

where the function s_r is π-periodic in φ. Since $\ell_{12} = O(\mu^2)$ we get $\lambda_R = O(\mu)$ as Lipschitz constant for the fibers $W^s(\varphi, r)$.

We conclude that the van der Pol equation (12.3) admits a μ-weakly attractive limit cycle which is point symmetric with respect to the origin. In Figure 12.1 the limit cycle is shown for several μ-values. It is nicely seen that the limit cycle is point symmetric

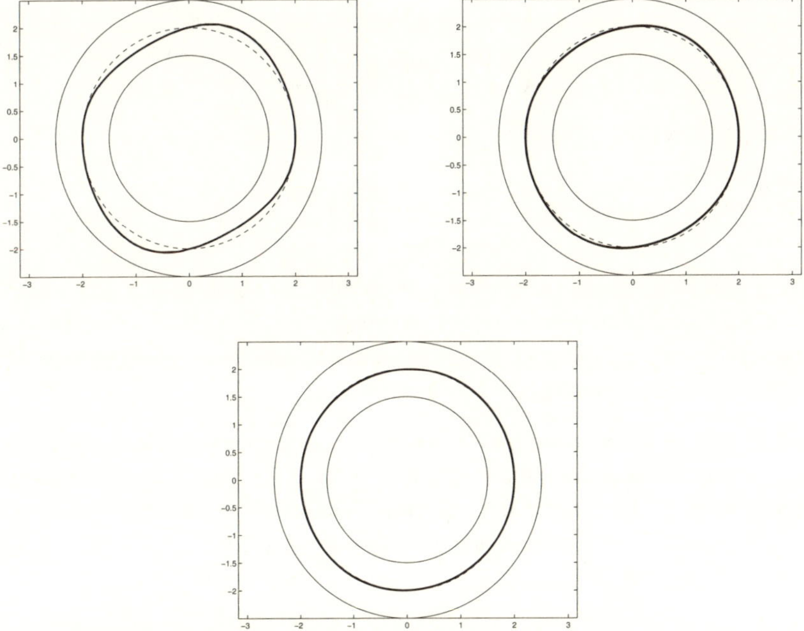

Figure 12.1. The limit cycle of the regularly perturbed van der Pol equation for the values $\mu = 0.25, 0.1, 0.03$.

with respect to the origin and that it converges to the circle with radius 2 for $\mu \to 0$. In polar coordinates this limit cycle has the form

$$M_r = \{(\varphi, r) \mid \varphi \in \mathbb{R},\ r = s_r(\varphi, \mu)\}$$

with the π-periodic function s_r given in (12.17).

12.2 The symplectic Euler method

In this section we consider numerical integration methods applied to the ODE (12.3) with step size h which are area preserving for $\mu = 0$. For the sake of simplicity we take the symplectic Euler method

$$\begin{aligned} \bar{x} &= x + hy, \\ \bar{y} &= -hx + (1 - h^2)y + \mu hy\big(1 - (x + hy)^2\big). \end{aligned} \tag{12.18}$$

We will show that this discrete system admits a weakly attractive invariant curve close to the limit cycle of the ODE (12.3) for μ, h small enough and independent of each other.

We first consider the case $\mu = 0$. In this case the map (12.18) is linear and takes the form

$$\begin{pmatrix} \bar{x} \\ \bar{y} \end{pmatrix} = A \begin{pmatrix} x \\ y \end{pmatrix} \quad \text{with } A = \begin{pmatrix} 1 & h \\ -h & 1-h^2 \end{pmatrix}. \tag{12.19}$$

This map preserves the quadratic form

$$Q(x,y) := x^2 + hxy + y^2 = (x,y)\, S \begin{pmatrix} x \\ y \end{pmatrix}, \quad S = \begin{pmatrix} 1 & h/2 \\ h/2 & 1 \end{pmatrix}. \tag{12.20}$$

For every $c > 0$ the curve $Q(x,y) = c$ is an ellipse. The matrix S has eigenvalues $\lambda_{1,2} = 1 \mp h/2$ and corresponding eigenvectors $q_{1,2} = (\mp 1, 1)^T$. The semi-major axis and the semi-minor axis of the ellipse, respectively, is $a = \sqrt{c/\lambda_1}$ and $b = \sqrt{c/\lambda_2}$, respectively, with directions q_1 and q_2, respectively.

As for the continuous system we want to introduce polar coordinates in (12.18). It is convenient to first transform the ellipses to circles. This is equivalent to transforming the matrix A in (12.19) to real normal form. The matrix A has eigenvalues $\lambda_{1,2} = 1 - h^2/2 \pm ih\sqrt{1-h^2/4}$ and corresponding eigenvectors $p_{1,2} = (1, -h/2)^T \pm i(0, \sqrt{1-h^2/4})^T$. Thus the coordinate transformation

$$\begin{pmatrix} x \\ y \end{pmatrix} = T \begin{pmatrix} u \\ v \end{pmatrix} \quad \text{with } T = \begin{pmatrix} 1 & 0 \\ -h/2 & d \end{pmatrix}, \; T^{-1} = \begin{pmatrix} 1 & 0 \\ h/(2d) & 1/d \end{pmatrix} \tag{12.21}$$

where $d := \sqrt{1-h^2/4}$ takes the map (12.19) to

$$\begin{pmatrix} \bar{u} \\ \bar{v} \end{pmatrix} = \Omega \begin{pmatrix} u \\ v \end{pmatrix}, \quad \text{with } \Omega = \begin{pmatrix} 1-h^2/2 & hd \\ -hd & 1-h^2/2 \end{pmatrix} =: \begin{pmatrix} \cos\omega & \sin\omega \\ -\sin\omega & \cos\omega \end{pmatrix}$$

where $w := \arcsin(hd) = 2\arcsin(h/2)$.

For $\mu > 0$ the transformation (12.21) takes the map (12.18) to

$$\begin{pmatrix} \bar{u} \\ \bar{v} \end{pmatrix} = \Omega \begin{pmatrix} u \\ v \end{pmatrix} + \mu h\, q(u,v,h)\, T^{-1} \begin{pmatrix} 0 \\ 1 \end{pmatrix} \tag{12.22}$$

with $q(u,v,h) = \big(-(h/2)u + dv\big)\big[1 - (u\cos\omega + v\sin\omega)^2\big]$.

We introduce polar coordinates (β, b) by

$$\begin{pmatrix} u \\ v \end{pmatrix} = \begin{pmatrix} b\cos\beta \\ -b\sin\beta \end{pmatrix} =: \Phi(\beta, b). \tag{12.23}$$

The linear part of (12.22) is transformed to

$$\Omega\Phi(\beta,b) = \begin{pmatrix} \cos\omega & \sin\omega \\ -\sin\omega & \cos\omega \end{pmatrix} \begin{pmatrix} b\cos\beta \\ -b\sin\beta \end{pmatrix} = \Phi(\beta+\omega, b)$$

and the map (12.22) is taken to

$$\Phi(\bar{\beta},\bar{b}) = \Phi(\beta+\omega,b) + \frac{\mu h}{d}\, q\big(\Phi(\beta,b),h\big)\begin{pmatrix}0\\1\end{pmatrix}.$$

In order to solve for $\bar{\beta},\bar{b}$ we apply Φ^{-1} and expand the right-hand side with respect to μh. Using

$$D\Phi^{-1}\big(\Phi(\beta+\omega,b)\big) = \begin{pmatrix} -\frac{1}{b}\sin(\beta+w) & -\frac{1}{b}\cos(\beta+\omega) \\ \cos(\beta+\omega) & -\sin(\beta+\omega)\end{pmatrix}$$

and expressing $q(\Phi(\beta,b),h)$ in terms of $\beta+\omega$ we obtain

$$\begin{aligned}\bar{\beta} &= \beta+\omega - \frac{\mu h}{bd}\, q\big(\Phi(\beta,b),h\big)\cos(\beta+\omega) + O(\mu^2h^2),\\ \bar{b} &= b - \frac{\mu h}{d} q\big(\Phi(\beta,b),h\big)\sin(\beta+\omega) + O(\mu^2h^2)\end{aligned} \tag{12.24}$$

with

$$q\big(\Phi(\beta,b),h\big) = b\Big(\frac{h}{2}\cos(\beta+\omega) - d\sin(\beta+\omega)\Big)\big[1-b^2\cos^2(\beta+\omega)\big].$$

If we apply the symplectic Euler method to the more general ODE (12.6) we get a smooth map, 2π-periodic in β, of the form

$$\bar{z} = z + \begin{pmatrix}\omega\\0\end{pmatrix} + \mu h\, f^1(z,h) + \mu^2 h\, f^2(z) + O(\mu^2h^2), \tag{12.25}$$

where $z = (\beta,b)^T$, with

$$f^1(\beta,b,h) = \begin{pmatrix} a_0(b,h)/2 + \sum_{j=1}^{k} [a_j(b,h)\cos j\beta + b_j(b,h)\sin j\beta] \\ c_0(b,h)/2 + \sum_{j=1}^{k} [c_j(b,h)\cos j\beta + d_j(b,h)\sin j\beta]\end{pmatrix} \tag{12.26}$$

and with analogous f^2. Note that $f^1(z,0)$ coincides with the vector field $f^1(z)$ in the ODE (12.6).

Similarly as in Subsection 12.1.1 we transform the map (12.25) to a normal form by the method of averaging. This is done in Subsection 12.2.1. In Subsection 12.2.2 we apply invariant manifold theorems for maps, cf. Part I. In Subsection 12.2.3 we apply the results obtained to the map (12.18).

12.2.1 The method of averaging for the map

We apply a transformation of the form

$$z = \zeta + \mu g^1(\zeta, h) + \mu^2 g^2(\zeta) \tag{12.27}$$

with $\zeta = (\alpha, a)^T$ to the map (12.25). This allows to simplify the μh- and the $\mu^2 h$-terms in (12.25) to terms independent of α. In a first step we have

$$\begin{aligned}
&\bar{\zeta} - \zeta - \binom{\omega}{0} + \mu g^1(\bar{\zeta}, h) + \mu^2 g^2(\bar{\zeta}) - \mu g^1(\zeta, h) - \mu^2 g^2(\zeta) \\
&\quad = \mu h\, f^1(\zeta, h) + \mu^2 h\big[Df^1(\zeta, h) g^1(\zeta, h) + f^2(\zeta)\big] + O(\mu^2 h^2) + O(\mu^3 h).
\end{aligned}$$

As in Subsection 12.1.1 we introduce the notation $\hat{f}^1(\cdot, h)$ for the average of $f^1(\alpha, \cdot, h)$, $\alpha \in [0, 2\pi]$. Adding the terms

$$-\mu h\, \hat{f}^1(a, h)$$

and

$$\begin{aligned}
&\mu\Big[g^1\Big(\zeta + \binom{\omega}{0}, h\Big) - g^1\Big(\zeta + \binom{\omega}{0} + \mu h\, \hat{f}^1(a, h), h\Big)\Big] \\
&\quad = -\mu^2 h\, Dg^1\Big(\zeta + \binom{\omega}{0}, h\Big)\, \hat{f}^1(a, h) + O(\mu^3 h^2)
\end{aligned}$$

and

$$-\mu^2\Big[g^2\Big(\zeta + \binom{\omega}{0} + \mu h\, \hat{f}^1(a, h)\Big) - g^2\Big(\zeta + \binom{\omega}{0}\Big)\Big] = O(\mu^3 h)$$

on both sides we get

$$\begin{aligned}
\bar{\zeta} - \zeta - \binom{\omega}{0} - \mu h\, \hat{f}^1(a, h) &+ \mu\Big[g^1(\bar{\zeta}, h) - g^1\Big(\zeta + \binom{\omega}{0} + \mu h\, \hat{f}^1(a, h), h\Big)\Big] \\
&+ \mu^2\Big[g^2(\bar{\zeta}) - g^2\Big(\zeta + \binom{\omega}{0} + \mu h\, \hat{f}^1(a, h)\Big)\Big] \\
&+ \mu\Big[g^1\Big(\zeta + \binom{\omega}{0}, h\Big) - g^1(\zeta, h)\Big] \\
&+ \mu^2\Big[g^2\Big(\zeta + \binom{\omega}{0}\Big) - g^2(\zeta)\Big] \\
= \mu h\big[f^1(\zeta, h) - \hat{f}^1(a, h)\big]& \\
+\mu^2 h\Big[Df^1(\zeta, 0)\, g^1(\zeta, 0) &- Dg^1\Big(\zeta + \binom{\omega}{0}, 0\Big)\, \hat{f}^1(a, 0) + f^2(\zeta)\Big] \\
+O(\mu^2 h^2) + O(\mu^3 h).&
\end{aligned}$$

We denote the last bracket by $F^2(\zeta)$. Adding $-\mu^2 h\,\hat{F}^2(a)$ on both sides and replacing $g^j(\zeta+\binom{\omega}{0}+\mu h\,\hat{f}^1(a,h),h)$ by $g^j(\zeta+\binom{\omega}{0}+\mu h\,\hat{f}^1(a,h)+\mu^2 h\,\hat{F}^2(a),h)+O(\mu^2 h)$, $j=1,2$, we obtain

$$\begin{aligned}(I+O(\mu)+O(\mu^2))\Big[\bar{\zeta}-\Big(\zeta+\binom{\omega}{0}+\mu h\,\hat{f}^1(a,h)+\mu^2 h\,\hat{F}^2(a,h)\Big)\Big]&\\ +\mu\Big[g^1\Big(\zeta+\binom{\omega}{0},h\Big)-g^1(\zeta,h)\Big]+\mu^2\Big[g^2\Big(\zeta+\binom{\omega}{0}\Big)-g^2(\zeta)\Big]&\qquad(12.28)\\ =\mu h\big[f^1(\zeta,h)-\hat{f}^1(a,h)\big]+\mu^2 h\big[F^2(\zeta)-\hat{F}^2(a)\big]+O(\mu^2h^2)+O(\mu^3 h).&\end{aligned}$$

We first look for a function g^1 of the form

$$g^1(\alpha,a,h)=\begin{pmatrix}\sum_{j=1}^{k}\big[A_j(a,h)\cos j\alpha+B_j(a,h)\sin j\alpha\big]\\ \sum_{j=1}^{k}\big[C_j(a,h)\cos j\alpha+D_j(a,h)\sin j\alpha\big]\end{pmatrix}\tag{12.29}$$

such that

$$g^1\Big(\zeta+\binom{\omega}{0},h\Big)-g^1(\zeta,h)=h\big(f^1(\zeta,h)-\hat{f}^1(a,h)\big).$$

After some calculations we find

$$\begin{pmatrix}A_j & B_j\\ C_j & D_j\end{pmatrix}=\frac{h}{2\sin\frac{j\omega}{2}}\begin{pmatrix}a_j & b_j\\ c_j & d_j\end{pmatrix}\begin{pmatrix}-\sin\dfrac{j\omega}{2} & \cos\dfrac{j\omega}{2}\\ -\cos\dfrac{j\omega}{2} & -\sin\dfrac{j\omega}{2}\end{pmatrix},\tag{12.30}$$

where we have suppressed the dependence of the coefficients on a and h. Since $\sin(\omega/2)=h/2$ this equation simplifies to

$$\begin{pmatrix}A_j & B_j\\ C_j & D_j\end{pmatrix}=\frac{1}{j}\begin{pmatrix}-b_j & a_j\\ -d_j & c_j\end{pmatrix}\tag{12.31}$$

in the limit $h\to 0$, cf. (12.11). It follows that $g^1(\zeta,0)$ is equal to $g^1(\zeta)$ in the ODE case. Second, we choose g^2 such that

$$g^2\Big(\zeta+\binom{\omega}{0}\Big)-g^2(\zeta)=h\big(F^2(\zeta)-\hat{F}^2(a)\big)+O(h^2)$$

in (12.28). By this choice of g^1 and g^2 the transformation (12.27) takes the map (12.25) to the averaged map

$$\begin{pmatrix}\bar{\alpha}\\ \bar{a}\end{pmatrix}=\begin{pmatrix}\alpha+\omega\\ a\end{pmatrix}+\mu h\,\hat{f}^1(a,h)+\mu^2 h\,\hat{F}^2(a)+O(\mu^2h^2)+O(\mu^3 h).\tag{12.32}$$

Remark 12.2. Note that $\hat{f}^1(a,0)$ and $\hat{F}^2(a)$, respectively, is equal to the μ- and μ^2-term of the ODE (12.12), respectively. In particular, the two terms are independent of α and both O-terms are 2π-periodic in α.

12.2.2 The invariant manifold of the map

In order to apply the invariant manifold Theorems 3.6 and 4.1 to the map (12.32) we make the same assumption as in Subsection 12.1.2, i.e., we assume Assumption ARA to hold. This means (cf. Remark 12.2) that the second component $\hat{f}_2^1$ of the function $\hat{f}^1(\cdot,0)$ in (12.32) satisfies $\hat{f}_2^1(\tilde{a},0)=0$ and $\frac{d}{da}\hat{f}_2^1(a,0)\le -\ell$ for $a\in[a^-,a^+]$.

We verify the assumptions of Theorems 3.6 and 4.1.

Hypothesis HM a) is satisfied for μ and h sufficiently small since

$$|\bar{a}-\tilde{a}|\le(1-\ell\mu h)\,|a-\tilde{a}|+O\big(\mu h(\mu+h)\big),\quad a\in[a^-,a^+].$$

Hypothesis HM b) holds since $X=\mathbb{R}$.

Hypothesis HM c) holds with

$$\Gamma_{11}=1+O\big(\mu^2h(\mu+h)\big),\quad L_{12}=O(\mu h),$$
$$L_{21}=O\big(\mu^2h(\mu+h)\big),\qquad L_{22}=1-\mu h\,\ell/2$$

for μ and h small enough.

Hypothesis HMA is satisfied for $y^*=\tilde{a}$.

It follows that for $k\in\mathbb{N}$, Conditions CM, CMA and CMA(k) are satisfied for μ and h sufficiently small. Hence, Theorems 3.6 and 4.1 imply the existence of a smooth attractive invariant manifold of the map (12.32) described by a 2π-periodic function $(\alpha,\mu,h)\mapsto\sigma_a(\alpha,\mu,h)$ as well as the foliation of the space adjacent to the manifold.

We estimate σ_a. By the contraction principle the equation

$$\hat{f}_2^1(a,h)+\mu\hat{F}_2^1(a)=0 \tag{12.33}$$

has a unique solution $a^*(\mu,h)=\tilde{a}+O(\mu+h)$. Note that $a^*(\mu,0)$ is equal to the solution $a^*(\mu)$ of (12.13). We apply Theorem 2.3 and find $\rho=O\big(\mu^2h(\mu+h)\big)$ implying

$$\sigma_a(\alpha,\mu,h)=a^*(\mu,h)+O\big(\mu(\mu+h)\big).$$

We formulate these results in the variables (β,b) of the map (12.25).

Theorem 12.3. *Let the map* (12.25) *be of class C^k, $k\in\mathbb{N}$, and let the average $\hat{f}^1(\cdot,0)$ of the function $f^1(\alpha,\cdot,0)$ satisfy Assumption ARA.*

Then there are $\mu_0,h_0,c>0$ and a C_b^k-function $\sigma_b\colon\mathbb{R}\times(0,\mu_0]\times(0,h_0]\to\mathbb{R}$, 2π-periodic with respect to the first argument, such that the following assertions hold for $\mu\in(0,\mu_0]$, $h\in(0,h_0]$.

i) *The set* $M_b := \{(\beta, b) \mid \beta \in \mathbb{R},\, b = \sigma_b(\beta,\mu,h)\}$ *is an invariant manifold of* (12.25). *The function* σ_b *satisfies*

$$\sigma_b(\beta,\mu,h) = a^*(\mu,h) + \mu\, g_2^1(\beta,\tilde{a},0) + O\big(\mu(\mu+h)\big)$$

with a^* *being the unique solution of* (12.33) *and* g^1 *being defined by* (12.29), (12.30).

ii) *The manifold* M_b *is attractive in the sense that for* $\chi_A = 1 - \mu h\ell/4$ *the estimate*

$$\big|b_j - \sigma_b(\beta_j,\mu,h)\big| \le \big(1 + O(\mu)\big)\, \chi_A{}^j\, \big|b_0 - \sigma_b(\beta_0,\mu,h)\big|$$

holds for every orbit (β_j, b_j), $j \ge 0$, *of the map* (12.25) *with* $(\beta_0, b_0) \in \mathbb{R} \times [a^- + c\mu_0,\, a^+ - c\mu_0]$.

iii) *The space* $\mathbb{R} \times [a^- + c\mu_0,\, a^+ - c\mu_0]$ *is foliated by stable fibers* $W_h^s(\beta,b) = \{(\xi,\eta) \mid \eta \in [a^- + c\mu_0,\, a^+ - c\mu_0],\ \xi = w_h^s(\beta,b,\eta)\}$. *The function* $w_h^s(\beta,b,\cdot)$ *is of class* C_b^k *and* λ_R*-Lipschitz continuous with* $\lambda_R = 2L_{12}/(\mu h\,\ell) + O(\mu)$.

12.2.3 Application to the van der Pol equation

We apply Theorem 12.3 to the map (12.24). This map is of the form (12.25) with

$$\begin{aligned} f_1^1(\beta,b,h) = &-\frac{h}{16d}\,(4-3b^2) - \frac{h}{4d}\,(1-b^2)\cos 2(\beta+\omega) \\ &+ \frac{1}{4}\,(2-b^2)\sin 2(\beta+\omega) \\ &+ \frac{hb^2}{16d}\cos 4(\beta+\omega) - \frac{b^2}{8}\,\sin 4(\beta+\omega) \end{aligned}$$

$$\begin{aligned} f_2^1(\beta,b,h) = &\frac{b}{8}\,(4-b^2) - \frac{b}{2}\,\cos 2(\beta+\omega) - \frac{hb}{8d}\,(2-b^2)\sin 2(\beta+\omega) \\ &+ \frac{b^3}{8}\cos 4(\beta+\omega) + \frac{hb^3}{16d}\,\sin 4(\beta+\omega), \end{aligned}$$

where $d := \sqrt{1-h^2/4}$ and with $f^2(\beta,b) = 0$.

We determine the averaged map (12.32) for our example. By Remark 12.2 we know that $\widehat{F}^2(a)$ is equal to $\overline{Df^1(\alpha,a)\cdot g^1(\alpha,a)}$ in (12.15). Hence, the averaged map of (12.24) is the map

$$\bar{\alpha} = \alpha + \omega - \frac{\mu h^2}{16d}(4-a^2) - \frac{\mu^2 h}{8}\Big(1 - \frac{3}{2}a^2 + \frac{11}{32}a^4\Big) + O\big(\mu^2 h(\mu+h)\big),$$

$$\bar{a} = a + \frac{\mu h a}{8}(4-a^2) + O\big(\mu^2 h(\mu+h)\big),$$

where the O-terms are π-periodic with respect to α. We have

$$\hat{f}^1(\cdot,0) = (0, a(4-a^2)/8)^T.$$

Assumption ARA is satisfied with $\tilde{a} = 2$, and with the choice $a^- = 3/2$, $a^+ = 5/2$ and $\ell = 1/3$. We get the equation $a(4-a^2)/8 = 0$ for a^*, cf. (12.33). It follows that $a^*(\mu,h) = 2$, independent of μ and h. Since $g^1(\alpha,a,0)$ is equal to $g^1(\alpha,a)$ in (12.14), cf. (12.31), the weakly attractive invariant manifold M_b of the map (12.24) is described by the π-periodic function

$$\sigma_b(\beta,\mu,h) = 2 + \mu\Big(-\frac{1}{2}\sin 2\beta + \frac{1}{4}\sin 4\beta + O(h)\Big) + O(\mu^2). \tag{12.34}$$

We discuss the relation between the invariant manifolds of the continuous and the discrete system. In particular, we inspect the limits $h \to 0$ and $\mu \to 0$.

In the case $h \to 0$ the coordinates (β, b) of the discrete system (12.24) tend to the coordinates (φ, r) of the continuous system (12.5) since

$$\begin{pmatrix} r\cos\varphi \\ -r\sin\varphi \end{pmatrix} = \begin{pmatrix} x \\ y \end{pmatrix} = T\begin{pmatrix} u \\ v \end{pmatrix} = T\begin{pmatrix} b\cos\beta \\ -b\sin\beta \end{pmatrix}$$

with T tending to the identity, cf. (12.21). Hence, in the limit $h \to 0$ the two leading terms in the μ-expansion of the function σ_b in (12.34) tend to the corresponding terms of the function s_r in (12.17).

Next we investigate the limit $\mu \to 0$. From (12.34) we have

$$\lim_{\mu\to 0} \sigma_b(\beta,\mu,h) = 2.$$

This limit function describes the circle $(u,v) = 2(\cos\beta, -\sin\beta)$ in the (u,v)-coordinates. By the coordinate change (12.21) the limit curve is an ellipse in the original coordinates (x,y). By means of (12.20) this ellipse has semi-major axis $2d/\sqrt{1-h/2}$, $d = \sqrt{1-h^2/4}$, in the direction of $(-1,1)^T$ and semi-minor axis $2d/\sqrt{1+h/2}$ in the direction of $(1,1)^T$. Its area is $4d\pi$ compared to 4π, the area of the limit circle of the ODE. In the polar coordinates (φ, r) of (12.4) this ellipse may be described by a function R as $\{(\varphi,r) \mid \varphi \in [0,2\pi],\ r = 2 + hR(\varphi,h)\}$. The function R has the symmetries

$$\begin{aligned} R(\varphi+\pi,h) &= R(\varphi,h), \\ R(\pi/4+\varphi,h) &= R(\pi/4-\varphi,h). \end{aligned} \tag{12.35}$$

The invariant manifold M_b given by σ_b, cf. (12.34), implies the existence of an invariant manifold of the map (12.18). We express this manifold in the polar coordinates (φ, r). Since the connection between the coordinates (β, b) of (12.23) and (φ, r) of (12.4) is

$$\begin{pmatrix} \varphi \\ r \end{pmatrix} = \Phi^{-1} \circ T \circ \Phi \begin{pmatrix} \beta \\ b \end{pmatrix} = (I + O(h)) \begin{pmatrix} \beta \\ b \end{pmatrix},$$

the function σ_b is transformed to

$$\begin{aligned}\sigma_r(\varphi,\mu,h) &= 2 + hR(\varphi,h) + \mu\Big(-\frac{1}{2}\sin 2\varphi + \frac{1}{4}\sin 4\varphi + O(h)\Big) + O(\mu^2)\\ &= s_r(\varphi,\mu) + hR(\varphi,h) + \mu O(\mu + h),\end{aligned}$$

where σ_r is π-periodic with respect to φ. This relates the invariant curve of the discrete system (12.18) described by σ_r to the limit cycle of the continuous system (12.3) described by s_r. In Figure 12.2 the invariant curve of the symplectic Euler method is shown for μ fixed and various h-values. For large h the term hR dominates and the symmetries (12.35) can be observed. For small h it is seen that the invariant curve of the symplectic Euler method approaches the limit cycle of the van der Pol equation.

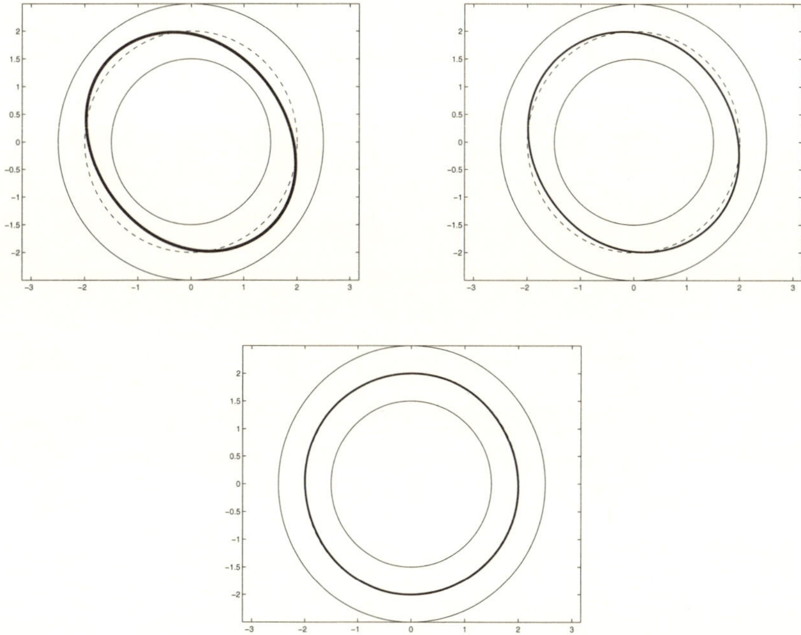

Figure 12.2. The invariant curve of the symplectic Euler method for $\mu = 0.03$ and $h = 0.4, 0.25, 0.05$.

12.3 The Euler method

In this section we apply an integration method to the van der Pol equation (12.3) which is not area preserving for $\mu = 0$. For simplicity we restrict ourselves to the explicit Euler method

$$\begin{pmatrix}\bar{x}\\ \bar{y}\end{pmatrix} = A\begin{pmatrix}x\\ y\end{pmatrix} + \mu h\begin{pmatrix}0\\ y(1-x^2)\end{pmatrix}, \quad A = \begin{pmatrix}1 & h\\ -h & 1\end{pmatrix}. \tag{12.36}$$

We will show that the map (12.36) admits an invariant curve in the domain $D = \{(x,y) \mid 3/2 \le \sqrt{x^2+y^2} \le 5/2\}$ if the step size h is small compared to the perturbation parameter μ. And we will show that if μ is small compared to h then every orbit starting in D eventually leaves D.

We transform the map (12.36) to polar coordinates by

$$\begin{pmatrix}x\\ y\end{pmatrix} = \begin{pmatrix}r\cos\varphi\\ -r\sin\varphi\end{pmatrix} =: \Phi(\varphi, r). \tag{12.37}$$

For $\mu = 0$ the right-hand side of (12.36) is taken to

$$A\Phi(\varphi,r) = rd\begin{pmatrix}\cos\omega & \sin\omega\\ -\sin\omega & \cos\omega\end{pmatrix}\begin{pmatrix}\cos\varphi\\ -\sin\varphi\end{pmatrix} = \Phi(\varphi+\omega, rd)$$

where $d := \sqrt{1+h^2}$ and $\omega = \arcsin(h/d)$. For $\mu > 0$ the map (12.36) is transformed to

$$\begin{aligned}\bar{\varphi} &= \varphi + \omega + \mu h\, f_1^1(\varphi, r, h) + O(\mu^2 h^2),\\ \bar{r} &= rd + \mu h\, f_2^1(\varphi, r, h) + O(\mu^2 h^2)\end{aligned} \tag{12.38}$$

with

$$\begin{aligned} f_1^1(\varphi,r,h) = \frac{1}{2d^2}\Big[&-h\Big(1-\frac{r^2}{4}\Big) + h\cos 2\varphi \\ &+ \Big(1-\frac{r^2}{2}\Big)\sin 2\varphi - \frac{hr^2}{4}\cos 4\varphi - \frac{r^2}{4}\sin 4\varphi\Big],\end{aligned}$$

$$f_2^1(\varphi,r,h) = \frac{r}{2d}\Big[1-\frac{r^2}{4} - \cos 2\varphi + h\Big(1-\frac{r^2}{2}\Big)\sin 2\varphi + \frac{r^2}{4}\cos 4\varphi - \frac{hr^2}{4}\sin 4\varphi\Big].$$

This is derived in an analogous way as (12.24) in Section 12.2.

Theorem 12.4. *For $(\varphi, r) \in \mathbb{R} \times [3/2, 5/2]$ consider the map (12.38), i.e., the explicit Euler method applied to the van der Pol equation (12.3) expressed in polar coordinates.*

Then there are positive constants μ_0, h_0, c and a C_b^1-function $\sigma_r\colon \mathbb{R} \times (0,\mu_0] \times (0,h_0] \to \mathbb{R}$, π-periodic with respect to the first argument, such that the following assertions hold for $\mu \in (0,\mu_0]$, $h \in (0,h_0]$.

i) *If* $h \leq \mu/2$ *the set* $M_r = \{(\varphi, r) \mid \varphi \in \mathbb{R},\ r = \sigma_r(\varphi, \mu, h)\}$ *is an invariant manifold of* (12.38). *The function* σ_r *satisfies*

$$\begin{aligned}\sigma_r(\varphi, \mu, h) = a^*(\mu, h) + \mu\Big[&-\frac{a^*(\mu, h)}{4}\sin 2\varphi \\ &+ \frac{a^*(\mu, h)^3}{32}\sin 4\varphi + O(h)\Big] + O(\mu^2)\end{aligned}$$

with $a^*(\mu, h) = 2\sqrt{1 + \frac{h}{\mu} \cdot \frac{2d}{d+1}} \in (2, 5/2)$ *and* $d = \sqrt{1 + h^2}$.

The estimates

$$\begin{aligned}\sigma_r(\varphi, \mu, h) &= s_r(\varphi, \mu) + a^*(\mu, h) - 2 + O\big(\mu(\mu + h)\big), \\ a^*(\mu, h) - 2 &= \frac{h}{\mu} \cdot \frac{2d}{d+1} \cdot q(\mu, h), \quad \frac{8}{9} < q(\mu, h) < 1,\end{aligned} \tag{12.39}$$

hold where s_r *defines the invariant manifold of the van der Pol equation* (12.3), *cf.* (12.17) *in Subsection* 12.1.3.

The manifold M_r *is attractive in the sense that for* $\chi_A = 1 - \mu h/100$ *the estimate*

$$|r_j - \sigma_r(\varphi_j, \mu, h)| \leq \big(1 + O(\mu)\big)\, {\chi_A}^j\, |r_0 - \sigma_r(\varphi_0, \mu, h)|$$

holds for every orbit (φ_j, r_j), $j \geq 0$, *of the map* (12.38) *with* $(\varphi_0, r_0) \in \mathbb{R} \times [3/2 + c\mu_0,\ 5/2 - c\mu_0]$.

The space $\mathbb{R} \times [3/2 + c\mu_0,\ 5/2 - c\mu_0]$ *is foliated by stable fibers.*

ii) *If* $h \geq 3\mu/5$ *every orbit* (φ_j, r_j), $j \geq 0$, *of the map* (12.38) *starting in the domain* $\mathbb{R} \times [3/2 + c\mu_0,\ 5/2 - c\mu_0]$ *leaves this domain.*

Remark 12.5. For fixed $\mu \leq \mu_0$ and for $h \to 0$ the invariant manifold of the Euler method approaches the invariant manifold of the van der Pol equation, i.e., $\sigma_r(\varphi, \mu, h) \to s_r(\varphi, \mu)$ as $h \to 0$, cf. (12.39).

Proof. i) Assume $h \leq \mu/2$. As described in Subsection 12.2.1 we apply the averaging procedure to the map (12.38). We find

$$\hat{f}^{\,1}(a, h) = \frac{4 - a^2}{8d^2}\begin{pmatrix} -h \\ da \end{pmatrix},$$

$$Df^{\,1}(\alpha, a, 0) = \begin{pmatrix} \left(1 - \frac{a^2}{2}\right)\cos 2\alpha - \frac{a^2}{2}\cos 4\alpha & -\frac{a}{2}\sin 2\alpha - \frac{a}{4}\sin 4\alpha \\ a \sin 2\alpha - \frac{a^3}{2}\sin 4\alpha & \frac{1}{2} - \frac{3}{8}a^2 - \frac{1}{2}\cos 2\alpha + \frac{3}{8}a^2\cos 4\alpha \end{pmatrix},$$

$$g^1(\alpha, a, 0) = \begin{pmatrix} -\frac{1}{4}\left(1 - \frac{a^2}{2}\right)\cos 2\alpha + \frac{a^2}{32}\cos 4\alpha \\ -\frac{a}{4}\sin 2\alpha + \frac{a^3}{32}\sin 4\alpha \end{pmatrix},$$

$$\hat{F}^2(a) = \overline{Df^1(\alpha, a, 0) \cdot g^1(\alpha, a, 0)} = \frac{1}{8}\begin{pmatrix} \frac{a^2}{2} - \left(1 - \frac{a^2}{2}\right)^2 - \frac{3a^4}{32} \\ 0 \end{pmatrix}.$$

The averaged map is, cf. (12.32),

$$\begin{pmatrix} \bar{\alpha} \\ \bar{a} \end{pmatrix} = \begin{pmatrix} \mathcal{F}(\alpha, a, \mu, h) \\ \mathcal{G}(\alpha, a, \mu, h) \end{pmatrix} = \begin{pmatrix} \alpha + \omega \\ ad \end{pmatrix} + \mu h \hat{f}^1(a, h) + \mu^2 h \hat{F}^2(a) + \mu^2 h O(\mu + h). \tag{12.40}$$

The equation

$$a = ad + \mu h \hat{f}^1_2(a, h) + \mu^2 h \hat{F}^2_2(a)$$

has the solution

$$a^*(\mu, h) = 2\sqrt{1 + \frac{h}{\mu} \cdot \frac{2d}{d+1}}.$$

We apply Theorem 3.6 for $k = 1$ and Theorem 4.1 to the map (12.40). We verify Hypotheses HM, HMA and Conditions CM and CMA.

Hyptohesis HM a): For h small enough we find $a^*(\mu, h) \in (2, 2.475)$. The map (12.40) is inflowing with respect to $(3/2, 5/2)$ for μ and h small enough since

$$\begin{aligned} |a^- - a^*| &= |\mathcal{G}(\alpha, a, \mu, h) - \mathcal{G}(\alpha, a^*(\mu, h), \mu, h)| + \mu^2 h O(\mu + h) \\ &\le p|a - a^*(\mu, h)| + \mu^2 h O(\mu + h) \end{aligned}$$

with $p = 1 - \mu h/4$.

Hypothesis HM b) holds since $\alpha \in \mathbb{R}$.

Hypothesis HM c) holds with

$$\Gamma_{11} = 1 + \mu^2 h O(\mu + h), \quad L_{12} = \mu h O(\mu + h),$$

$$L_{21} = \mu^2 h O(\mu + h), \qquad L_{22} = 1 - \mu h/60$$

for μ and h sufficiently small.

Hypothesis HMA is satisfied for $y^* = 2$.

Conditions CM and CMA are satisfied for μ small enough.

We conclude that the map (12.40) admits an invariant manifold given by $a = \sigma_a(\alpha, \mu, h)$, σ_a being π-periodic with respect to α. Applying Theorem 2.3 with approximation $a^*(\mu, h)$, $\rho = \mu^2 h O(\mu + h)$ and $\chi_A = 1 - \mu h/100$ yields

$$\sigma_a(\alpha, \mu, h) = a^*(\mu, h) + \mu O(\mu + h).$$

Going back to the variables (φ, r) we obtain

$$\sigma_r(\varphi, \mu, h) = a^*(\mu, h) + \mu\Big[-\frac{a^*(\mu, h)}{4}\sin 2\varphi + \frac{a^*(\mu, h)^3}{32}\sin 4\varphi + O(h)\Big] + O(\mu^2),$$

σ_r being π-periodic with respect to φ.

ii) It is easy to verify that $\bar{a} > a$, $a \in (3/2, 5/2)$, for the map (12.40) if $\mu \leq 5h/3$ and if h is sufficiently small. □

In Figure 12.3 we illustrate the behaviour of the invariant manifold of the Euler method for $\mu = 0.05$ and for $h = 0.01, 0.025, 0.03, 0.05$. In these plots it is observed that the distance of the invariant manifold of the Euler method to the limit cycle of the van der Pol equation is approximately h/μ as indicated in Theorem 12.4 i). In particular, it is seen that the invariant manifold of the discrete system lies within the annulus $3/2 \leq r \leq 5/2$ for $h \leq \mu/2$ and outside for $h \geq 3\mu/5$.

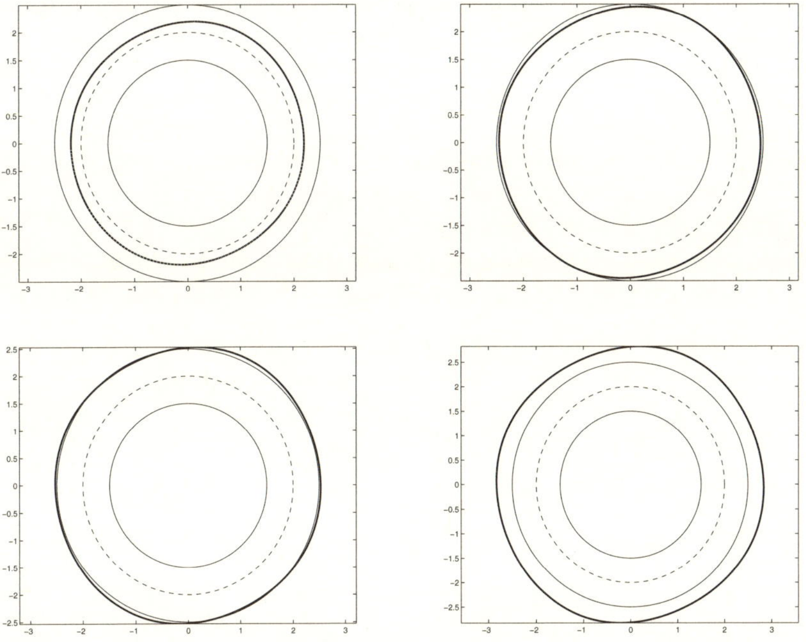

Figure 12.3. The invariant manifold of the Euler method for the values $\mu = 0.05$ and $h = 0.01$, 0.025, 0.03, 0.05.

Chapter 13
Blow-up in singular perturbations

13.1 Introduction

We consider the stiff van der Pol differential equation, cf. (12.2) in Section 12.1,

$$\begin{aligned} \dot{x} &= 1 + y, \\ \varepsilon^3 \dot{y} &= -x - y^2 - y^3/3, \end{aligned} \tag{13.1}$$

and its discretisation by a one-step method. In (13.1) the fold point $(2/3, 1)$ of (12.2) is shifted to the origin and the parameter ε is replaced by ε^3 since in our investigations some functions depend smoothly on the third root of this parameter. The differential equation (13.1) admits a so-called reduced manifold which in this case is a curve $\mathcal{G}$ given by the algebraic equation $-x - y^2 - y^3/3 = 0$. The upper branch $\mathcal{G}_+$ and the lower branch $\mathcal{G}_-$ of the curve $\mathcal{G}$, respectively, may be described as the graph of a function $y = s_+^0(x)$, $x \in (-\infty, 0]$ and $y = s_-^0(x)$, $x \in [-4/3, \infty)$, respectively, cf. Figure 13.1.

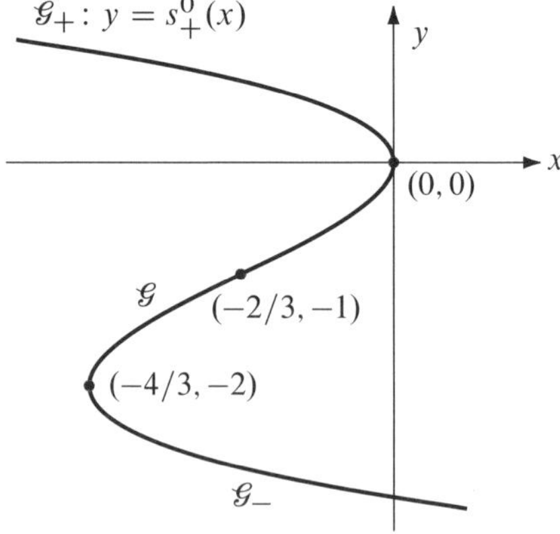

Figure 13.1. The reduced manifold of the van der Pol equation (13.1).

Away from the origin, in an ε^3-neighbourhood of $\mathcal{G}_+$, there is a highly attractive (negatively) invariant manifold $M_{+,\varepsilon}^{\infty}$ described as the graph of a function $y = s_+^{\infty}(x, \varepsilon)$. The existence of this manifold is proved in Theorem 10.3. For the efficient numerical integration of (13.1) near $M^{\infty}_{+,\varepsilon}$ stiff integration methods with a step size large compared to ε are needed. For details we refer to Chapter 11. Analogous statements hold true for the lower branch $\mathcal{G}_-$. There is a broad literature on stiff integration methods, cf. Hairer, Wanner [53] and references therein.

In the vicinity of the origin, the so-called fold point, the behaviour of the solutions of (13.1) is more complex. Such solutions have thoroughly been investigated by the method of matched asymptotic expansions, cf. O'Malley [102], Lagerstrom, Casten [75], Mishchenko, Rozov [89], Nipp [91] and [93]–[95]. More recently, a more geometric approach using invariant manifold techniques was developped. For this approach named geometric singular perturbation theory we refer to Jones [62], Dumortier [35] and Krupa, Szmolyan [73]. In [35] and [73] the so-called blow-up was introduced to get local coordinates in charts independent of the perturbation parameter. In [73] this was applied to extend the (negatively) invariant manifold $M^{\infty}_{+,\varepsilon}$ past the fold point of the differential equation (13.1). First results on using the blow-up approach for the discrete dynamical system obtained by applying the Euler method to (13.1) were presented at the ICNAAM Conference 2009 in Rethymno, Crete, by Nipp, Stoffer, Szmolyan.

In this chapter we consider a map which is a one-step approximation of the time-h map of the differential equation (13.1). For our approach it is essential that h is small compared to the parameter ε^3. For this map we prove the existence of a negatively invariant manifold near the fold point using the blow-up approach. This is done in several charts and allows to describe the manifold as the graph of a function in every chart. Our theoretical concept for the graph transform in several charts is described in Section 1.4.

The remaining part of this chapter is organized as follows. In Section 13.2 we state the main result. In Section 13.3 we sketch the blow-up of the fold point of (13.1) and establish a reference manifold which is used to introduce appropriate local coordinates. In Section 13.4 we prove the main result.

13.2 The main result

Since the function s^0_+ is not bounded we modify the x-equation by multiplying the right-hand side by a monotone function $q\colon [-3,\infty) \to [0,1]$ of class C^{k+1}_b, $k \geq 1$, with $q(-3) = 0, q(x) \equiv 1$ for $x \geq -2$. Moreover, we transform the slow time variable t to the fast time variable τ by $t = \varepsilon^3\tau$ and obtain the continuous dynamical system

$$\begin{aligned} \frac{dx}{d\tau} &= \varepsilon^3 q(x)(1+y) =: \varepsilon^3 f(x,y), \\ \frac{dy}{d\tau} &= -x - y^2 - y^3/3 =: g(x,y). \end{aligned} \tag{13.2}$$

As the corresponding discrete dynamical system we consider the following map from $\mathbb{R}^4$ to $\mathbb{R}^4$. Note that the parameters ε, h will be treated as variables below.

$$P: \begin{pmatrix} x \\ y \\ \varepsilon \\ h \end{pmatrix} \longmapsto \begin{pmatrix} \bar{x} \\ \bar{y} \\ \bar{\varepsilon} \\ \bar{h} \end{pmatrix} = \begin{pmatrix} x + h\varepsilon^3 f(x,y) + h^2\varepsilon^3 \hat{f}(x,y,\varepsilon,h) \\ y + hg(x,y) + h^2 \hat{g}(x,y,\varepsilon,h) \\ \varepsilon \\ h \end{pmatrix}. \tag{13.3}$$

For an appropriate choice of $\hat{f}$ and $\hat{g}$ this map describes a one-step method applied to (13.2) with step size h. We restrict ourselves to one-step methods which can be expressed as B-series, cf. Hairer, Nørsett, Wanner [52], as, e.g., Runge–Kutta methods. In particular, for the explicit Euler method $\hat{f} = \hat{g} = 0$.

We show that the continuous dynamical system (13.2) and the discrete dynamical system (13.3), respectively, admits a family of attractive (negatively) invariant curves around the fold point depending on the parameter ε and on the parameters ε and h, respectively.

Theorem 13.1 (Main result). *There exist positive numbers δ_0, ν_0 such that for every $(\varepsilon, h) \in (0, \delta_0] \times (0, \nu_0]$ the map P given in* (13.3) *admits an attractive negatively invariant curve. The union of these negatively invariant curves form a* 3*-dimensional attractive negatively invariant manifold $\mathcal{M}$. The manifold $\mathcal{M}$ is described in coordinates relative to a reference manifold $\mathcal{M}^r$. The reference manifold $\mathcal{M}^r$ is a negatively invariant manifold of the differential equation* (13.2) *and it is piecewise defined as the graph of some smooth functions $s_1, \dots, s_6$ in six charts $\Phi_1, \dots, \Phi_6$ given in Section* 13.4. *The manifold $\mathcal{M}$ is piecewise defined as the graph of some Lipschitz continuous functions $\sigma_1, \dots, \sigma_6$.*

Expressing the manifolds $\mathcal{M}^r$ and $\mathcal{M}$ in the original variables yields functions $s_1^*, \dots, s_6^*$ and $\sigma_1^*, \dots, \sigma_6^*$. The pieces $\mathcal{M}_i^r$ are given as

$$\begin{aligned} \mathcal{M}_i^r &= \{(x,y) \mid x \in X_i,\ y = s_i^*(x,\varepsilon)\}, \quad i = 1,2,3, \\ \mathcal{M}_i^r &= \{(x,y) \mid y \in Y_i,\ x = s_i^*(y,\varepsilon)\}, \quad i = 4,5,6, \end{aligned}$$

for some intervals $X_1, X_2, X_3 \subset \mathbb{R}$ and $Y_4, Y_5, Y_6 \subset \mathbb{R}$. The functions s_i^* satisfy the estimates

$$\begin{aligned} s_1^*(x,\varepsilon) &= s_+^0(x) + \varepsilon^3 \frac{q(x)(1 + s_+^0(x))}{[s_+^0(x)(2 + s_+^0(x))]^2} + O(\varepsilon^6), \\ s_2^*(x,\varepsilon) &= s_+^0(x) + O((-x)^{3/2} + \varepsilon), \\ s_3^*(x,\varepsilon) &= \varepsilon A(x/\varepsilon^2) + O(\varepsilon^2), \\ s_4^*(y,\varepsilon) &= \varepsilon^2 A^{\mathrm{inv}}(y/\varepsilon) + O(\varepsilon^3), \\ s_5^*(y,\varepsilon) &= O(\varepsilon^2), \\ s_6^*(y,\varepsilon) &= O(\varepsilon^2), \end{aligned}$$

where the function A is defined in (13.13). *The pieces* $\mathcal{M}_i$ *are given as*

$$\mathcal{M}_i = \{(x,y) \mid x \in X_i,\ y = s_i^*(x,\varepsilon) + \sigma_i^*(x,h,\varepsilon)\}, \quad i = 1,2,3,$$
$$\mathcal{M}_i = \{(x,y) \mid y \in Y_i,\ x = s_i^*(y,\varepsilon) + \sigma_i^*(y,h,\varepsilon)\}, \quad i = 4,5,6.$$

The functions σ_i^* *satisfy the estimates*

$$\begin{aligned} \sigma_1^*(x,h,\varepsilon) &= O(\varepsilon^3 h), \\ \sigma_2^*(x,h,\varepsilon) &= O(\varepsilon^2 h), \\ \sigma_3^*(x,h,\varepsilon) &= O(\varepsilon^2 h), \\ \sigma_4^*(y,h,\varepsilon) &= O(\varepsilon^3 h), \\ \sigma_5^*(y,h,\varepsilon) &= O(\varepsilon^2 h), \\ \sigma_6^*(y,h,\varepsilon) &= O(\varepsilon^2 h). \end{aligned}$$

Remark 13.2. (1) Our construction of the reference manifold of the differential equation (13.2) proves the corresponding result in Krupa, Szmolyan [73] in a different way and in addition describes the manifold as the graph of a function in every chart.

(2) Applying Theorem 11.1 to the situation of chart Φ_1 gives an improved estimate $\sigma_1^* = O(\varepsilon^6 h^p)$ for Runge–Kutta methods of order p. We expect that a refined investigation would show that also in the estimates for σ_i^*, $i = 2,\dots,6$, the factor h may be replaced by h^p. We did not work out the details.

13.3 Preliminaries

13.3.1 The blow-up

We briefly describe how we choose the charts Φ_2, Φ_3, Φ_4, Φ_5 in the vicinity of the fold point (0,0) of (13.2). Krupa, Szmolyan [73] investigate a blow-up of this fold point. It magnifies the critical vicinity and thus allows to introduce appropriate local coordinates. We apply such a blow-up to the map (13.3) near the fold point. The blow-up transformation

$$x = \tilde{r}^2 \tilde{x}, \quad y = \tilde{r}\tilde{y}, \quad \varepsilon = \tilde{r}\tilde{\varepsilon}, \quad h = \tilde{h}/\tilde{r}$$

takes $(\tilde{x},\tilde{y},\tilde{\varepsilon},\tilde{h},\tilde{r}) \in B := S^2 \times (0,H] \times (0,R]$ to $(x,y,\varepsilon,h) \in \mathbb{R}^4$. In order to define charts near the fold point we consider half-spheres of S^2 with centers $\tilde{x} = -1$, $\tilde{\varepsilon} = 1$, $\tilde{y} = -1$, respectively, cf. Figure 13.2. The choice $\tilde{x} = -1$ leads to chart Φ_2. The charts Φ_3, Φ_4 are obtained by setting $\tilde{\varepsilon} = 1$, chart Φ_5 by setting $\tilde{y} = -1$. To exemplify the definition of our blow-up charts we discuss Φ_2. In order to simplify the transformation we replace the half sphere with center $\tilde{x} = -1$ by the tangent plane of S^2 at $(-1,0,0)$ and use $\tilde{\varepsilon}$, $\tilde{y}$ as local coordinates. The final coordinates r_2, y_2, ε_2 and

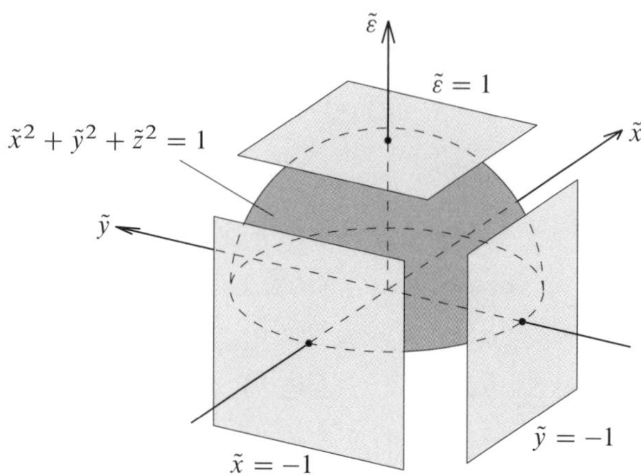

Figure 13.2. The blow-up charts Φ_2 where $\tilde{x} = -1$, Φ_3 and Φ_4 where $\tilde{\varepsilon} = 1$ and Φ_5 where $\tilde{y} = -1$ without the h-variable.

h_2 are introduced as follows:

$$x = -r_2^2, \quad y = r_2 y_2, \quad \varepsilon = r_2 \varepsilon_2, \quad h = \frac{h_2}{r_2} \tag{13.4}$$

with $(r_2, y_2, \varepsilon_2, h_2)$ in an appropriate domain.

13.3.2 The reference manifold

In every chart the variable in the contracting direction is taken relative to a reference manifold $\mathcal{M}^r$. We build the reference manifold $\mathcal{M}^r$ as follows. In chart Φ_1, for $x \in [-3, -\xi]$, $\varepsilon \in (0, \delta]$, $\xi, \delta > 0$, we take the unique (negatively) invariant manifold M_1^r of the differential equation (13.2) which is the graph of some function $(x, \varepsilon) \mapsto s_1(x, \varepsilon)$ as established in Lemma 13.3 below. In charts $\Phi_2, \dots, \Phi_6$ the manifold M_1^r is extended under the flow of (13.2) to the manifolds $M_2^r, \dots, M_6^r$.

Lemma 13.3. *For every $\xi \in (0, 3)$ there is $\delta > 0$ such that there exists a function $s_1 \colon [-3, -\xi] \times (0, \delta] \to \mathbb{R}^+$ satisfying the following assertions.*

i) *The set $M_1^r = \{(x, y) \mid x \in [-3, -\xi],\ y = s_1(x, \varepsilon)\}$ is a negatively invariant manifold of* (13.2).

ii) *The function s_1 satisfies the invariance equation*

$$\varepsilon^3 \Big[\frac{\partial}{\partial x} s_1(x, \varepsilon)\Big] f(x, s_1(x, \varepsilon)) = g(x, s_1(x, \varepsilon)).$$

iii) *The function* s_1 *is of class* C_b^k.

iv) *The function* s_1 *satisfies*

$$s_1(x,\varepsilon) = s_+^0(x) + \varepsilon^3 s_+^1(x,\varepsilon), \quad x \in [-3,-\xi],$$

where s_+^0 *is defined by* $g(x, s_+^0(x)) = 0$ *and where* s_+^1 *is of class* C_b^k *and satisfies*

$$s_+^1(x,\varepsilon) = \frac{q(x)(1+s_+^0(x))}{[s_+^0(x)(2+s_+^0(x))]^2} + O(\varepsilon^3).$$

Proof. We consider the differential equation (13.2) in the slow time variable t and we introduce the new variable z by $y = s_+^0(x) + \varepsilon^3 z$. We obtain the transformed differential equation

$$\begin{aligned}
\dot{x} &= q(x)(1+s_+^0(x)+\varepsilon^3 z),\\
\varepsilon^3 \dot{z} &= -z\big[2s_+^0(x) + s_+^0(x)^2 + \varepsilon^3 z(1+s_+^0(x)+\varepsilon^3 z/3)\big]\\
&\quad - \frac{ds_+^0(x)}{dx} q(x)(1+s_+^0(x)+\varepsilon^3 z),
\end{aligned}$$

where

$$\frac{ds_+^0(x)}{dx} = -\frac{1}{s_+^0(x)(2+s_+^0(x))}.$$

We apply Theorem 10.1. Assumption ASA$(k+1)$ is satisfied with $s_A^0(x) = q(x)(1+s_+^0(x))/[s_+^0(x)(2+s_+^0(x))]^2$, $d_0 > 0$ and $b_A = 2s_+^0(\xi)$ for ε small enough. □

13.4 Proof of the main result

We first describe the structure of the proof of the main result without the details to be done in every chart. These details are given in Subsections 13.4.1–13.4.6 below.

The main result is proved by applying Theorem 1.10. In each chart $\Phi_i, i = 1,\dots,6$, we verify Hypothesis HMAK and Condition CMAK and introduce coordinates relative to the reference manifold M_i^r. The existence of M_1^r is established in Subsection 13.3.2.

We prove that there are $\delta_0, \nu_0 > 0$ such that for all $\delta \le \delta_0$, $\nu \le \nu_0$ the assertions of Theorem 13.1 hold for $(\varepsilon, h) \in [\delta/2, \delta] \times [\nu/2, \nu]$. We need to bound ε from below to get a bounded norm factor in chart Φ_6 and we need to bound h from below to get an attrictivity constant $\chi_i < 1$ in chart $\Phi_i, i = 1,\dots,6$. Then by uniqueness the manifold $\mathcal{M}$ is established for $(\varepsilon, h) \in (0,\delta] \times (0,\nu]$. If, e.g., $\varepsilon < \delta_0/2$ we set $\delta := 4\varepsilon/3 < \delta_0$. Then $\varepsilon \in [\delta/2, \delta]$. The extension for $h < \nu_0/2$ is done analogously.

To apply Theorem 1.10 we have to compute Lipschitz constants of maps which have to satisfy certain conditions. It is crucial that in charts Φ_1, Φ_3, Φ_4, Φ_6 the

Lipschitz constants may depend on the parameters ε_i and h_i, $i = 1, 3, 4, 6$, and they are supposed to be *local* if considered with respect to ε_i or h_i. More precisely, for a function $F\colon (x, \varepsilon, h) \mapsto F(x, \varepsilon, h)$,

$$\operatorname{Lip}_x F := \sup_{x^1, x^2 \in X,\ x^1 \neq x^2} \frac{|F(x^1, \varepsilon, h) - F(x^2, \varepsilon, h)|}{|x^1 - x^2|},$$

$$\operatorname{Lip}_\varepsilon F := \sup_{x \in X} \limsup_{\varepsilon^* \to \varepsilon} \frac{|F(x, \varepsilon, h) - F(x, \varepsilon^*, h)|}{|\varepsilon - \varepsilon^*|},$$

$$\operatorname{Lip}_h F := \sup_{x \in X} \limsup_{h^* \to h} \frac{|F(x, \varepsilon, h) - F(x, \varepsilon, h^*)|}{|h - h^*|}.$$

In each chart Φ_i, $i = 1, \dots, 6$, small positive constants have to be chosen appropriately. The constants in chart Φ_i may depend on the constants of chart Φ_{i-1}. As a consequence one might have to adapt subsequently the constants in charts $\Phi_{i-1}, \dots, \Phi_1$. This induces chains of dependencies on certain constants which must not contain loops. E.g., there is the chain $\eta_6 \to \eta_5 \to \eta_4 \to \eta_3 \to \eta_2 \to \eta_1 \to \nu$ where η_i defines the neighbourhood considered of the reference manifold M_i^r in chart Φ_i.

According to Hypothesis HMAK d), for every chart Φ_i a function space C_i has to be introduced to define the function space Σ. The space Σ is nonempty since $\sigma = (\sigma_1, \dots, \sigma_6)$ with $\sigma_i \equiv 0$, $i = 1, \dots, 6$, is contained in Σ. This follows from $0 \in C_i$. The function $\sigma \equiv 0$ describes the reference manifold $\mathcal{M}^r$. For Hypothesis HMAK d) to hold it remains to verify that the operator $\mathcal{F}$ takes Σ into itself. The details of the proof of Theorem 13.1 in the charts $\Phi_1, \dots, \Phi_6$ are given in the following subsections.

The estimates for s_i^*, $i = 1, \dots, 6$, follow directly from the construction of the reference manifold $\mathcal{M}^r$. The estimates for σ_i^*, $i = 1, \dots, 6$, are obtained as follows. An argument similar to the one used to prove Assertion vi) of Theorem 1.5 shows that in chart Φ_i, $i = 1, \dots, 6$, the set $Z_{c_i} := \{z_i \mid |z_i| \le c_i h_i\}$ is invariant under the map P_i for a sufficiently large constant c_i. For $i = 2, \dots, 6$, the function $\bar{\sigma}_{i,i-1}$ is contained in Z_{c_i} for an appropriate c_i. It follows that $\sigma_i = O(h_i)$. Going back to the original variables gives the estimates for σ_i^*.

13.4.1 The chart Φ_1

Verification of Hypothesis HMAK a), b), d) and Condition CMAK a)

Reference manifold

We consider the coordinates

$$x = x_1,$$
$$y = y_1.$$

For small positive constants $\xi_1, \check{\xi}_1$ with $\xi_1 < \check{\xi}_1$ we put

$$X_1 := [-3, -\xi_1], \quad \widetilde{X}_1 := [-3, 0],$$
$$\check{X}_1 := [-\check{\xi}_1, -\xi_1], \quad E_1 := [\delta/2, \delta].$$

We take the graph M_1^r of the function $s_1 : X_1 \times E_1 \to \mathbb{R}$ established in Lemma 13.3 as reference manifold for Chart Φ_1.

Local coordinates

The chart Φ_1 is given by the coordinate transformation

$$\Phi_1 : \quad \begin{aligned} x &= x_1, \\ y &= s_1(x_1, \varepsilon_1) + \varepsilon_1^3 z_1, \\ \varepsilon &= \varepsilon_1, \\ h &= h_1. \end{aligned}$$

Domains

For a small positive constants η_1 we set

$$Z_1 := [-\eta_1, \eta_1], \quad H_1 := [\nu/2, \nu].$$

In this chart ε_1 and h_1 are parameters and we have $V_1 = Z_1$ and $U_1 = \hat{U}_1 = X_1 \times E_1 \times H_1$, similarly $\widetilde{U}_1 = \widetilde{X}_1 \times E_1 \times H_1$, $\check{U}_1 = \check{X}_1 \times E_1 \times H_1$.

Map

The change to local coordinates takes (13.3) to (for simplicity we write s instead of $s_1(x_1, \varepsilon_1)$ and $\bar{s}$ instead of $s_1(\bar{x}_1, \varepsilon_1)$)

$$\begin{aligned} \bar{x}_1 &= x_1 + h\varepsilon_1^3 f(x_1, s + \varepsilon_1^3 z_1) + h_1^2 \varepsilon_1^3 \hat{f}(x_1, s + \varepsilon_1^3 z_1, \varepsilon_1, h_1), \\ \varepsilon_1^3 \bar{z}_1 &= -(\bar{s} - s) + \varepsilon_1^3 z_1 + h_1 g(x_1, s + \varepsilon_1^3 z_1) + h_1^2 \hat{g}(x_1, s + \varepsilon_1^3 z_1, \varepsilon_1, h_1). \end{aligned} \tag{13.5}$$

The function s satisfies the invariance equation in Lemma 13.3 ii), i.e.,

$$\varepsilon_1^3 s_x f(x_1, s) = g(x_1, s). \tag{13.6}$$

To show that the function $\hat{g}$ is of order ε_1^3 we use that the map P may be expressed as a B-series and that $g(x_1, s + \varepsilon_1^3 z_1) = O(\varepsilon_1^3)$ by the invariance equation. By the mean value theorem we have that

$$-(\bar{s} - s) = -\int_0^1 s_x(x_1 + t(\bar{x}_1 - x_1), \varepsilon_1) dt \cdot (\bar{x}_1 - x_1) =: -I_1 \cdot (\bar{x}_1 - x_1).$$

Hence, using (13.6) and once more applying the mean value theorem we get with $\bar{x}_1 - x_1$ from (13.5)

$$\begin{aligned}
&-(\bar{s}-s)+h_1 g(x_1,s)\\
&\quad=(\bar{s}-s)+h_1\varepsilon_1^3 s_x f(x_1,s)\\
&\quad=-h_1\varepsilon_1^3 f(x_1,s)\int_0^1\int_0^1 t s_{xx}(x_1+rt(\bar{x}_1-x_1),\varepsilon_1)dt\,dr\cdot(\bar{x}_1-x_1)\\
&\qquad-h_1\varepsilon_1^3 I_1\cdot(\varepsilon_1^3 z_1+O(h_1)).
\end{aligned}$$

Introducing this into (13.5) we obtain the map

$$P_1:\quad \begin{aligned}
\bar{x}_1&=x_1+h_1\varepsilon_1^3 f(x_1,s+\varepsilon_1^3 z_1)+h_1^2\varepsilon_1^3\hat{f}(x_1,s+\varepsilon_1^3 z_1,\varepsilon_1,h_1),\\
\bar{z}_1&=\big[1+h_1\big(g_y(x_1,s)+O(h_1)+O(\varepsilon_1^3)\big)\big]z_1+O(h_1^2),
\end{aligned}$$

or, more explicitely,

$$P_1:\quad \begin{aligned}
\bar{x}_1&=x_1+h_1\varepsilon_1^3 q(x_1)(1+s+\varepsilon_1^3 z_1)+h_1^2\varepsilon_1^3\hat{f}(x_1,s+\varepsilon_1^3 z_1,\varepsilon_1,h_1),\\
\bar{z}_1&=\big[1+h_1\big(-2s-s^2+O(h_1)+O(\varepsilon_1^3)\big)\big]z_1+O(h_1^2).
\end{aligned}$$

Note that the h_1^2-term of the z_1-equation is independent of z_1.

Graph transform

We apply the graph transform result Theorem 1.11. For ε_1, h_1 and η_1 small enough Hypothesis HMAG is satisfied with

$$\begin{gathered}
\Gamma_{11}=1+h_1 O(\delta^3),\quad L_{12}=h_1 O(\delta^6),\quad L_{13}=h_1 O(\delta^2),\\
L_{10}=O(\delta^3),\\
L_{21}=h_1 O(\eta_1+h_1),\quad L_{22}=1-h_1 b,\quad L_{23}=h_1 O(\eta_1+h_1),\\
L_{20}=O(\eta_1+h_1),
\end{gathered}$$

where b is a positive constant smaller than $2s_+^0(-\xi_1)+s_+^0(-\xi_1)^2$, cf. Lemma 13.3 iv).

Since Condition CM holds as $h_1 O(\delta^3) < h_1(b+O(\delta^3))$, for δ small enough, we may choose $\alpha_1 := \alpha_{\min} = O(\eta_1+h_1)$. We choose $\nu \le \eta_1$ implying $\alpha_1 = O(\eta_1)$. Condition CMAG takes the form $\eta_1 O(\delta^6) < b$ which is satisfied for small δ. Since $h_1 \in [\nu/2,\nu]$ the assertions of Theorem 1.11 hold with

$$\alpha_1=O(\eta_1),\quad \beta_1=O(\eta_1),\quad \gamma_1=O(\eta_1)/\nu$$

and $\chi_1 \le 1-b\nu/4$.

Function space

We define the space of functions

$$\begin{aligned}C_1(U_1, V_1) := \{\sigma_1 \mid \sigma_1 \colon U_1 \to V_1,\ &\sigma_1 \text{ is } \alpha_1\text{-Lipschitz continuous with respect to } x_1,\\ &\beta_1\text{-Lipschitz continuous with respect to } \varepsilon_1,\\ &\gamma_1\text{-Lipschitz continuous with respect to } h_1\}.\end{aligned}$$

Theorem 1.11 states that the operator $\mathcal{F}_1$ induced by P_1 takes $\sigma_1 \in C_1(U_1, V_1)$ to $\hat{\sigma}_1 \in C_1(U_1, V_1)$.

13.4.2 The chart Φ_2

Verification of Hypothesis HMAK a), b) and Condition CMAK a)

Reference manifold

We consider the coordinates

$$\begin{aligned} x &= -r_2^2,\\ y &= r_2 y_2,\\ \varepsilon &= r_2 \varepsilon_2, \end{aligned}$$

cf. (13.4). For positive constants $\hat{\rho}_2, \rho_2, \check{\delta}_2, \delta_2, \tilde{\delta}_2$ with $\xi_1 < \hat{\rho}_2^2 < \rho_2^2 < \check{\xi}_1$ and $\check{\delta}_2 < \delta_2 < \tilde{\delta}_2$ we put

$$\begin{aligned} &X_2 := (0, \rho_2], \quad \hat{X}_2 := (0, \hat{\rho}_2],\\ &E_2 := (0, \delta_2], \quad \tilde{E}_2 := (0, \tilde{\delta}_2], \quad \check{E}_2 := [\check{\delta}_2, \delta_2]. \end{aligned}$$

Rescaling the time by $t_2 = r_2(1 + r_2 y_2)\tau$, equation (13.2) expressed in the new coordinates is taken to

$$\begin{aligned} dr_2/dt_2 &= -\frac{r_2\varepsilon_2^3}{2},\\ d\varepsilon_2/dt_2 &= \frac{\varepsilon_2^4}{2},\\ dy_2/dt_2 &= \frac{1 - y_2^2 - r_2 y_2^3/3}{1 + r_2 y_2} + \frac{\varepsilon_2^3 y_2}{2}. \end{aligned} \tag{13.7}$$

In these coordinates we describe the reference manifold $\mathcal{M}^r$ as the graph M_2^r of a function $y_2 = s_2(r_2, \varepsilon_2)$ for $r_2 \in X_2, \varepsilon_2 \in E_2, r_2\varepsilon_2 \in [\delta/2, \delta]$. Expressing the manifold M_1^r in the new coordinates for $r_2 = \rho_2, \varepsilon_2 \in [\delta/2, \delta]/\rho_2$ leads to $y_2 = s_1(-\rho_2^2, \rho_2\varepsilon_2)/\rho_2$. Let (r_2^0, ε_2^0) be given in the domain considered. We take the solution of the r_2- and the ε_2-equation of (13.7) with initial conditions (r_2^0, ε_2^0) and choose $t_2^0 < 0$

such that $r_2(t_2^0) = \rho_2$. Set $\varepsilon_2^* := \varepsilon_2(t_2^0)$. Note that $\varepsilon_2^* \in [\delta/2, \delta]/\rho_2$. Now take the solution of (13.7) with initial conditions

$$r_2(t_2^0) = \rho_2, \quad \varepsilon_2(t_2^0) = \varepsilon_2^*, \quad y_2(t_2^0) = s_1(-\rho_2^2, \rho_2\varepsilon_2^*)/\rho_2$$

and define $s_2(r_2^0, \varepsilon_2^0) := y_2(0)$, cf. Figure 13.3. The existence, uniqueness and regularity theorem for ordinary differential equations implies that s_2 is at least two times

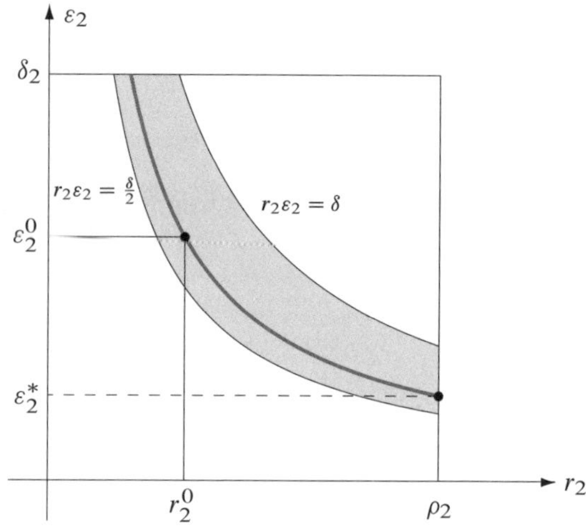

Figure 13.3. The domain of (r_2, ε_2).

differentiable with continuous and bounded derivatives. From (13.7) we get that s_2 satisfies the invariance equation

$$-\frac{\partial s_2}{\partial r_2}\frac{r_2\varepsilon_2^3}{2} + \frac{\partial s_2}{\partial \varepsilon_2}\frac{\varepsilon_2^4}{2} = \frac{1 - s_2^2 - r_2 s_2^3/3}{(1 + r_2 s_2)} + \frac{\varepsilon_2^3 s_2}{2}.$$

In the coordinates r_2, ε_2, y_2 the reference manifold is given as the graph M_2^r of the function s_2.

Local coordinates

The chart Φ_2 is given by the coordinate transformation

$$\Phi_2: \quad \begin{aligned} x &= -r_2^2, \\ y &= r_2(s_2(r_2, \varepsilon_2) + \varepsilon_2^3 z_2), \\ \varepsilon &= r_2\varepsilon_2, \\ h &= h_2/r_2, \end{aligned} \tag{13.8}$$

cf. (13.4). By the invariance equation and Theorem 7.8 we conclude that $s_2(r_2, \varepsilon_2) = 1 - r_2/6 + O(r_2^2 + \varepsilon_2^3)$ and that s_2 is of class C_b^2 for $r_2 \in X_2, \varepsilon_2 \in E_2, r_2\varepsilon_2 \in [\delta/2, \delta]$.

Domains

For positive constants η_2, ν_2 we put

$$Z_2 := [-\eta_2, \eta_2], \quad H_2 := (0, \nu_2]$$

and set

$$\begin{aligned} &V_2 := Z_2, \\ &U_2 := \{(r_2, \varepsilon_2, h_2) \in X_2 \times E_2 \times H_2 \mid \delta/2 \le r_2\varepsilon_2 \le \delta,\ \nu/2 \le h_2/r_2 \le \nu\} \end{aligned}$$

with

$$\delta < \hat{\rho}_2 \check{\delta}_2, \quad \nu < \nu_2/\rho_2, \tag{13.9}$$

cf. Figure 13.4. Moreover, we set

$$\begin{aligned} \widetilde{U}_2 &:= X_2 \times \widetilde{E}_2 \times H_2, \\ \widehat{U}_2 &:= \{(r_2, \varepsilon_2, h_2) \in U_2 \mid r_2 \in \widehat{X}_2\}, \\ \check{U}_2 &:= \{(r_2, \varepsilon_2, h_2) \in U_2 \mid \varepsilon_2 \in \check{E}_2\}. \end{aligned}$$

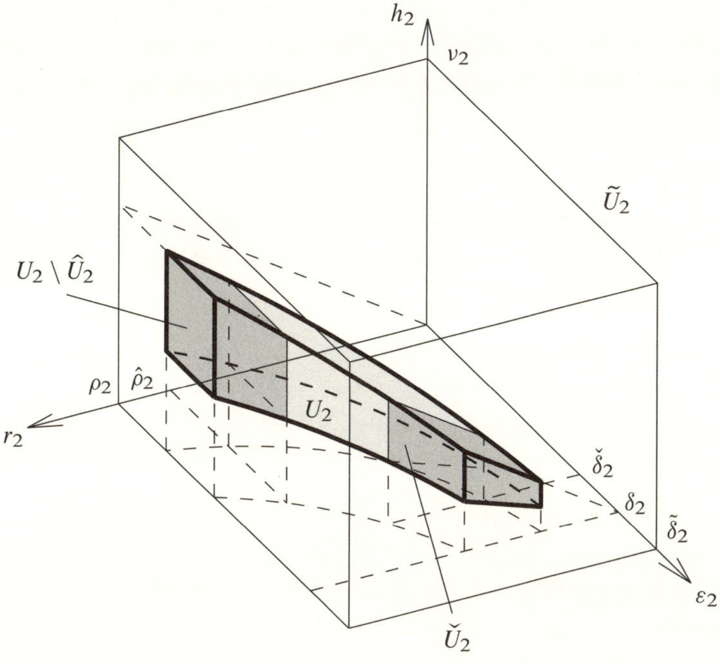

Figure 13.4. The sets U_2, $\widetilde{U}_2$, $U_2 \setminus \widehat{U}_2$ and $\check{U}_2$.

The condition $\delta < \hat{\rho}_2\check{\delta}_2$ guarantees that the sets $U_2 \setminus \hat{U}_2$ and $\check{U}_2$ are disjoint and the condition $\nu < \nu_2/\rho_2$ guarantees that the upper boundary of U_2 is contained in the plane $\nu_2 = \nu\rho_2$, cf. Figure 13.4.

Map

The coordinate change (13.8) takes (13.3) to

$$P_2\colon \quad \begin{aligned} \bar{h}_2 &= h_2/\mu(r_2, \varepsilon_2, z_2, h_2), \\ \bar{r}_2 &= r_2/\mu(r_2, \varepsilon_2, z_2, h_2), \\ \bar{\varepsilon}_2 &= \varepsilon_2\mu(r_2, \varepsilon_2, z_2, h_2), \\ \bar{z}_2 &= \big[1 + h_2 B(r_2, \varepsilon_2, z_2, h_2)\big] z_2 + h_2^2 G(r_2, \varepsilon_2, h_2), \end{aligned}$$

where

$$\begin{aligned} \mu(r_2, \varepsilon_2, z_2, h_2) &= 1 + h_2 A(r_2, \varepsilon_2, z_2, h_2), \\ A(r_2, \varepsilon_2, z_2, h_2) &= \frac{\varepsilon_2^3}{2}\big[1 + r_2(s_2 + \varepsilon_2^3 z_2) + O(h_2)\big], \\ B(r_2, \varepsilon_2, z_2, h_2) &= -2s_2 + O(r_2 + |z_2| + \varepsilon_2 + h_2), \\ G(r_2, \varepsilon_2, z_2, h_2) &= O(1). \end{aligned}$$

All functions are Lipschitz continuous. The form of the map P_2 is obtained in an analogous way as P_1 in chart Φ_1 by applying twice the mean value theorem to $\bar{s}_2 - s_2$ (here $\bar{s}_2$ stands for $s_2(\bar{r}_2, \bar{\varepsilon}_2)$). In addition we have used that the functions $\hat{f}$ and $\hat{g}$ of (13.3) satisfy $\hat{f} = O(r_2)$ and $\hat{g} = O\big(r_2^3\varepsilon_2^3\big)$, respectively. This is a consequence of the fact that the map (13.3) may be expressed as a B-series.

Graph transform

We apply Theorem 1.12. Hypothesis HMAV is satisfied with

$$\begin{aligned} \ell_{10} &= O(1), & \ell_{11} &= O(\delta_2^2), & \ell_{12} &= O(\delta_2^6\rho_2), \\ \ell_{20} &= O(1), & \ell_{21} &= O(1), & \ell_{22} &= O(1), \\ L_{20} &= O(1), & L_{21} &= O(1), & L_{22} &= 0. \end{aligned}$$

We verify Condition CMAV. We choose $b = 3/2$, $a = 1/2$, $c = 1$ and ρ_2, η_2, δ_2, ν_2 small enough such that $B < -b$ and $|A| < a = c/2$.

Thus, for η_2 and ν_2 sufficiently small the assertions of Theorem 1.12 hold. The upper bound of h_2 is ν_2 and the lower bound is $\nu\delta/(4\delta_2)$. Hence, Theorem 1.12 yields

$$\alpha_2 = O(\eta_2), \quad \gamma_2 = \frac{O(\eta_2)}{\nu\delta}, \quad \chi_2 \le 1 - 3\nu/4. \tag{13.10}$$

Note that in chart Φ_3 again δ_2 has to satisfy a smallness condition.

Function space

We define the space of functions in Φ_2 as $C_2(U_2, V_2) := \{\sigma_2 \mid \sigma_2\colon U_2 \to V_2,\ \sigma_2$ is α_2-Lipschitz continuous with respect to (r_2, ε_2) and γ_2-Lipschitz continuous with respect to $h_2\}$.

Verification of Hypothesis HMAK c), d) and Condition CMAK b)

Connection to Φ_1

$\Phi_2 \circ \Phi_1^{-1}$ is given by

$$\begin{aligned} r_2 &= \sqrt{-x_1}, & x_1 &= -r_2^2, \\ \varepsilon_2 &= \varepsilon_1/\sqrt{-x_1}, & z_1 &= z_2/r_2^2, \\ h_2 &= h_1\sqrt{-x_1}, & \varepsilon_1 &= r_2\varepsilon_2, \\ z_2 &= -x_1 z_1, & h_1 &= h_2/r_2. \end{aligned}$$

Given $\check{\sigma}_1 \in C_1(\check{U}_1, V_1)$ and $u_2 = (r_2, \varepsilon_2, h_2) \in U_2 \setminus \hat{U}_2$ we have

$$\begin{aligned} u_1 &= (x_1, \varepsilon_1, h_1) := (-r_2^2, r_2\varepsilon_2, h_2/r_2) \\ &\in [-\rho_2^2, -\hat{\rho}_2^2] \times [\delta/2, \delta] \times [\nu/2, \nu] \subset [-\check{\xi}_1, -\xi_1] \times [\delta/2, \delta] \times [\nu/2, \nu] = \check{U}_1 \end{aligned}$$

and $\Phi_2 \circ \Phi_1^{-1}(u_1, \check{\sigma}_1(u_1)) = (u_2, v_2)$ with

$$\begin{aligned} v_2 = z_2 &= -x_1\check{\sigma}_1(x_1, \varepsilon_1, h_1) = r_2^2\check{\sigma}_1(-r_2^2, r_2\varepsilon_2, h_2/r_2) \\ &=: \bar{\sigma}_{2,1}(r_2, \varepsilon_2, h_2). \end{aligned} \tag{13.11}$$

We find $|v_2| \le \eta_1\check{\xi}_1$ implying $v_2 \in V_2$ if one sets $\eta_1 = \eta_2/\check{\xi}_1$. The function $\bar{\sigma}_{2,1}$ is $\alpha_{2,1}$-Lipschitz continuous with respect to (r_2, ε_2) and $\gamma_{2,1}$-Lipschitz continuous with respect to h_2 with

$$\begin{aligned} \alpha_{2,1} &= 2\rho_2\eta_1 + \rho_2^2(2\rho_2\alpha_1 + \beta_1(\rho_2 + \delta_2)) + \gamma_1 h_2 = O(\eta_1), \\ \gamma_{2,1} &= \rho_2\gamma_1 = O(\eta_1)/\nu_2. \end{aligned}$$

It holds that $\alpha_{2,1} \le \alpha_2$ and $\gamma_{2,1} \le \gamma_2$ holds for η_1 small enough.

Norm factor

Let $\check{\sigma}_1^{(j)} \in C_1(\check{U}_1, V_1)$, $j = 1, 2$. According to (13.11) we have for $(r_2, \varepsilon_2, h_2) \in U_2 \setminus \hat{U}_2$

$$\left|\bar{\sigma}_{2,1}^{(1)}(r_2, \varepsilon_2, h_2) - \bar{\sigma}_{2,1}^{(2)}(r_2, \varepsilon_2, h_2)\right| \le -x_1\left|\check{\sigma}_1^{(1)}(x_1, \varepsilon_1, h_1) - \check{\sigma}_1^{(2)}(x_1, \varepsilon_1, h_1)\right|$$

with $(x_1, \varepsilon_1, h_1) \in \check{U}_1$. Taking first the supremum on the right-hand side and then on the left-hand side and multiplying by N_2 we obtain

$$|\bar{\sigma}_{2,1}^{(1)} - \bar{\sigma}_{2,1}^{(2)}|_{U_2\setminus\hat{U}_2} \le \frac{3N_2}{N_1}|\check{\sigma}_1^{(1)} - \check{\sigma}_1^{(2)}|_{\check{U}_1}$$

leading to the choice $N_2 = N_1/3$.

13.4.3 The chart Φ_3

Verification of Hypothesis HMAK a), b) and Condition CMAK a)

Reference manifold

We consider the coordinates

$$
\begin{aligned}
x &= r_3^2 x_3,\\
y &= r_3 y_3,\\
\varepsilon &= r_3.
\end{aligned}
$$

For positive constants $\xi_3, \hat{\xi}_3$ with $\check{\delta}_2^2 < \xi_3 < \hat{\xi}_3 < \delta_2^2$ and $\delta \le \rho_2 \sqrt{\xi_3}$ we put

$$
\begin{aligned}
&X_3 := [-1/\xi_3, 1/2], \quad \widetilde{X}_3 := [-1/\xi_3, 1], \quad \widehat{X}_3 := [-1/\hat{\xi}_3, 1/2],\\
&\check{X}_3 := [1/4, 1/2], \qquad E_3 := [\delta/2, \delta].
\end{aligned}
$$

Rescaling the time by $t_3 = r_3 \tau$, equation (13.2) expressed in the new coordinates is taken to

$$
\begin{aligned}
dx_3/dt_3 &= 1 + r_3 y_3,\\
dy_3/dt_3 &= -x_3 - y_3^2(1 + r_3 y_3/3).
\end{aligned} \tag{13.12}
$$

We derive the reference manifold M_3^r. Expressing the manifold M_2^r in the new coordinates for $x_3 = -1/\xi_3, r_3 \in [\delta/2, \delta]$ leads to $y_3 = s_2\big(r_3/\sqrt{\xi_3}, \sqrt{\xi_3}\big)/\sqrt{\xi_3}$. Let $(x_3^0, r_3^0) \in X_3 \times E_3$ be given. We take the solution of (13.12) with initial conditions

$$
x_3(0) = -1/\xi_3, \quad y_3(0) = s_2\big(r_3^0/\sqrt{\xi_3}, \sqrt{\xi_3}\big)/\sqrt{\xi_3}.
$$

Choose $t_3^0 > 0$ such that $x_3(t_3^0) = x_3^0$ and define $s_3(x_3^0, r_3^0) := y_3(t_3^0)$. The existence, uniqueness and regularity theorem for ordinary differential equations implies that s_3 is at least two times differentiable with continuous and bounded derivatives. From (13.12) we get that s_3 satisfies the invariance equation

$$
\frac{\partial s_3}{\partial x_3}(1 + r_3 s_3) = -x_3 - s_3^2\,\big(1 + r_3 s_3/3\big).
$$

We give an approximation of s_3. For $r_3 = 0$ the differential equation (13.12) takes the form

$$
dv/du = -u - v^2 \tag{13.13}
$$

admitting a unique solution $v = A(u)$ with the property $\lim_{u \to -\infty}(A(u) - \sqrt{-u}) = 0$. It is well known that

$$
A(u) = \frac{-\mathrm{Ai}'(-u)}{\mathrm{Ai}(-u)},
$$

where Ai is the Airy function. Moreover, $A(\omega_0) = 0$ and $\lim_{u\to\Omega_0} A(u) = -\infty$ holds for $\omega_0 = 1.0187\ldots$, $\Omega_0 = 2.3381\ldots$. We will also need

$$A'(u) = -u - A^2(u), \tag{13.14}$$

$$A^{\mathrm{inv}\prime}(v) = -\frac{1}{A^{\mathrm{inv}}(v) + v^2}, \tag{13.15}$$

where A^{inv} denotes the inverse function of A. By the invariance equation and Theorem 7.8 we have $s_3(x_3, r_3) = A(x_3) + O(r_3)$ for $r_3 > 0$.

Local coordinates

$$\Phi_3\colon \quad \begin{aligned} x &= r_3^2 x_3, \\ y &= r_3(s_3(x_3, r_3) + z_3), \\ \varepsilon &= r_3, \\ h &= h_3/r_3. \end{aligned} \tag{13.16}$$

Domains

For $\nu\delta \le \nu_2\sqrt{\xi_3}$ and for a positive constant η_3 we put

$$Z_3 := [-\eta_3, \eta_3], \quad H_3 := [\nu\delta/4, \nu\delta]$$

and set, cf. Figure 13.5,

$$\begin{aligned} V_3 &:= Z_3, \\ U_3 &:= \{(x_3, r_3, h_3) \in X_3 \times E_3 \times H_3 \mid \nu/2 \le h_3/r_3 \le \nu\}, \\ \widetilde{U}_3 &:= \widetilde{X}_3 \times E_3 \times H_3, \\ \widehat{U}_3 &:= \{(x_3, r_3, h_3) \in U_3 \mid x_3 \in \widehat{X}_3\}, \\ \check{U}_3 &:= \{(x_3, r_3, h_3) \in U_3 \mid x_3 \in \check{X}_3\}. \end{aligned}$$

Map

$$P_3\colon \quad \begin{aligned} \bar{x}_3 &= x_3 + h_3\big[1 + r_3\big(s_3(x_3, r_3) + z_3\big)\big] + O(h_3^2), \\ \bar{z}_3 &= \big[1 + h_3(-2s_3(x_3, r_3) + O(r_3) + O(|z_3|))\big]z_3 + O(h_3^2). \end{aligned}$$

Graph transform

We apply Theorem 1.11. Hypothesis HMAG a) holds for η_3, δ small enough. Hypothesis HMAG b) holds for δ small enough. For η_3, δ small enough Hypothesis HMAG c) is satisfied with

$$\Gamma_{11} = 1 + h_3 O(\delta), \quad L_{12} = h_3 O(\delta), \quad L_{13} = O(h_3),$$
$$L_{10} = 1 + O(\delta),$$

$$L_{21} = h_3\big[O(\eta_3) + O(h_3)\big], \quad L_{22} = 1 - h_3 b, \quad L_{23} = O(h_3),$$
$$L_{20} = O(\eta_3) + O(h_3).$$

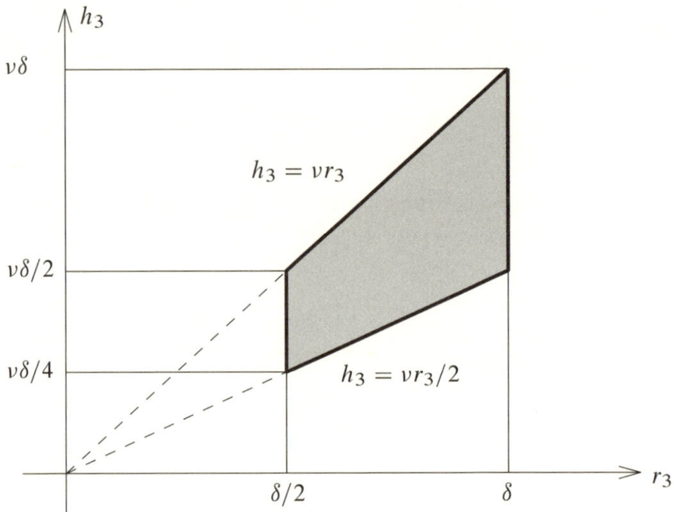

Figure 13.5. The projection of the domain U_3 into the (r_3, h_3)-plane.

where b is a positive constant. Hence, Condition CM,

$$2\sqrt{L_{12}L_{21}} = h_3\sqrt{O(\delta)} < h_3(b + O(\delta)) = \Gamma_{11} - L_{22},$$

holds for δ sufficiently small. According to (1.36) we choose

$$\alpha_3 := \alpha_{\min} = O(\eta_3 + \nu\delta) = O(\eta_3)$$

if δ is sufficiently small. Condition CMAG,

$$\chi_3 = L_{22} + L_{12}\alpha_3 \leq 1 - b\nu\delta/8 < 1,$$

holds for δ sufficiently small and we choose $\beta_3 = \beta_{\min} = O(1), \gamma_3 = \gamma_{\min} = O(\eta_3)/(\nu\delta)$ for δ sufficiently small.

Function space

We define the space of functions in chart Φ_3 as $C_3(U_3, V_3) := \{\sigma_3 \mid \sigma_3 \colon U_3 \to V_3$, σ_3 is α_3-Lipschitz continuous with respect to x_3, β_3-Lipschitz continuous with respect to r_3 and γ_3-Lipschitz continuous with respect to $h_3\}$.

Verification of Hypothesis HMAK c), d) and Condition CMAK b)

Connection to Φ_2

$\Phi_3 \circ \Phi_2^{-1}$ is given by

$$\begin{aligned} x_3 &= -1/\varepsilon_2^2, & r_2 &= r_3\sqrt{-x_3}, \\ z_3 &= \varepsilon_2^2 z_2, & z_2 &= -x_3 z_3, \\ r_3 &= r_2\varepsilon_2, & \varepsilon_2 &= 1/\sqrt{-x_3}, \\ h_3 &= h_2\varepsilon_2, & h_2 &= h_3\sqrt{-x_3}. \end{aligned}$$

Given $\check{\sigma}_2 \in C_2(\check{U}_2, V_2)$ and $u_3 = (x_3, r_3, h_3) \in U_3 \setminus \hat{U}_3$ we have

$$\begin{aligned} u_2 = (r_2, \varepsilon_2, h_2) &:= (r_3\sqrt{-x_3}, 1/\sqrt{-x_3}, h_3\sqrt{-x_3}) \\ &\in \{(r_2, \varepsilon_2, h_2) \in X_2 \times \check{E}_2 \times H_2 \mid \delta/2 \le r_2\varepsilon_2 \le \delta,\ \nu/2 \le h_2/r_2 \le \nu\} = \check{U}_2 \end{aligned}$$

and $\Phi_3 \circ \Phi_2^{-1}(u_2, \check{\sigma}_2(u_2)) = (u_3, v_3)$ with

$$\begin{aligned} v_3 = z_3 = \varepsilon_2^2\check{\sigma}_2(u_2) &= \frac{1}{-x_3}\check{\sigma}_2\Big(r_3\sqrt{-x_3}, \frac{1}{\sqrt{-x_3}}, h_3\sqrt{-x_3}\Big) \\ &=: \bar{\sigma}_{3,2}(x_3, r_3, h_3) \end{aligned}$$

implying that $v_3 \in V_3$ if the condition $\hat{\xi}_3\eta_2 < \eta_3$ is satisfied. The function $\bar{\sigma}_{3,2}$ is $\alpha_{3,2}$-Lipschitz continuous with respect to x_3, $\beta_{3,2}$-Lipschitz continuous with respect to r_3 and $\gamma_{3,2}$-Lipschitz continuous with respect to h_3 with

$$\begin{aligned} \alpha_{3,2} &= O(\eta_2), \\ \beta_{3,2} &= O(\eta_2), \\ \gamma_{3,2} &= O(\eta_2)/\nu. \end{aligned}$$

The function $\bar{\sigma}_{3,2}$ is in $C_3(U_3 \setminus \hat{U}_3, V_3)$ for η_2 small enough (i.e., $\alpha_{3,2} \le \alpha_3$, $\beta_{3,2} \le \beta_3$ and $\gamma_{3,2} \le \gamma_3$ holds).

Norm factor

Let $\check{\sigma}_2^{(j)} \in C_2(\check{U}_2, V_2)$, $j = 1, 2$. We have for $(x_3, r_3, h_3) \in U_3 \setminus \hat{U}_3$

$$\big|\bar{\sigma}_{3,2}^{(1)}(x_3, r_3, h_3) - \bar{\sigma}_{3,2}^{(2)}(x_3, r_3, h_3)\big| \le \varepsilon_2^2\big|\check{\sigma}_2^{(1)}(r_2, \varepsilon_2, h_2) - \check{\sigma}_2^{(2)}(r_2, \varepsilon_2, h_2)\big|$$

with $(r_2, \varepsilon_2, h_2) \in \check{U}_2$. Taking first the supremum on the right-hand side and then on the left-hand side and multiplying by N_3 we obtain

$$\big|\bar{\sigma}_{3,2}^{(1)} - \bar{\sigma}_{3,2}^{(2)}\big|_{U_3 \setminus \hat{U}_3} \le \frac{N_3\delta_2^2}{N_2}\big|\check{\sigma}_2^{(1)} - \check{\sigma}_2^{(2)}\big|_{\check{U}_2}$$

leading to the choice $N_3 = N_2/\delta_2^2$.

13.4.4 The chart Φ_4

Verification of Hypothesis HMAK a), b) and Condition CMAK a)

Reference manifold

We consider the coordinates

$$
\begin{aligned}
x &= r_4^2 x_4, \\
y &= r_4 A(y_4), \\
\varepsilon &= r_4.
\end{aligned}
$$

Note that the use of the function A, cf. (13.13) in Subsection 13.4.3, is motivated by the wish that in the map P_4 the variable y_4 grows essentially uniformly. In this chart the reference manifold is the graph of a function of y_4 and r_4.

For small positive constants $\xi_4, \check{\xi}_4, \hat{\xi}_4$ with $\xi_4 < \check{\xi}_4$ we put

$$
\begin{aligned}
Y_4 &:= [3/8, \Omega_0 - \xi_4], & \widetilde{Y}_4 &:= [3/8, \Omega_0], \\
\widehat{Y}_4 &:= [3/8 + \hat{\xi}_4, \Omega_0 - \xi_4], & \check{Y}_4 &:= [\Omega_0 - \check{\xi}_4, \Omega_0 - \xi_4], \quad E_4 := [\delta/2, \delta],
\end{aligned}
$$

cf. Figure 13.6. Rescaling the time by $t_4 = r_4 \tau$ and using (13.14), equation (13.2)

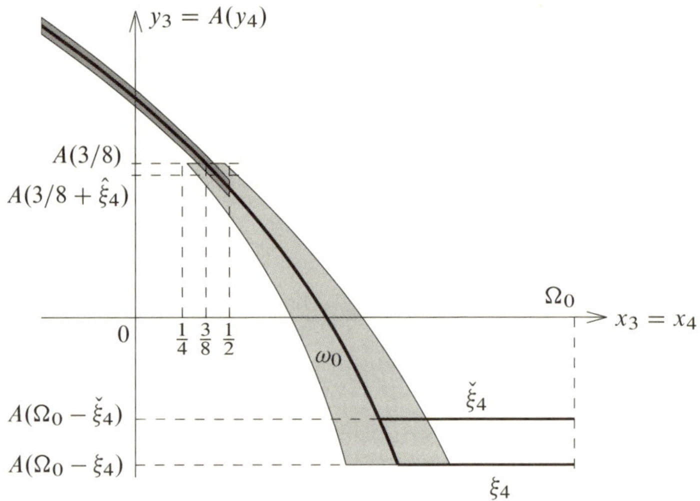

Figure 13.6. The charts Φ_3 and Φ_4 in the coordinates x_3, y_3.

expressed in the new coordinates is taken to

$$\begin{aligned} dy_4/dt_4 &= \frac{x_4 + A(y_4)^2(1 + r_4 A(y_4)/3)}{y_4 + A(y_4)^2}, \\ dx_4/dt_4 &= 1 + r_4 A(y_4). \end{aligned} \tag{13.17}$$

Let $(y_4^0, r_4^0) \in Y_4 \times E_4$ be given. To express the manifold M_3^r in the new coordinates for $y_4 = 3/8$ and $r_4 = r_4^0$ we solve $A(3/8) = s_3(x_4, r_4^0)$ for x_4. Since $s_3(x_4, r_4) = A(x_4) + O(r_4)$ there is a unique solution $x_4^* = 3/8 + O(r_4^0)$. We take the solution of (13.17) with initial conditions

$$y_4(0) = 3/8, \quad x_4(0) = x_4^*.$$

Choose $t_4^0 > 0$ such that $y_4(t_4^0) = y_4^0$ and define $s_4(y_4^0, r_4^0) := x_4(t_4^0)$. The existence, uniqueness and regularity theorem for ordinary differential equations implies that s_4 is at least two times differentiable with continuous and bounded derivatives. The function s_4 satisfies the invariance equation

$$\frac{\partial s_4}{\partial y_4} \frac{s_4 + A(y_4)^2(1 + r_4 A(y_4)/3)}{y_4 + A(y_4)^2} = 1 + r_4 A(y_4).$$

Note that by the invariance equation $s_4(y_4, r_4) = y_4 + O(r_4)$ holds.

Local coordinates

This allows to introduce the chart

$$\Phi_4: \quad \begin{aligned} x &= r_4^2[s_4(y_4, r_4) + z_4 e^{y_4}], \\ y &= r_4 A(y_4)), \\ \varepsilon &= r_4, \\ h &= h_4/r_4. \end{aligned}$$

The factor e^{y_4} is introduced to guarantee attractivity of the map P_4.

Domains

For a positive constant η_4 we put

$$Z_4 := [-\eta_4, \eta_4], \quad H_4 := (\nu\delta/4, \nu\delta],$$

and we define

$$\begin{aligned} V_4 &:= Z_4, \\ U_4 &:= \{(y_4, r_4, h_4) \in Y_4 \times E_4 \times H_4 \mid \nu/2 \le h_4/r_4 \le \nu\}, \\ \widetilde{U}_4 &:= \widetilde{Y}_4 \times E_4 \times H_4, \\ \widehat{U}_4 &:= \{(y_4, r_4, h_4) \in U_4 \mid y_4 \in \widehat{Y}_4\}, \\ \check{U}_4 &:= \{(y_4, r_4, h_4) \in U_4 \mid y_4 \in \check{Y}_4\}. \end{aligned}$$

Map
Using (13.15) and $s_4(y_4, r_4) = y_4 + O(r_4)$ the map 13.3 is transformed to

$$P_4\colon \quad \begin{aligned} \bar{y}_4 &= y_4 + h_4[1 + O(r_4) + O(|z_4|)] + O(h_4^2), \\ \bar{z}_4 &= \big[1 + h_4\big(-1 + O(r_4) + O(|z_4|) + O(h_4)\big)\big]z_4 + O(h_4^2). \end{aligned}$$

Graph transform
We apply Theorem 1.11. For η_4, δ small enough Hypothesis HMAG is satisfied with

$$\begin{gathered} \Gamma_{11} = 1 + h_4 O(\eta_4 + \delta), \quad L_{12} = O(h_4), \quad L_{13} = O(h_4), \\ L_{10} = O(1), \\ L_{21} = h_4 O(\eta_4 + \delta), \quad L_{22} = 1 - h_4/2, \quad L_{23} = h_4 O(\eta_4 + \delta), \\ L_{20} = O(\eta_4 + \delta). \end{gathered}$$

Hence, Condition CM,

$$2\sqrt{L_{12}L_{21}} = h_4\sqrt{O(\eta_4 + \delta)} < \frac{h_4}{4} < \Gamma_{11} - L_{22},$$

holds for η_4, δ sufficiently small. We choose $\alpha_4 := \alpha_{\min} = O(\eta_4 + \delta)$. Thus

$$\alpha_4 = O(\eta_4)$$

if δ is sufficiently small. Condition CMAG,

$$\chi_4 = L_{22} + L_{12}\alpha_4 \le \nu\delta/10 < 1,$$

holds for η_4 sufficiently small and we get $\beta_{\min} = O(\eta_4), \gamma_{\min} = O(\eta_4)/(\nu\delta)$. We choose

$$\beta_4 = \max\{\beta_{\min}, \beta_{4,3}\}, \quad \gamma_4 = \gamma_{\min},$$

where $\beta_{4,3}$ is determined in (13.18) below.

Function space
We define the space of functions in Φ_4 as $C_4(U_4, V_4) := \{\sigma_4 \mid \sigma_4\colon U_4 \to V_4,\ \sigma_4$ is α_4-Lipschitz continuous with respect to y_4, β_4-Lipschitz continuous with respect to r_4 and γ_4-Lipschitz continuous with respect to $h_4\}$.

Verification of Hypothesis HMAK c), d) and Condition CMAK b)

Connection to Φ_3

$\Phi_4 \circ \Phi_3^{-1}$ is given by

$$\begin{aligned} z_4 &= e^{-A^{\text{inv}}(s_3(x_3,r_3)+z_3)}\big[x_3 - s_4(A^{\text{inv}}(s_3(x_3,r_3)+z_3), r_3)\big],\\ y_4 &= A^{\text{inv}}(s_3(x_3,r_3)+z_3),\\ r_4 &= r_3,\\ h_4 &= h_3,\\ x_3 &= s_4(y_4,r_4) + z_4 e^{y_4},\\ z_3 &= A(y_4) - s_3(s_4(y_4,r_4)+z_4e^{y_4}, r_4),\\ r_3 &= r_4,\\ h_3 &= h_4, \end{aligned}$$

where A^{inv} denotes the inverse of A, cf. (13.15).

Given $\check{\sigma}_3 \in C_3(\check{U}_3, V_3)$ and $u_4 = (y_4, r_4, h_4) \in U_4 \setminus \hat{U}_4$ and taking $\hat{\xi}_4$, η_4 and δ small enough we have that for $u_3 = (x_3, r_3, h_3) := (s_4(y_4, r_4) + z_4 e^{y_4}, r_4, h_4)$ with $z_4 \in V_4$ the inclusion $u_3 \in \check{U}_3$ holds. Moreover, if in

$$z_4 = e^{-A^{\text{inv}}(s_3(x_3,r_3)+\check{\sigma}_3(x_3,r_3,h_3))}\big[\underbrace{x_3 - s_4(A^{\text{inv}}(s_3(x_3,r_3)+\check{\sigma}_3(x_3,r_3,h_3)), r_3)}_{=O(\eta_3)}\big]$$

x_3 is replaced by $s_4(y_4,r_4)+z_4e^{y_4}$, r_3 by r_4 and h_3 by h_4, an equation for z_4 depending on y_4, r_4, h_4 is obtained. By a contraction argument one can show that this equation has a unique solution $z_4 = \bar{\sigma}_{4,3}(y_4, r_4, h_4)$. For η_3 small enough $z_4 \in V_4$. One obtains that $\bar{\sigma}_{4,3}$ is $\alpha_{4,3}$-Lipschitz continuous with respect to y_4, $\beta_{4,3}$-Lipschitz continuous with respect to r_4 and $\gamma_{4,3}$-Lipschitz continuous with respect to h_4 with

$$\begin{aligned} \alpha_{4,3} &= O(\eta_3),\\ \beta_{4,3} &= O(\beta_3), \\ \gamma_{4,3} &= O(\gamma_3) = O(\eta_3)/(\nu\delta). \end{aligned} \tag{13.18}$$

The function $\bar{\sigma}_{4,3}$ is in $C_4(U_4 \setminus \hat{U}_4, V_4)$ for η_3 small enough (i.e., $\alpha_{4,3} \le \alpha_4$, $\beta_{4,3} \le \beta_4$ and $\gamma_{4,3} \le \gamma_4$ holds).

Norm factor

Let $\check{\sigma}_3^{(j)} \in C_3(\check{U}_3, V_3)$, $j = 1, 2$. We have for $(y_4, r_4, h_4) \in U_4 \setminus \hat{U}_4$

$$\begin{aligned} \bar{\sigma}_{4,3}^{(j)}(y_4, r_4, h_4) = e^{-A^{\text{inv}}\left(s_3(x_3^{(j)},r_3)+\check{\sigma}_3^{(j)}(x_3^{(j)},r_3,h_3)\right)}\big[x_3^{(j)} - s_4\big(A^{\text{inv}}(s_3(x_3^{(j)}, r_3) \\ + \check{\sigma}_3^{(j)}(x_3^{(j)}, r_3, h_3)), r_3\big)\big] \end{aligned}$$

with $(x_3^{(j)}, r_3, h_3) \in \check{U}_3$. We conclude that there are constants K_3 and K_4 such that

$$|x_3^{(1)} - x_3^{(2)}| \le K_3 |\check{\sigma}_3^{(1)} - \check{\sigma}_3^{(2)}| \quad \text{and} \quad |\bar{\sigma}_{4,3}^{(1)} - \bar{\sigma}_{4,3}^{(2)}| \le K_4 |\check{\sigma}_3^{(1)} - \check{\sigma}_3^{(2)}|.$$

Therefore, we may choose the norm factor $N_4 = N_3/K_4$.

13.4.5 The chart Φ_5

Verification of Hypothesis HMAK a), b) and Condition CMAK a)

Reference manifold

We consider the coordinates

$$\begin{aligned} x &= r_5^2 x_5, \\ y &= -r_5, \\ \varepsilon &= r_5 \varepsilon_5. \end{aligned}$$

For small positive constants $\check{\rho}_5, \rho_5, \tilde{\rho}_5, \hat{\delta}_5, \delta_5$ with $\check{\rho}_5 < \rho_5 < \tilde{\rho}_5$ and

$$\frac{1}{-A(\Omega_0 - \xi_4)} < \hat{\delta}_5 < \delta_5 < \frac{1}{-A(\Omega_0 - \check{\xi}_4)}$$

we put

$$\begin{aligned} Y_5 &:= (0, \rho_5], \quad \widetilde{Y}_5 := (0, \tilde{\rho}_5], \quad \check{Y}_5 := [\check{\rho}_5, \rho_5], \\ E_5 &:= (0, \delta_5], \quad \widehat{E}_5 := (0, \hat{\delta}_5], \end{aligned}$$

Rescaling the time by $t_5 = r_5(1 + x_5 - r_5/3)\tau$, equation (13.2) expressed in the new coordinates is taken to

$$\begin{aligned} dr_5/dt_5 &= r_5, \\ d\varepsilon_5/dt_5 &= -\varepsilon_5, \\ dx_5/dt_5 &= -2x_5 + \frac{\varepsilon_5^3(1 - r_5)}{1 + x_5 - r_5/3}. \end{aligned} \tag{13.19}$$

We describe the reference manifold M_5^r as the graph of a function $x_5 = s_5(r_5, \varepsilon_5)$ for $r_5 \in Y_5, \varepsilon_5 \in E_5, r_5\varepsilon_5 \in [\delta/2, \delta]$. We proceed as in chart Φ_2. Expressing the manifold M_4^r in the new coordinates for $\varepsilon_5 = \delta_5, r_5 \in [\delta/2, \delta]/\delta_5$ leads to $x_5 = \varepsilon_5^2 s_4(A^{\text{inv}}(-1/\delta_5), r_5\delta_5)$. Let (r_5^0, ε_5^0) be given in the domain considered. We take the solution of the r_5- and the ε_5-equation of (13.19) with initial conditions (r_5^0, ε_5^0) and choose $t_5^0 < 0$ such that $\varepsilon_5(t_5^0) = \delta_5$. Set $r_5^* := r_5(t_5^0)$. Note that $r_5^* \in [\delta/2, \delta]/\delta_5$. Now take the solution of (13.19) with initial conditions

$$r_5(t_5^0) = r_5^*, \quad \varepsilon_5(t_5^0) = \delta_5, \quad x_5(t_5^0) = \delta_5^2 s_4(A^{\text{inv}}(-1/\delta_5), r_5^*\delta_5)$$

and define $s_5(r_5^0, \varepsilon_5^0) := x_5(0)$. The existence, uniqueness and regularity theorem for ordinary differential equations implies that s_5 is at least two times differentiable with continuous and bounded derivatives. From (13.19) we get that s_5 satisfies the invariance equation

$$\frac{\partial s_5}{\partial r_5} r_5 - \frac{\partial s_5}{\partial \varepsilon_5} \varepsilon_5 = -2 s_5 + \frac{\varepsilon_5^3 (1 - r_5)}{1 + s_5 - r_5/3} \tag{13.20}$$

implying that $s_5(r_5, \varepsilon_5) = O(r_5 + \varepsilon_5)$. For ρ_5, δ_5 small enough we get from (13.19) that $\varepsilon_5(t_5) = \delta_5 e^{-(t_5 - t_5^0)}$ and $\dot{x}_5(t_5) \le -2x_5(t_5) + \varepsilon_5(t_5)^3$. It follows that

$$s_5(r_5, \varepsilon_5) \le {\varepsilon_5}^2 \big(s_4(A^{\mathrm{inv}}(-1/\delta_5), r_5 \varepsilon_5) + \delta_5\big).$$

Local coordinates

$$\Phi_5 \colon \quad \begin{aligned} x &= r_5^2 \big(s_5(r_5, \varepsilon_5) + z_5\big), \\ y &= -r_5, \\ \varepsilon &= r_5 \varepsilon_5, \\ h &= h_5 / r_5. \end{aligned} \tag{13.21}$$

Domains

For positive constants η_5, ν_5, $\tilde{\nu}_5$ with $\nu_5 < \tilde{\nu}_5$ we put

$$H_5 := (0, \nu_5], \quad \widetilde{H}_5 := (0, \tilde{\nu}_5], \quad Z_5 := [-\eta_5, \eta_5]$$

and set

$$\begin{aligned} &V_5 := Z_5, \\ &U_5 := \{(r_5, \varepsilon_5, h_5) \in Y_5 \times E_5 \times H_5 \mid \delta/2 \le r_5 \varepsilon_5 \le \delta,\ \nu/2 \le h_5 / r_5 \le \nu\} \end{aligned}$$

with

$$\delta < \check{\rho}_5 \hat{\delta}_5, \quad \nu < \nu_5 / \tilde{\rho}_5, \tag{13.22}$$

cf. Figure 13.7. Moreover, we set

$$\begin{aligned} \widetilde{U}_5 &:= \widetilde{Y}_5 \times E_5 \times \widetilde{H}_5, \\ \widehat{U}_5 &:= \{(r_5, \varepsilon_5, h_5) \in U_5 \mid \varepsilon_5 \in \widehat{E}_5\}, \\ \check{U}_5 &:= \{(r_5, \varepsilon_5, h_5) \in U_5 \mid r_5 \in \check{Y}_5\}. \end{aligned}$$

The condition $\delta < \check{\rho}_5 \hat{\delta}_5$ guarantees that the sets $U_5 \setminus \widehat{U}_5$ and $\check{U}_5$ are disjoint. The condition $\nu < \nu_5 / \tilde{\rho}_5$ guarantees that the upper boundary of U_5 is contained in the plane $\nu_5 = \nu \tilde{\rho}_5$ and that U_5 is mapped to $\widetilde{U}_5$, cf. Figure 13.7.

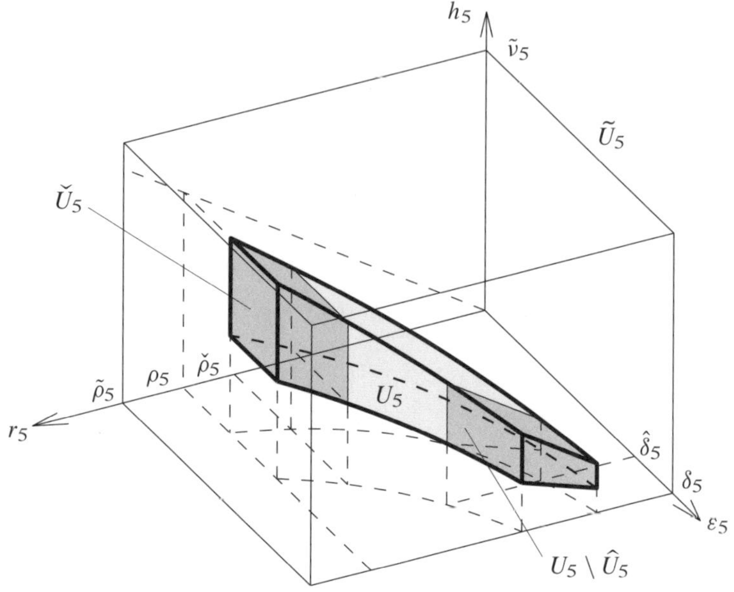

Figure 13.7. The sets U_5, $\widetilde{U}_5$, $U_5 \setminus \hat{U}_5$ and $\check{U}_5$.

Map

The coordinate change (13.21) takes (13.3) to

$$P_5\colon \quad \begin{aligned} \bar{h}_5 &= h_5/\mu(r_5, \varepsilon_5, z_5, h_5), \\ \bar{r}_5 &= r_5/\mu(r_5, \varepsilon_5, z_5, h_5), \\ \bar{\varepsilon}_5 &= \varepsilon_5 \mu(r_5, \varepsilon_5, z_5, h_5), \\ \bar{z}_5 &= \big[1 + h_5 B(r_5, \varepsilon_5, z_5, h_5)\big] z_5 + h_5^2 G(r_5, \varepsilon_5, h_5), \end{aligned}$$

where

$$\begin{aligned} \mu(r_5, \varepsilon_5, z_5, h_5) &= 1 + h_5 A(r_5, \varepsilon_5, z_5, h_5), \\ A(r_5, \varepsilon_5, z_5, h_5) &= -\big[1 + s_5(r_5, \varepsilon_5) + O(r_5 + z_5 + h_5)\big], \\ B(r_5, \varepsilon_5, z_5, h_5) &= -2\big[1 + s_5(r_5, \varepsilon_5) + O(r_5 + z_5 + h_5)\big], \\ G(r_5, \varepsilon_5, z_5, h_5) &= O(1). \end{aligned}$$

All functions are Lipschitz continuous. In addition we have used that the functions $\hat{f}$ and $\hat{g}$ in (13.3) satisfy $\hat{f} = O(r_5)$ and $\hat{g} = O(r_5^3)$, respectively. This is a consequence of the fact that the map (13.3) may be expressed as a B-series.

Graph transform

We apply Theorem 1.12. Hypothesis HMAV is satisfied with

$$\begin{aligned}
\ell_{10} &= O(1), & \ell_{11} &= O(1), & \ell_{12} &= O(1),\\
\ell_{20} &= O(1), & \ell_{21} &= O(1), & \ell_{22} &= O(1),\\
L_{20} &= O(1), & L_{21} &= O(1), & L_{22} &= 0.
\end{aligned}$$

We verify Condition CMAV. We choose $b = 5/3, a = 4/3, c = 0$ and ρ_5 and ν_5 small enough such that $B < -b$, $|A| < a$ and $A < c/2 = 0$.

Thus, for ν_5 sufficiently small the assertions of Theorem 1.12 hold with

$$\alpha_5 = O(\eta_5), \quad \gamma_5 = O(\eta_5)/(\nu\delta), \quad \chi_5 \le 1 - 5\nu\delta/(24\delta_5). \tag{13.23}$$

Function space

We define the space of functions in Φ_5 as $C_5(U_5, V_5) := \{\sigma_5 \mid \sigma_5\colon U_5 \to V_5$, σ_5 is α_5-Lipschitz continuous with respect to (r_5, ε_5) and γ_5-Lipschitz continuous with respect to h_5 and $|\sigma_5(r_5, \varepsilon_5, h_5)| \le (c_5 - 3\varepsilon_5)\varepsilon_5^2$ with $c_5 = \eta_5/\delta_5^2 + 3\delta_5\}$. A closer look at P_5 shows that

$$\bar{z}_5 = \mu^2\big[z_5 + h_5\varepsilon_5^3(1 - r_5) + (s_5 - \bar{s}_5/\mu^2) + O(h_5^2\varepsilon_5^3)\big],$$

where $\mu = \mu(r_5, \varepsilon_5, z_5, h_5)$, $s_5 = s_5(r_5, \varepsilon_5)$ and $\bar{s}_5 = s_5(\bar{r}_5, \bar{\varepsilon}_5)$. Using the differential equation (13.19) we may estimate $|s_5 - \bar{s}_5/\mu^2| \le h_5\varepsilon_5^3$. This allows to show that the property $|\sigma_5(r_5, \varepsilon_5, h_5)| \le (c_5 - 3\varepsilon_5)\varepsilon_5^2$ is invariant under the map P_5.

Verification of Hypothesis HMAK c), d) and Condition CMAK b)

Connection to Φ_4

$\Phi_5 \circ \Phi_4^{-1}$ is given by

$$\begin{aligned}
r_5 &= -r_4 A(y_4), & z_4 &= z_5 e^{-A^{\mathrm{inv}}(-1/\varepsilon_5)}/\varepsilon_5^2,\\
\varepsilon_5 &= -1/A(y_4), & y_4 &= A^{\mathrm{inv}}(-1/\varepsilon_5),\\
h_5 &= -h_4 A(y_4), & r_4 &= r_5\varepsilon_5,\\
z_5 &= z_4 e^{y_4}/A(y_4)^2, & h_4 &= h_5\varepsilon_5.
\end{aligned}$$

Here again A and A^{inv} denote the functions defined in (13.14) and (13.15). Given $\breve{\sigma}_4 \in C_4(\breve{U}_4, V_4)$ and $u_5 = (r_5, \varepsilon_5, h_5) \in U_5 \setminus \hat{U}_5$ we have that for $u_4 = (y_4, r_4, h_4) := (A^{\mathrm{inv}}(-1/\varepsilon_5), r_5\varepsilon_5, h_5\varepsilon_5)$ the inclusion $u_4 \in \breve{U}_4$ holds. Moreover, we have $\Phi_5 \circ \Phi_4^{-1}(u_4, \breve{\sigma}_4(u_4)) = (u_5, v_5)$ with

$$\begin{aligned}
v_5 = z_5 = \frac{\breve{\sigma}_4(y_4, r_4, h_4)e^{y_4}}{A(y_4)^2} &= \varepsilon_5^2\breve{\sigma}_4(A^{\mathrm{inv}}(-1/\varepsilon_5), r_5\varepsilon_5, h_5\varepsilon_5)e^{A^{\mathrm{inv}}(-1/\varepsilon_5)}\\
&=: \bar{\sigma}_{5,4}(r_5, \varepsilon_5, h_5).
\end{aligned}$$

We find $v_5 = O(\delta_5^2 \eta_4)$ implying $v_5 \in V_5$ for η_4 small enough. The function $\bar{\sigma}_{5,4}$ is $\alpha_{5,4}$-Lipschitz continuous with respect to (r_5, ε_5) and $\gamma_{5,4}$-Lipschitz continuous with respect to h_5 with

$$\alpha_{5,4} = O(\eta_4),$$
$$\gamma_{5,4} = O(\eta_4)/(\nu\delta).$$

In view of (13.23), $\alpha_{5,4} \leq \alpha_5$ and $\gamma_{5,4} \leq \gamma_5$ holds for η_4 small enough.

Norm factor

Let $\check{\sigma}_4^{(j)} \in C_4(\check{U}_4, V_4)$, $j = 1, 2$. We have for $(r_5, \varepsilon_5, h_5) \in U_5 \setminus \hat{U}_5$

$$|\bar{\sigma}_{5,4}^{(1)}(r_5, \varepsilon_5, h_5) - \bar{\sigma}_{5,4}^{(2)}(r_5, \varepsilon_5, h_5)| \leq \varepsilon_5^2 e^{A^{\text{inv}}(-1/\varepsilon_5)} |\check{\sigma}_4^{(1)}(y_4, r_4, h_4) - \check{\sigma}_4^{(2)}(y_4, r_4, h_4)|$$

with $(y_4, r_4, h_4) \in \check{U}_4$. Taking first the supremum on the right-hand side and then on the left-hand side and multiplying by N_5 we obtain

$$|\bar{\sigma}_{5,4}^{(1)} - \bar{\sigma}_{5,4}^{(2)}|_{U_5 \setminus \hat{U}_5} \leq \frac{N_5 \delta_5^2 e^{\Omega_0}}{N_4} |\check{\sigma}_4^{(1)} - \check{\sigma}_4^{(2)}|_{\check{U}_4}$$

leading to the choice $N_5 = N_4/(\delta_5^2 e^{\Omega_0})$.

13.4.6 The chart Φ_6

Verification of Hypothesis HMAK a), b) and Condition CMAK a)

Reference manifold

We define $J : \mathbb{R} \to \mathbb{R}$ as the solution of the differential equation

$$\frac{d}{dr} J = J' = J^2(1 - J/3)$$

with initial condition $J(0) = p_6$ where p_6 is a small positive constant given below. This solution satisfies $\lim_{r \to -\infty} J(r) = 0$ and $\lim_{r \to \infty} J(r) = 3$, cf. Figure 13.8. We will also need

$$J^{\text{inv}\prime}(v) = \frac{1}{v^2(1 - v/3)},$$

where J^{inv} denotes the inverse of J. We consider the coordinates

$$x = \varepsilon_6^2 x_6,$$
$$y = -J(r_6),$$
$$\varepsilon = \varepsilon_6.$$

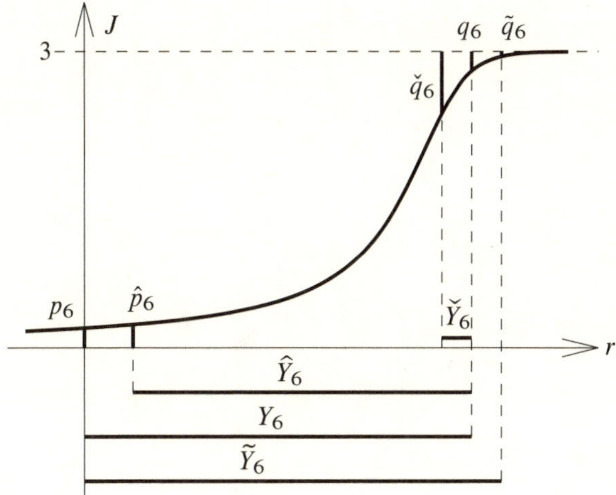

Figure 13.8. Sketch of the function J.

For positive constants p_6, $\hat{p}_6$, $\tilde{q}_6$, q_6 with $J^{\text{inv}}(\check{\rho}_5) < p_6 < \hat{p}_6 < J^{\text{inv}}(\rho_5)$ and $\tilde{q}_6 < q_6$ we put

$$
\begin{aligned}
Y_6 &:= [0, J^{\text{inv}}(3 - q_6)], & \widetilde{Y}_6 &:= [0, J^{\text{inv}}(3 - \tilde{q}_6)],\\
\widehat{Y}_6 &:= [J^{\text{inv}}(\hat{p}_6), J^{\text{inv}}(3 - q_6)], & \check{Y}_6 &:= [J^{\text{inv}}(3 - \check{q}_6), J^{\text{inv}}(3 - q_6)],\\
E_6 &:= [\delta/2, \delta],
\end{aligned}
$$

cf. Figure 13.8. Rescaling the time by

$$
t_6 = \left(1 + \frac{\varepsilon_6^2 x_6}{J(r_6)^2(1 - J(r_6)/3)}\right)\tau,
$$

the differential equation (13.2) expressed in the new coordinates is taken to

$$
\begin{aligned}
dr_6/dt_6 &= 1,\\
dx_6/dt_6 &= \varepsilon_6(1 - J(r_6))\Big/\left[1 + \frac{\varepsilon_6^2 x_6}{J(r_6)^2(1 - J(r_6)/3)}\right].
\end{aligned}
$$

The function $x_6 = s_6(r_6, \varepsilon_6)$ describing the manifold M_6^r is obtained similarly as s_3 is obtained in chart Φ_3. The function s_6 satisfies the invariance equation

$$
\frac{\partial s_6}{\partial r_6} = \varepsilon_6(1 - J(r_6))\Big/\left[1 + \frac{\varepsilon_6^2 s_6}{J(r_6)^2(1 - J(r_6)/3)}\right].
$$

Note that with Theorem 7.8 it follows that $s_6(r_6, \varepsilon_6) = 1 + O(\varepsilon_6)$.

Local coordinates

$$\Phi_6\colon \quad \begin{aligned} x &= \varepsilon_6^2\big(s_6(r_6, \varepsilon_6) + z_6 e^{r_6}\big), \\ y &= -J(r_6), \\ \varepsilon &= \varepsilon_6, \\ h &= h_6. \end{aligned} \tag{13.24}$$

Here again the factor e^{r_6} is introduced to guarantee attractivity of the map P_6.

Domains

For the positive constant η_6 we put

$$Z_6 := [-\eta_6, \eta_6], \quad H_6 := [\nu/2, \nu]$$

and set

$$\begin{aligned} &V_6 := Z_6, \\ &U_6 := Y_6 \times E_6 \times H_6, \quad \widetilde{U}_6 := \widetilde{Y}_6 \times E_6 \times H_6, \\ &\widehat{U}_6 := \widehat{Y}_6 \times E_6 \times H_6, \quad \check{U}_6 := \check{Y}_6 \times E_6 \times H_6. \end{aligned}$$

Map

The coordinate change (13.24) takes (13.3) to

$$P_6\colon \quad \begin{aligned} \bar{r}_6 &= r_6 + h_6\big[1 + O(\varepsilon_6^2) + O(h_6)\big], \\ \bar{z}_6 &= \big[1 - h_6\big(1 + O(\varepsilon_6^2) + O(h_6)\big)\big]z_6 + O(h_6^2). \end{aligned}$$

Graph transform

We apply Theorem 1.11. Hypothesis HMAG a) holds for δ, ν small enough. HMAG b) holds for ν small enough. For δ, ν small enough Hypothesis HMAG c) is satisfied with

$$\begin{gathered} \Gamma_{11} = 1 + h_6 O(\delta^2 + \nu), \quad L_{12} = h_6\delta^2 c, \quad L_{13} = h_6 O(\delta + \nu), \\ L_{10} = O(1), \\ L_{21} = h_6 O(\delta^2 + \nu), \quad L_{22} = 1 - h_6/2, \quad L_{23} = h_6 O(\delta + \nu), \\ L_{20} = O(\eta_6 + \nu). \end{gathered}$$

There is such a positive constant c in L_{12} since in the r_6-equation the variable z_6 allways has a factor ε_6^2. It follows that Condition CM,

$$2\sqrt{L_{12}L_{21}} = h_6 O(\delta) < h_6(1/2 + O(\delta^2 + \nu)) = \Gamma_{11} - L_{22},$$

holds for δ and ν sufficiently small. We have

$$\alpha_{\min} = O(\delta^2 + \nu), \quad \alpha_{\max} = \frac{1 + O(\delta^2 + \nu)}{2c\delta^2}$$

and for small δ and ν we choose

$$\alpha_6 := \frac{1}{4c\delta^2} \in [\alpha_{\min}, \alpha_{\max}].$$

Condition CMAG is satisfied if δ and ν are sufficiently small:

$$\chi_6 = L_{22} + L_{12}\alpha_6 \le 1 - \nu/8 < 1.$$

We have $\beta_{\min} = \delta^{-2}O(\delta + \nu)$ and $\gamma_{\min} = \delta^{-2}O(1)/\nu$. We choose

$$\beta_6 = \max\{\beta_{\min}, \beta_{6,5}\}, \quad \gamma_6 = \max\{\gamma_{\min}, \gamma_{6,5}\},$$

where $\beta_{6,5}$ and $\gamma_{6,5}$ are determined in (13.25) below.

Function space

We define the space of functions in chart Φ_6 as $C_6(U_6, V_6) := \{\sigma_6 \mid \sigma_6\colon U_6 \to V_6$, σ_6 is α_6-Lipschitz continuous with respect to z_6, β_6-Lipschitz continuous with respect to r_6 and γ_6-Lipschitz continuous with respect to $h_6\}$.

Verification of Hypothesis HMAK c), d) and Condition CMAK b)

Connection to Φ_5

$\Phi_6 \circ \Phi_5^{-1}$ is given by

$$\begin{aligned}
z_6 &= \varepsilon_5^{-2} z_5 e^{-J^{\mathrm{inv}}(r_5)}, & z_5 &= \varepsilon_6^2 z_6 e^{r_6}/J(r_6)^2,\\
r_6 &= J^{\mathrm{inv}}(r_5), & r_5 &= J(r_6),\\
\varepsilon_6 &= r_5\varepsilon_5, & \varepsilon_5 &= \varepsilon_6/J(r_6),\\
h_6 &= h_5/r_5, & h_5 &= h_6 J(r_6).
\end{aligned}$$

Given $\check{\sigma}_5 \in C_5(\check{U}_5, V_5)$ and $u_6 = (r_6, \varepsilon_6, h_6) \in U_6 \setminus \hat{U}_6$ we have that for $u_5 = (r_5, \varepsilon_5, h_5) := (J(r_6), \varepsilon_6/J(r_6), h_6 J(r_6))$ the inclusion $u_5 \in \check{U}_5$ holds. Moreover, we have $\Phi_6 \circ \Phi_5^{-1}(u_5, \check{\sigma}_5(u_5)) = (u_6, v_6)$ with

$$\begin{aligned}
v_6 = z_6 &= \varepsilon_5^{-2} e^{-J^{\mathrm{inv}}(r_5)} \check{\sigma}_5(r_5, \varepsilon_5, h_5),\\
&= \varepsilon_6^{-2} J(r_6)^2 e^{-r_6} \check{\sigma}_5(J(r_6), \varepsilon_6/J(r_6), h_6 J(r_6)),\\
&=: \bar{\sigma}_{6,5}(r_6, \varepsilon_6, h_6).
\end{aligned}$$

Using $\check{\sigma}_5(r_5, \varepsilon_5, h_5) \le (c_5 - 3\varepsilon_5)\varepsilon_5^2$ we get $v_6 \le c_5$. Setting $\eta_6 = c_5$ implies $v_6 \in V_6$. The function $\bar{\sigma}_{6,5}$ is $\alpha_{6,5}$-Lipschitz continuous with respect to r_6, $\beta_{6,5}$-Lipschitz continuous with respect to ε_6 and $\gamma_{6,5}$-Lipschitz continuous with respect to h_6 with

$$\begin{aligned}
\alpha_{6,5} &\le \delta^{-2} O(\eta_5 + \alpha_5 + h_6\gamma_5) = \delta^{-2} O(\eta_5),\\
\beta_{6,5} &\le \delta^{-2} O(\eta_5),\\
\gamma_{6,5} &\le \delta^{-2} O(\eta_5)/(\nu\delta).
\end{aligned} \tag{13.25}$$

The function $\bar{\sigma}_{6,5}$ is in $C_6(U_6 \setminus \hat{U}_6, V_6)$ for η_5 small enough (i.e., $\alpha_{6,5} \le \alpha_6$, $\beta_{6,5} \le \beta_6$ and $\gamma_{6,5} \le \gamma_6$ holds).

Norm factor

Let $\check{\sigma}_5^{(j)} \in C_5(\check{U}_5, V_5)$, $j = 1, 2$. We have for $(z_6, r_6, h_6) \in U_6 \setminus \hat{U}_6$ that

$$\left|\bar{\sigma}_{6,5}^{(1)}(r_6, \varepsilon_6, h_6) - \bar{\sigma}_{6,5}^{(2)}(r_6, \varepsilon_6, h_6)\right| \le \frac{9}{{\varepsilon_6}^2}\left|\check{\sigma}_5^{(1)}(r_5, \varepsilon_5, h_5) - \check{\sigma}_5^{(2)}(r_5, \varepsilon_5, h_5)\right|$$

with $(r_5, \varepsilon_5, h_5) \in \check{U}_5$. Taking first the supremum on the right-hand side and then on the left-hand side and multiplying by N_6 we obtain

$$\left|\bar{\sigma}_{6,5}^{(1)} - \bar{\sigma}_{6,5}^{(2)}\right|_{U_6 \setminus \hat{U}_6} \le \frac{9N_6}{N_5\delta^2/4}\left|\check{\sigma}_5^{(1)} - \check{\sigma}_5^{(2)}\right|_{\check{U}_5},$$

leading to the choice $N_6 = N_5\delta^2/36$.

Chapter 14

Application of Runge–Kutta methods to differential-algebraic equations

The dynamics of a differential-algebraic equation (DAE) takes place on a lower dimensional manifold in the phase space. We ask the question: When applying a numerical integration scheme to a DAE does the discrete dynamical system preserve this geometric property of the continuous dynamical system?

For index-1 problems of the form

$$\begin{aligned} \dot{x} &= f(x, y), \\ 0 &= g(x, y), \end{aligned} \tag{14.1}$$

where the algebraic equation has a locally unique solution $y = s^0(x)$, all solutions lie on the manifold $M_0 = \{(x, y) \mid y = s^0(x)\}$. Applying stiffly accurate Runge–Kutta methods (RKMs) to (14.1) all results may be obtained via the singularly perturbed case given in Section 11.2 by just taking the limit of the perturbation parameter $\varepsilon \to 0$. Indeed there is a commuting diagram:

$$\begin{array}{ccc}
\begin{array}{c}\text{ODE (11.1);}\\ M_\varepsilon, s_A(x,\varepsilon)\end{array} & \begin{array}{c}\xrightarrow{\ \varepsilon=0\ }\\ \xleftarrow[\text{0 replaced by } \varepsilon\dot{y}]{}\end{array} & \begin{array}{c}\text{DAE (14.1);}\\ M_0, s^0(x) = s_A(x,0)\end{array} \\
\text{RKM}\Big\downarrow & & \Big\downarrow\text{RKM} \\
\begin{array}{c}\text{map (11.11);}\\ M_{h,\varepsilon}, \sigma_A(x,\varepsilon,h);\ (\text{GE})_{h,\varepsilon}\end{array} & \xrightarrow{\ \varepsilon=0\ } & \begin{array}{c}\text{map (14.6);}\\ M_{h,0}, \sigma^0(x,h) = \sigma_A(x,0,h);\ (\text{GE})_{h,0}\end{array}
\end{array}$$

In this diagram the stages of the map (14.6) satisfy $g(X, Y) = 0$, the letter M stands for invariant manifold and (GE) denotes the global error of the RKM, cf. Nipp [97].

Index-2 problems are more involved and are also discussed in Nipp [97] for RKMs satisfying $a_{si} = b_i, i = 1, \dots, s$. Here we investigate index-2 problems approximated by general RKMs satisfying Assumption ASARK below, cf. also Section 11.2. We consider the index-2 DAE of Hessenberg form, cf. Hairer, Lubich, Roche [49] and Hairer, Wanner [53],

$$\begin{aligned} \dot{x} &= f(x, y), \\ 0 &= G(x) \end{aligned} \tag{14.2}$$

satisfying

Assumption ADAE($k+1$)

a) *The function* f *is bounded and* $f \in C_b^{k+1}(\mathbb{R}^m \times \mathbb{R}^n, \mathbb{R}^m)$, $G \in C_b^{k+1}(\mathbb{R}^m, \mathbb{R}^n)$, $k \geq 1$.

b) *There is a function* $s^0 \in C_b^k(\mathbb{R}^m, \mathbb{R}^n)$ *such that* $G_x(x) f(x, s^0(x)) = 0$ *for* $x \in \mathbb{R}^m$.

c) *The matrix* $G_x(x) f_y(x, s^0(x))$ *is invertible and has a uniformly bounded inverse for* $x \in \mathbb{R}^m$.

Under these assumptions, equation (14.2) is of index 2 since differentiating $0 = G(x)$ with respect to t yields $0 = G_x(x) f(x, y)$ which together with $\dot{x} = f(x, y)$ is an index-1 problem by Assumption ADAE($k+1$) b), c). The equation $0 = G_x(x) f(x, y)$ has a unique solution $y = s^0(x)$ for $(x, y) \in \Omega_d := \{(x, y) \mid x \in \mathbb{R}^m,\ |y - s^0(x)| < d\}$, d small enough. All solutions $(x(t), y(t))$ in Ω_d of the index-1 DAE (the so-called index-1 formulation of the index-2 problem (14.2))

$$\begin{aligned} \dot{x} &= f(x, y), \\ 0 &= g(x, y) := G_x(x) f(x, y) \end{aligned}$$

lie in the m-dimensional surface

$$M = \{(x, y) \mid x \in \mathbb{R}^m,\ y = s^0(x)\} \in \mathbb{R}^m \times \mathbb{R}^n.$$

Moreover, the function G is a first integral since

$$\frac{d}{dt} G(x(t)) = G_x(x(t)) f(x(t), s^0(x(t))) = 0.$$

Hence, all solutions of (14.2) (in Ω_d) lie in the submanifold

$$K = \{(x, y) \mid G(x) = 0,\ y = s^0(x)\} \subset M, \tag{14.3}$$

cf. Figure 14.1.

As in Section 11.2 for singularly perturbed ODEs we apply an RKM to (14.2) satisfying

Assumption ASARK

a) *The RKM has order* p *and stage order* q *with* $1 \leq q \leq p$.

b) *The RK-matrix* A *is invertible.*

c) *The stability function* $R(z) := 1 + zb^T(I_s - zA)^{-1}\mathbb{1}$, $z \in \mathbb{C}$, *where* $\mathbb{1} = (1, \dots, 1)^T \in \mathbb{R}^s$, *satisfies* $|R(\infty)| < 1$.

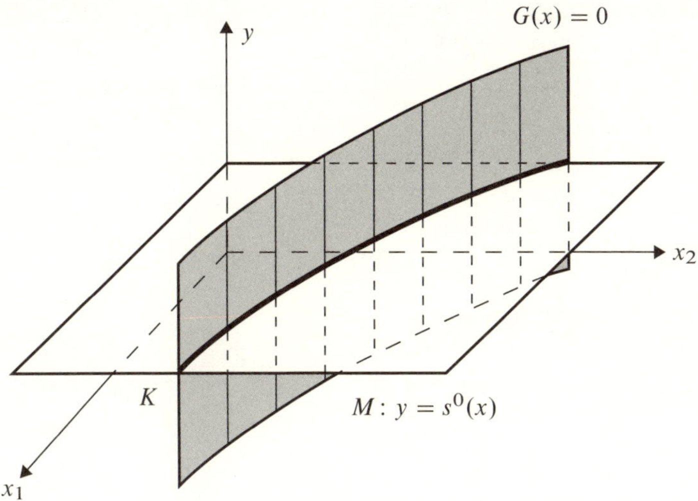

Figure 14.1. The invariant manifold K of the index-2 DAE (14.2) and the manifold M.

The RKM applied to the DAE (14.2) is defined as

$$\begin{aligned} \bar{x} &= x + h(b^T \otimes I_m) f(X, Y), \\ \bar{y} &= y + h(b^T \otimes I_n) Y', \end{aligned} \tag{14.4}$$

where the stages satisfy

$$\begin{aligned} X &= \mathbb{1} \otimes x + h(A \otimes I_m) f(X, Y), \\ 0 &= G(X), \\ Y &= \mathbb{1} \otimes y + h(A \otimes I_n) Y'. \end{aligned} \tag{14.5}$$

Since by assumption the RK-matrix A is invertible the third equation of (14.5) may uniquely be solved for Y'. Inserting the solution into (14.4) and using $R(\infty) = 1 - b^T A^{-1} \mathbb{1}$ yields the RKM

$$P: \quad \begin{aligned} \bar{x} &= x + h(b^T \otimes I_m) f(X, Y), \\ \bar{y} &= R(\infty) y + (b^T A^{-1} \otimes I_n) Y, \end{aligned} \tag{14.6}$$

where the stages satisfy

$$\begin{aligned} X &= \mathbb{1} \otimes x + h(A \otimes I_m) f(X, Y) \\ 0 &= G(X). \end{aligned} \tag{14.7}$$

In order to show that (14.6) defines a smooth map $P : (x, y) \mapsto (\bar{x}, \bar{y})$ one has to show that the equation (14.7) has a unique solution X, Y depending smoothly on x. This is done in Lemma 14.1 for a generalised equation.

Lemma 14.1. *There are positive constants ν, ρ, c such that for $\Delta \in \mathbb{R}^n$ with $|\Delta| < \rho$, $x \in \mathbb{R}^m$ and $h \in [\nu/2, \nu]$ the nonlinear system*

$$\begin{aligned} X^\Delta &= \mathbb{1} \otimes x + h(A \otimes I_m) f(X^\Delta, Y^\Delta), \\ G(X^\Delta) &= \mathbb{1} \otimes (G(x) - h\Delta) \end{aligned} \tag{14.8}$$

has a unique solution $(X^\Delta, Y^\Delta)(x,h)$ of class C_b^k. Moreover, it holds that

$$|X^\Delta(x,h) - \mathbb{1} \otimes x| \le ch, \quad |Y^\Delta(x,h) - s^0(X^\Delta)| \le c(h + |\Delta|).$$

Proof. We write the second equation of (14.8) as

$$\begin{aligned} G(\mathbb{1} \otimes x) - G(X^\Delta) &= \int_0^1 \operatorname{diag}\big[G_x(X^\Delta + \tau(\mathbb{1} \otimes x - X^\Delta))\big] d\tau (\mathbb{1} \otimes x - X^\Delta) \\ &= h(\mathbb{1} \otimes \Delta), \end{aligned}$$

where $\operatorname{diag}\big[G_x(X^\Delta)\big]$ denotes the $sn \times sn$-block-diagonal matrix with $n \times n$-blocks $G_x(X_j^\Delta)$, $j = 1, \dots, s$. Using the first equation and dividing by h we get

$$\int_0^1 \big\{\operatorname{diag}\big[G_x(X^\Delta)\big] + \tau O(h)\big\} d\tau (A \otimes I_m) f(X^\Delta, Y^\Delta) = -\mathbb{1} \otimes \Delta.$$

Since

$$\begin{aligned} \operatorname{diag}\big[G_x(X^\Delta)\big](A \otimes I_m) &= (I_s \otimes G_x(x))(A \otimes I_m) + O(h) \\ &= (A \otimes I_n)(I_s \otimes G_x(x)) + O(h) \\ &= (A \otimes I_n) \operatorname{diag}\big[G_x(X^\Delta)\big] + O(h) \end{aligned} \tag{14.9}$$

we have

$$(A \otimes I_n) \operatorname{diag}\big[G_x(X^\Delta)\big] f(X^\Delta, Y^\Delta) + O(h) = -\mathbb{1} \otimes \Delta. \tag{14.10}$$

We introduce the variable $Z^\Delta := Y^\Delta - s^0(X^\Delta)$, expand f about $Y^\Delta = s^0(X^\Delta)$ and get

$$(A \otimes I_n) \operatorname{diag}\big[G_x(X^\Delta)\big] f_y(X^\Delta, s^0(X^\Delta)) Z^\Delta + O(|Z^\Delta|^2) + O(h) = -\mathbb{1} \otimes \Delta,$$

where we have used ADAE($k+1$) b). We define the block-diagonal matrix $C(X^\Delta) = \operatorname{diag}\big[G_x(X^\Delta) f_y(X^\Delta, s^0(X^\Delta))\big]$ which by Assumption ADAE($k+1$) c) is invertible with bounded inverse and get

$$\begin{aligned} X^\Delta &= \mathbb{1} \otimes x + h(A \otimes I_m) f(X^\Delta, s^0(X^\Delta) + Z^\Delta), \\ Z^\Delta &= C(X^\Delta)^{-1}(A^{-1} \otimes I_n)\big[-\mathbb{1} \otimes \Delta + O(|Z^\Delta|^2) + O(h)\big]. \end{aligned}$$

Considering the two equations as a fixed point equation the contraction principle implies the existence of a unique solution $(X^\Delta, Z^\Delta)(x, \Delta, h)$ for h and $|\Delta|$ sufficiently small. This solution satisfies $|X^\Delta - \mathbb{1} \otimes x| \le \text{const} \cdot h$, $|Z^\Delta| \le \text{const} \cdot (h + |\Delta|)$. The smoothness follows from the implicit function theorem. □

Note that the x-equation of (14.6) and the stages X, Y of (14.7) do not depend on y. This follows from Lemma 14.1 with the choice $h\Delta = G(x)$. Therefore we first investigate the x-part of the map P. Following an idea of A. Murua [90] we consider the auxiliary map from $\mathbb{R}^m \times \{z \in \mathbb{R}^n \mid |z| < \rho\}$ into $\mathbb{R}^m \times \mathbb{R}^n$

$$\begin{aligned} \bar{x} &= x + h(b^T \otimes I_m)\, f(X^\Delta, s^0(X^\Delta) + Z^\Delta) =: F(x, \Delta, h), \\ \bar{\Delta} &= \Delta + \frac{1}{h}\,[G(\bar{x}) - G(x)] =: H(x, \Delta, h), \end{aligned} \tag{14.11}$$

where $(X^\Delta, s^0(X^\Delta) + Z^\Delta)(x, \Delta, h)$ is the unique solution of equation (14.8).

Lemma 14.2. *There is a function $\delta\colon \mathbb{R}^m \times [\nu/2, \nu] \to \mathbb{R}^n$ of class C_b^k such that for $h \in [\nu/2, \nu]$ the set $V_h := \{(x, \Delta) \mid x \in \mathbb{R}^m,\ \Delta = \delta(x, h)\}$ is an invariant manifold of the auxiliary map* (14.11). *V_h is attractive with attractivity constant $\chi_x = |R(\infty)| + O(h) < 1$, i.e.,*

$$|\bar{\Delta} - \delta(\bar{x}, h)| \le \chi_x |\Delta - \delta(x, h)|.$$

The set $U_h := \{x \mid G(x) = h\delta(x, h)\}$ is invariant under the x-part of the RK-map P of (14.6).

Proof. We take a closer look at the function H. We estimate

$$\begin{aligned} \frac{1}{h}\big[G(\bar{x}) - G(x)\big] = \int_0^1 & G_x(x + \tau(\bar{x} - x))d\tau(b^T \otimes I_m) f(X^\Delta, s^0(X^\Delta) + Z^\Delta) \\ & - (b^T \otimes I_n)\,\mathrm{diag}\big[G_x(X^\Delta)\big] f(X^\Delta, s^0(X^\Delta) + Z^\Delta) \\ & - (b^T A^{-1} \otimes I_n)\big[(\mathbb{1} \otimes \Delta) + O(h)\big], \end{aligned}$$

where we have used (14.10). Similarly as for (14.9) it is shown that the sum of the first two terms is $O(h)$. Using $R(\infty) = 1 - b^T A^{-1}\mathbb{1}$ we conclude that $H(x, \Delta, h) = R(\infty)\Delta + O(h)$. We apply Theorem 3.6 to the map (14.11). Hypothesis HM is satisfied with

$$\begin{aligned} \Gamma_{11} &= 1 + O(h), & L_{12} &= O(h), \\ L_{21} &= O(h), & L_{22} &= |R(\infty)| + O(h) \end{aligned}$$

for h small enough. Hypothesis HMA is satisfied for $\Delta^* = 0$. Conditions CM, CMA and CMA(k) are satisfied for h sufficiently small. We conclude that there is a smooth function $\delta\colon \mathbb{R}^m \times [\nu_0, \nu] \to \mathbb{R}^n$ such that the map (14.11) admits an attractive invariant manifold $V_h = \{(x, \Delta) \mid x \in \mathbb{R}^m,\ \Delta = \delta(x, h)\}$ with attractivity constant $\chi_x = |R(\infty)| + O(h) < 1$. Restricting the map (14.11) to V_h one gets a map $x \mapsto \bar{x} = F(x, \delta(x, h), h)$, $\mathbb{R}^m \to \mathbb{R}^m$. Moreover, the set $U_h = \{x \mid G(x) = h\delta(x, h)\}$ is invariant under this map. This follows from the fact that on V_h the Δ-equation of (14.11) takes the form $h\delta(\bar{x}) = h\delta(x) + G(\bar{x}) - G(x)$. For $x \in U_h$ the systems (14.8) and (14.7) are identical. This proves Lemma 14.2. □

We now prove that the RK-map P given in (14.6) with stages (14.7) admits an invariant manifold K_h close to the invariant manifold K of the DAE (14.2), cf. Figure 14.2.

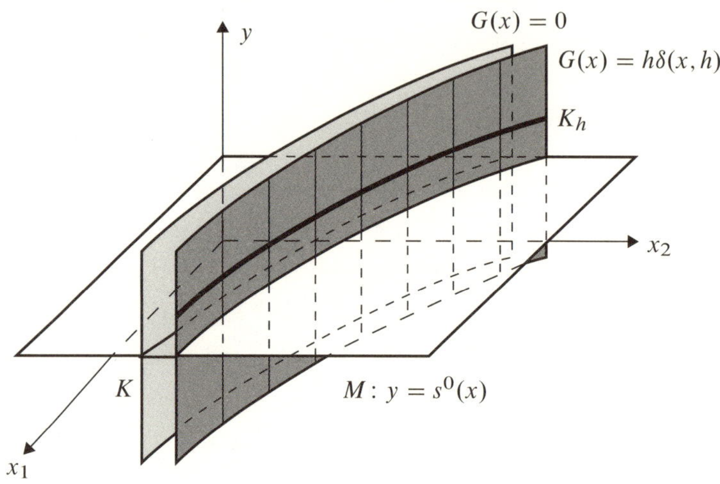

Figure 14.2. The invariant manifold K_h of the RKM (14.6) close to the invariant manifold K of the DAE (14.2) and the manifold M.

Theorem 14.3. *Let the DAE* (14.2) *satisfy Assumption ADAE*$(k+1)$, *assume that the RK-map P of* (14.6), (14.7) *satisfies Assumption ASARK and assume $k > p$.*

Then there exist positive constants ν_0, ν, d, c and functions δ and σ^0 of class C_b^k, $\delta\colon \mathbb{R}^m \times [\nu_0, \nu] \to \mathbb{R}^n$ and $\sigma^0\colon (x, h) \mapsto y = \sigma^0(x, h)$, $h \in [\nu_0, \nu]$, $x \in U_h = \{x \mid G(x) = h\delta(x, h)\}$ such that for $h \in [\nu_0, \nu]$ the following assertions hold.

i) *The set $K_h = \{(x, y) \mid x \in U_h,\ y = \sigma^0(x, h)\}$ is an invariant manifold of the RK-map P.*

ii) *The manifold K_h is x-attractive, i.e., if $|G(x_0)|/h$ is sufficiently small and if $|y_0 - s^0(x_0)| < d$ then the x-part of the RK-orbit (x_j, y_j), $j \geq 0$, satisfies*

$$|G(x_j) - h\delta(x_j, h)| \leq {\chi_x}^j |G(x_0) - h\delta(x_0, h)|$$

where $\chi_x = |R(\infty)| + O(h) < 1$.

iii) *The set $N_{h,d} := \{(x, y) \mid x \in U_h,\ |y - s^0(x)| < d\}$ is foliated by vertical stable fibers. Every RK-orbit (x_j, y_j), $j \geq 0$, with $(x_0, y_0) \in N_{h,d}$ converges to its "asymptotic phase orbit" $(x_j, \sigma^0(x_j, h))$ with*

$$|y_j - \sigma^0(x_j, h)| \leq {\chi_y}^j |y_0 - \sigma^0(x_0, h)|.$$

where $\chi_y = |R(\infty)| < 1$.

iv) *The manifold* K_h *is* $O(h^q)$*-close to the manifold* K *given in* (14.3)*, more precisely,*

$$|G(x)| \le ch^{q+1},$$
$$|\sigma^0(x,h) - s^0(x)| \le ch^q$$

holds for $x \in U_h$.

v) *If the RKM satisfies* $a_{si} = b_i$, $i = 1, \dots, s$, *then* $\delta(x,h) \equiv 0$ *and* $K_h = \{(x,y) \mid G(x) = 0,\ y = \sigma^0(x,h)\}$ *is infinitely attractive, i.e., if* $|G(x_0)|/h$ *is sufficiently small and if* $|y_0 - s^0(x_0)| < d$ *then the image* $(x_1, y_1) = P(x_0, y_0)$ *satisfies*

$$G(x_1) = 0, \quad y_1 = \sigma^0(x_1, h).$$

Proof. i), iii) Introducing z by $y = s^0(x) + z$ and Z by $Y = s^0(X) + Z$, the map P has the form

$$\begin{aligned}
\bar{x} &= x + h(b^T \otimes I_m)\, f(X, s^0(X) + Z),\\
\bar{z} &= R(\infty)\, z + (b^T A^{-1} \otimes I_n)\, Z\\
&\qquad + (b^T A^{-1} \otimes I_n)\big[s^0(X) - \mathbb{1} \otimes s^0(x))\big] - \big[s^0(\bar{x}) - s^0(x)\big], \qquad (14.12)\\
X &= \mathbb{1} \otimes x + h(A \otimes I_m) f(X, s^0(X) + Z),\\
0 &= G(X).
\end{aligned}$$

We consider this map for $(x,z) \in U_h \times \{z \mid |z| < d\}$.

In order to apply Theorems 3.6 and 4.1 we verify Hypotheses HM, HMA and Conditions CM, CMA, CMA(k). Since in (14.11) the function $H(x, \Delta, h) = R(\infty)\Delta + O(h)$, cf. the proof of Lemma 14.2, it follows from Theorem 1.5 iv) that $\delta(x,h) = O(h)$. Lemma 14.1 implies $Z = O(h)$. Hence, Hypothesis HM a) is satisfied for h small enough. Hypothesis HM b) is satisfied since the set U_h is invariant and Hypothesis HM c) is satisfied with

$$\Gamma_{11} = 1 + O(h), \quad L_{12} = 0,$$
$$L_{21} = O(1), \qquad L_{22} = |R(\infty)|.$$

Hypothesis HMA is satisfied for $z^* = 0$. Conditions CM, CMA and CMA(k) are satisfied for h small enough. We conclude that Theorems 3.6 and 4.1 apply. In Theorem 4.1 we take $\Omega := U_h \times \{z \mid |z| < d\}$. Going back to the (x,y)-variables, there is a function $\sigma^0\colon (x,h) \mapsto y = \sigma^0(x,h)$, $h \in [\nu_0, \nu]$, $x \in U_h$, such that the set $K_h = \{(x,y) \mid x \in U_h,\ y = \sigma^0(x,h)\}$ is an attractive invariant manifold of the RK-map P with attractivity constant $\chi_y := |R(\infty)|$, i.e., $(x,y) \in N_{h,d}$ implies $|\bar{y} - \sigma^0(\bar{x},h)| \le \chi_y |y - \sigma^0(x,h)|$. And the set $N_{h,d}$ is foliated by stable fibers which are vertical since $L_{12} = 0$.

ii) Consider the auxiliary map (14.11) for x with $G(x)/h = \Delta$, $|\Delta|$ sufficiently small. By Lemma 14.2 it holds that $|\bar{\Delta} - \delta(\bar{x}, h)| \leq \chi_x |\Delta - \delta(x, h)|$ with $\chi_x = |R(\infty)| + O(h) < 1$. Moreover, $\bar{\Delta} = G(\bar{x})/h$ holds by the Δ-equation of (14.11). It follows that

$$|G(\bar{x}) - h\delta(\bar{x}, h)| \leq \chi_x |G(x) - h\delta(x, h)|.$$

iv) We again consider the map P restricted to the set $N_{h,d}$ and estimate $\bar{z}$ in (14.12). We consider the auxiliary ODE

$$\begin{aligned} \dot{u} &= f(u, s^0(u)), \\ \dot{v} &= s^0_x(u) f(u, s^0(u)) \end{aligned} \tag{14.13}$$

with initial values $u = x$, $v = s^0(x)$. Note that $v(t) = s^0(u(t))$ holds for $t \in \mathbb{R}$. The RKM applied to (14.13) leads to the auxiliary RK-map

$$\begin{aligned} \bar{u} &= u + h(b^T \otimes I_m) f(U, s^0(U)), \\ \bar{v} &= v + h(b^T \otimes I_n) s^0_x(U) f(U, s^0(U)), \end{aligned} \tag{14.14}$$

where the stages satisfy

$$\begin{aligned} U &= \mathbb{1} \otimes u + h(A \otimes I_m) f(U, s^0(U)), \\ V &= \mathbb{1} \otimes v + h(A \otimes I_n) s^0_x(U) f(U, s^0(U)). \end{aligned}$$

We estimate the terms on the second line of the z-equation of (14.12) as

$$\begin{aligned} &(b^T A^{-1} \otimes I_n)\big[s^0(X) - \mathbb{1} \otimes s^0(x)\big] - \big[s^0(\bar{x}) - s^0(x)\big] \\ &\quad = (b^T A^{-1} \otimes I_n)\big[s^0(U) - s^0(\mathbb{1} \otimes u)\big] - \big[s^0(\bar{u}) - s^0(u)\big] \\ &\qquad + (b^T A^{-1} \otimes I_n)\big[s^0(X) - s^0(U)\big] - \big[s^0(\bar{x}) - s^0(\bar{u})\big]. \end{aligned} \tag{14.15}$$

First, we have

$$\begin{aligned} X - U &= h(A \otimes I_m)\big[f(X, s^0(X) + Z) - f(U, s^0(U) + Z)\big] \\ &\quad + h(A \otimes I_m)\big[f(U, s^0(U) + Z) - f(U, s^0(U))\big] \\ &= O(h)(X - U) + h(A \otimes I_m) \operatorname{diag}\big[f_y(U, s^0(U)) + O(|Z|)\big] Z \end{aligned}$$

leading to

$$X - U = h(A \otimes I_m) \operatorname{diag}\big[f_y(U, s^0(U)) + O(h) + O(|Z|)\big] Z. \tag{14.16}$$

Second, we get

$$\begin{aligned} \bar{x} - \bar{u} &= h(b^T \otimes I_m)\big[f(X, s^0(X) + Z) - f(U, s^0(U) + Z)\big] \\ &\quad + h(b^T \otimes I_m)\big[f(U, s^0(U) + Z) - f(U, s^0(U))\big] \\ &= O(h)(X - U) + O(h)Z = O(h)Z. \end{aligned} \tag{14.17}$$

It remains to estimate the first two terms on the right-hand side of (14.15). Since the RKM has order p and stage order q we get

$$\begin{aligned} s^0(\bar{u}) - s^0(u) &= s^0(u(h)) - s^0(u) + O(h^{p+1}) \\ &= v(h) - v + O(h^{p+1}) \\ &= \bar{v} - v + O(h^{p+1}) \\ &= h(b^T \otimes I_n) s_x^0(U) f(U, s^0(U)) + O(h^{p+1}) \end{aligned}$$

and analogously

$$s^0(U) - s^0(\mathbb{1} \otimes u) = h(A \otimes I_n) s_x^0(U) f(U, s^0(U)) + O(h^{q+1}).$$

Thus, the first two terms on the right-hand side of (14.15) add up to $O(h^{q+1}) + O(h^{p+1}) = O(h^{q+1})$. Hence, the z-equation of the RK-map (14.12) has the form

$$\bar{z} = R(\infty)z + (b^T A^{-1} \otimes I_n + O(h))Z + O(h^{q+1}).$$

We estimate Z. Using $G(X)=0$ and $(d/dt)G(u(t)) = G_x(u(t)) f(u(t), s^0(u(t))) = 0$ we obtain, since P is restricted to $N_{h,d}$,

$$\begin{aligned} G(X) - G(U) &= -G(U(ch)) + O(h^{q+1}) = -G(\mathbb{1} \otimes u) + O(h^{q+1}) \\ &= -h(\mathbb{1} \otimes \delta(u,h)) + O(h^{q+1}), \end{aligned}$$

where $U(ch) = (u(c_1 h)^T, \dots, u(c_s h)^T)^T$. On the other hand we get using (14.16)

$$\begin{aligned} G(X) - G(U) &= \big(\operatorname{diag}[G_x(U)] + O(|X-U|)\big)(X-U) \\ &= h \operatorname{diag}[G_x(U)](A \otimes I_m)\big(\operatorname{diag}[f_y(U, s^0(U))] + O(h) + O(|Z|)\big)Z. \end{aligned}$$

The argument used in (14.9) leads to

$$G(X) - G(U) = h\big(C(U) + O(h) + O(|Z|)\big)Z, \tag{14.18}$$

where the matrix $C(U) := (A \otimes I_n) \operatorname{diag}\big[G_x(U) f_y(U, s^0(U))\big]$ is invertible with uniformly bounded inverse. Combining these two estimates for $G(X) - G(U)$ we get $Z = O(|\delta|) + O(h^q)$ and thus

$$\bar{z} = R(\infty)z + O(|\delta|) + O(h^q). \tag{14.19}$$

Theorem 1.5 iv) implies that for $x \in U_h$ the function describing the invariant manifold in the z-variable is of order $O(|\delta|) + O(h^q)$.

We estimate the function δ. We apply Theorem 2.3 with the approximating function $\Delta = \sigma(x,h) \equiv 0$ to the auxiliary map (14.11) and obtain the estimate

$$\delta = O(|H(x,0,h)|).$$

We have

$$\begin{aligned} h\,H(x,0,h) &= G(\bar{x}) - G(x) \\ &= [G(\bar{x}) - G(\bar{u})] + [G(\bar{u}) - G(u(h))] + [G(u(h)) - G(u)] \\ &= [G(\bar{x}) - G(\bar{u})] + O(h^{p+1}) + 0, \end{aligned}$$

where again $u(t)$ is the solution of the u-equation of (14.13) with $u(0) = x$ and $\bar{u}$ is its RK-approximation, c.f. (14.14). We know from equations (14.16), (14.17) that $G(\bar{x}) - G(\bar{u}) = O(h)Z^0$ with $Z^0 = Y^0 - s^0(X^0)$, where X^0, Y^0 solve the equation (14.8) with $\Delta = 0$, and that (14.18) holds with X, Z replaced by X^0, Z^0. On the other hand we have

$$\begin{aligned} G(X^0) - G(U) &= [G(\mathbb{1} \otimes x) - G(U(ch))] + [G(U(ch)) - G(U)] \\ &= 0 + O(h^{q+1}) \end{aligned}$$

since the function G is a first integral of $\dot{u} = f(u, s^0(u))$ and since the RKM has stage order q. From 14.18 we conclude that $Z^0 = O(h^q)$ and hence $H(x,0,h) = (1/h)\,[G(\bar{x}) - G(\bar{u})] + O(h^p) = O(|Z^0|) + O(h^p) = O(h^q)$ implying $\delta = O(h^q)$. It follows that for $x \in U_h$ we have $G(x) = O(h^{q+1})$. Equation 14.19 implies the second estimate of iv).

v) Lemma 14.1 implies that for $\Delta \in \mathbb{R}^n$, small enough, and for a starting value x with $G(x) = h\Delta$ one has $G(X^{\Delta}) = 0$ implying $G(\bar{x}) = G(X_s^{\Delta}) = 0$. Moreover, for $\Delta = 0$ the set $\{x \mid G(x) = 0\}$ is invariant under (14.11). On the other hand, A invertible and $a_{si} = b_i$, $i = 1, \dots, s$, implies $R(\infty) = 0$, cf. Hairer, Wanner [53]. Hence, assertion iii) implies $\bar{y} = \sigma^0(\bar{x}, h)$. □

Part IV

Appendices

Appendix A
Hypotheses and conditions for maps

In this appendix we list all hypotheses and conditions used in this book.

Notation: We denote hypotheses and conditions by a sequence of letters defined as follows:

H for **H**ypothesis (first letter)
C for **C**ondition (first letter)
M for **M**ap (second letter)
D for **D**ifferential equation (used in Part B) (second letter)
A for **A**ttractive invariant manifold (third letter)
R for **R**epulsive invariant manifold (third letter)
(*k*) for smooth of class C^k
F for **F**oliation
B for smoothness with respect to the **B**ase point
K for manifolds described by several Charts (**K**arten)
G for **G**eneralised situation
V for **V**ariable parameter

A.1 Hypotheses

Hypothesis HM

The functions $F \in C^0(X \times Y \times E, \mathcal{B}_x)$, $G \in C^0(X \times Y \times E, \mathcal{B}_y)$ have the following properties.

a) *P_ϑ is inflowing with respect to Y, i.e., $G(x, y, \vartheta) \in Y$ holds for all $(x, y, \vartheta) \in X \times Y \times E$.*

b) *P_ϑ is outflowing with respect to X, i.e., for every $\bar{x} \in X$, $y \in Y$, $\vartheta \in E$ there is $x \in X$ such that $F(x, y, \vartheta) = \bar{x}$.*

c) *There are nonnegative constants Γ_{11}, L_{12}, L_{13}, L_{21}, L_{22} and L_{23} such that for $x, x_1, x_2 \in X$, $y, y_1, y_2 \in Y$, $\vartheta, \vartheta_1, \vartheta_2 \in E$ the functions F and G satisfy*

$$|F(x_1, y, \vartheta) - F(x_2, y, \vartheta)| \geq \Gamma_{11}|x_1 - x_2|,$$

$$|F(x, y_1, \vartheta_1) - F(x, y_2, \vartheta_2)| \leq L_{12}|y_1 - y_2| + L_{13}|\vartheta_1 - \vartheta_2|,$$

$$|G(x_1, y_1, \vartheta_1) - G(x_2, y_2, \vartheta_2)| \leq L_{21}|x_1 - x_2| + L_{22}|y_1 - y_2| + L_{23}|\vartheta_1 - \vartheta_2|.$$

Hypothesis HMA

There is $y^* \in Y$ *such that the function* $G(\cdot, y^*, \cdot)\colon X \times E \to Y$ *is bounded.*

Hypothesis HMAK

For $i = 1, \dots, \kappa$ *there are sets* $Q_i \subset \widetilde{Q}_i \subset \mathbb{R}^\ell$ *with* $\widetilde{Q}_i \supset P(Q_i)$, *sets* $\check{U}_i \subset \widehat{U}_i \subset U_i \subset \widetilde{U}_i \subset \mathbb{R}^m$ *with* $\widehat{U}_1 = U_1$ *and sets* $V_i \subset \mathbb{R}^n$ *and charts* Φ_i *with* $\Phi_i(Q_i) = W_i := U_i \times V_i$ *and* $\Phi_i(\widetilde{Q}_i) = \widetilde{W}_i := \widetilde{U}_i \times V_i$ *such that the following holds.*

a) *In the chart* Φ_i, $i = 1, \dots, \kappa$, *the map* P *induces a map*

$$P_i := \Phi_i \circ P \circ \Phi_i^{-1}\colon W_i = U_i \times V_i \longrightarrow \widetilde{W}_i = \widetilde{U}_i \times V_i,$$
$$\begin{pmatrix} u_i \\ v_i \end{pmatrix} \longmapsto \begin{pmatrix} \bar{u}_i \\ \bar{v}_i \end{pmatrix} = \begin{pmatrix} F_i(u_i, v_i) \\ G_i(u_i, v_i) \end{pmatrix}.$$

b) *For* $i = 1, \dots, \kappa$ *there is a complete function space* $C_i(U_i, V_i)$ *of bounded Lipschitz continuous functions such that for* $\sigma_i \in C_i(U_i, V_i)$ *the map* P_i *maps the set* $M_{\sigma_i} := \{(u_i, v_i) \mid u_i \in U_i,\ v_i = \sigma_i(u_i)\}$ *to the set* $P_i(M_{\sigma_i})$ *which is the graph of some function* $\tilde{\sigma}_i$. *The restriction* $\hat{\sigma}_i := \tilde{\sigma}_i|_{\widehat{U}_i}$ *lies in* $C_i(\widehat{U}_i, V_i)$, *i.e., the induced operator* $\mathcal{F}_i$ *takes* $\sigma_i \in C_i(U_i, V_i)$ *to* $\hat{\sigma}_i \in C_i(\widehat{U}_i, V_i)$.

c) *For every* $\check{\sigma}_{i-1} \in C_{i-1}(\check{U}_{i-1}, V_{i-1})$, $i = 2, \dots, \kappa$, *there is* $\bar{\sigma}_{i,i-1} \in C_i(U_i \setminus \widehat{U}_i, V_i)$ *such that for every* $u_i \in U_i \setminus \widehat{U}_i$ *there is* $u_{i-1} \in \check{U}_{i-1}$ *with* $\Phi_i \circ \Phi_{i-1}^{-1}(u_{i-1}, \check{\sigma}_{i-1}(u_{i-1})) = (u_i, \bar{\sigma}_{i,i-1}(u_i))$, *i.e., the induced operator* $\mathcal{T}_{i,i-1}$ *transforms* $\check{\sigma}_{i-1} \in C_{i-1}(\check{U}_{i-1}, V_{i-1})$ *to* $\bar{\sigma}_{i,i-1} \in C_i(U_i \setminus \widehat{U}_i, V_i)$.

d) *The function space*

$$\Sigma := \big\{\sigma = (\sigma_1, \dots, \sigma_\kappa) \mid \sigma_i \in C_i(U_i, V_i),\ i = 1, \dots, \kappa, \text{ and } \mathcal{T}_{i,i-1}(\sigma_{i-1}|_{\check{U}_{i-1}}) = \sigma_i|_{U_i \setminus \widehat{U}_i},\ i = 2, \dots, \kappa\big\}$$

is nonempty and the operator $\mathcal{F}$ *taking* $\sigma = (\sigma_1, \dots, \sigma_\kappa) \in \Sigma$ *to* $\bar{\sigma} = (\bar{\sigma}_1, \dots, \bar{\sigma}_\kappa)$ *with*

$$\bar{\sigma}_i(u_i) := \begin{cases} \bar{\sigma}_{i,i-1}(u_i) := \mathcal{T}_{i,i-1}(\mathcal{F}_{i-1}(\sigma_{i-1})|_{\check{U}_{i-1}})(u_i) & \text{for } u_i \in U_i \setminus \widehat{U}_i, \\ & i = 2, \dots, \kappa \\ \hat{\sigma}_i(u_i) = \mathcal{F}_i(\sigma_i)(u_i) & \text{for } u_i \in \widehat{U}_i, \\ & i = 2, \dots, \kappa \end{cases}$$

maps Σ *into itself, i.e.,* $\mathcal{F}\colon \Sigma \to \Sigma$.

Hypothesis HMAG

The functions $F \in C^0(X \times Y \times E, \mathcal{B}_x)$, $G \in C^0(X \times Y \times E, \mathcal{B}_y)$ *have the following properties.*

a) *Hypothesis HM* a).

b) *There is* $\hat{X} \subset \mathcal{B}_X$ *such that* P_ϑ *is flowing from* X *to* $\hat{X}$, *i.e., for every* $\bar{x} \in \hat{X}$, $y \in Y$, $\vartheta \in E$ *there is* $x \in X$ *such that* $F(x, y, \vartheta) = \bar{x}$.

c) *Hypothesis HM* c).

Hypothesis HMAV

For $\nu \in H_{\nu_0}$ *there are domains* $\hat{U}_\nu \subset U_\nu = X \times H_\nu$ *such that the following holds.*

a) *For all* $(\bar{x}, \bar{h}) \in \hat{U}_\nu$, $z \in Z_{d_0}$ *there is* $(x, h) \in U_\nu$ *such that* $F(x, z, h) = \bar{x}$ *and* $Q(x, z, h) = \bar{h}$.

b) *In* $X \times H_{\nu_0} \times Z_{d_0}$ *the functions* $\hat{Q}$, $\hat{F}$, $\hat{B}$, $\hat{G}$ *are bounded and Lipschitz continuous with Lipschitz constants*

$$
\begin{aligned}
&\operatorname{Lip}_h \hat{Q} = \hat{L}_{00}, \quad \operatorname{Lip}_x \hat{Q} = \hat{L}_{01}, \quad \operatorname{Lip}_z \hat{Q} = \hat{L}_{02},\\
&\operatorname{Lip}_h \hat{F} = \hat{L}_{10}, \quad \operatorname{Lip}_x \hat{F} = \hat{L}_{11}, \quad \operatorname{Lip}_z \hat{F} = \hat{L}_{12},\\
&\operatorname{Lip}_h \hat{B} = \hat{\ell}_{20}, \quad \operatorname{Lip}_x \hat{B} = \hat{\ell}_{21}, \quad \operatorname{Lip}_z \hat{B} = \hat{\ell}_{22},\\
&\operatorname{Lip}_h \hat{G} = \hat{L}_{20}, \quad \operatorname{Lip}_x \hat{G} = \hat{L}_{21}, \quad \operatorname{Lip}_z \hat{G} = \hat{L}_{22}.
\end{aligned}
$$

Hypothesis HMR

There is $x^* \in X$ *such that the function* $F(x^*, \cdot, \cdot)\colon Y \times E \to \mathcal{B}_X$ *is bounded.*

Hypothesis HMRF

There is a set $\Omega_\vartheta \subset X \times Y$ *which is negatively invariant under* P_ϑ. *The map* P_ϑ *is invertible on* Ω_ϑ, *i.e., for every* $(x, y) \in \Omega_\vartheta$ *there is a unique* $(\underline{x}, \underline{y}) \in \Omega_\vartheta$ *such that* $P_\vartheta(\underline{x}, \underline{y}) = (x, y)$.

Hypothesis HMB

a) *The sets* X *and* Y *are convex.*

b) *There is a constant* Γ_{22} *such that for* $x \in X$, $y_1, y_2 \in Y, \vartheta \in E$ *the function* G *satisfies*

$$|G(x, y_1, \vartheta) - G(x, y_2, \vartheta)| \geq \Gamma_{22}\, |y_1 - y_2|.$$

Hypothesis HMAB

There is a constant L_{11} *such that for* $x_1, x_2 \in X$, $y \in Y$, $\vartheta \in E$ *the function* F *satisfies*

$$|F(x_1, y, \vartheta) - F(x_2, y, \vartheta)| \leq L_{11}\, |x_1 - x_2|.$$

A.2 Conditions

Condition CM

$$2\sqrt{L_{12}L_{21}} < \Gamma_{11} - L_{22}.$$

Condition CMA

$$L_{22} + \Delta \;<\; 1,$$

where

$$\Delta = \frac{2L_{12}L_{21}}{\Gamma_{11} - L_{22} + \sqrt{(\Gamma_{11} - L_{22})^2 - 4L_{12}L_{21}}}.$$

Condition CMA(*k*)

$$L_{22} + \Delta < (\Gamma_{11} - \Delta)^k.$$

Condition CMAK

a) *For* $i = 1, \dots, \kappa$ *the operator* $\mathcal{F}_i$ *is* χ_i*-contracting, i.e., for* $\sigma_i^{(1)}, \sigma_i^{(2)} \in C_i(U_i, V_i)$

$$\big|\mathcal{F}_i(\sigma_i^{(1)}) - \mathcal{F}_i(\sigma_i^{(2)})\big|_{\hat{U}_i} \;\le\; \chi_i \,\big|\sigma_i^{(1)} - \sigma_i^{(2)}\big|_{U_i}$$

holds with $\chi_i < 1$.

b) *For* $i = 2, \dots, \kappa$ *the operator* $\mathcal{T}_{i,i-1}$ *is nonaugmenting, i.e.,*

$$\big|\mathcal{T}_{i,i-1}(\check{\sigma}_{i-1}^{(1)}) - \mathcal{T}_{i,i-1}(\check{\sigma}_{i-1}^{(1)})\big|_{U_i \setminus \hat{U}_i} \;\le\; \big|\check{\sigma}_{i-1}^{(1)} - \check{\sigma}_{i-1}^{(2)}\big|_{\check{U}_{i-1}}$$

holds for $\check{\sigma}_{i-1}^{(1)}, \check{\sigma}_{i-1}^{(2)} \in C_{i-1}(\check{U}_{i-1}, V_{i-1})$.

Condition CMAG

$$L_{22} + L_{12}\alpha < 1.$$

Condition CMAV

There are positive constants a, b *and* c *with* $b > a$ *and* $b > c$ *such that* $|I + h\hat{B}| < 1 - hb$, $\hat{L}_{11} < a$, $|1 + 2h\hat{Q}| > 1 - hc$ *holds for* $(x, h, z) \in X \times H_{\nu_0} \times Z_{d_0}$.

Condition CMR

$$1 < \Gamma_{11} - \Delta.$$

Condition CMR(*k*)

$$(L_{22} + \Delta)^k < \Gamma_{11} - \Delta.$$

Condition CMB

$$\Gamma_{22} - \Delta > 0.$$

Condition CMAB($k-1$)

$$\frac{\Gamma_{11} - \Delta}{L_{22} + \Delta} > (L_{11} + \Delta)^{k-1}.$$

Condition CMRB($k-1$)

$$\frac{L_{22} + \Delta}{\Gamma_{11} - \Delta} < (\Gamma_{22} - \Delta)^{k-1}.$$

Appendix B
Hypotheses and conditions for ODEs

B.1 Hypotheses

Hypothesis HD0

Let the closure of X satisfy $\bar{X} \subset X'$. Assume that there is $T > 0$ such that for all $(x, y, \vartheta) \in X \times Y \times E$ the solution $(\varphi(t; x, y, \vartheta), \psi(t; x, y, \vartheta))$ exists and remains in $X' \times Y$ for all $t \in [0, T]$ and assume that if x is on the boundary ∂X of X then $\varphi(t; x, y, \vartheta) \notin X$ for $t \in (0, T]$.

Hypothesis HD

Let $X \subset X' \subset \mathbb{R}^m$, $Y \subset \mathbb{R}^n$, $E \subset \mathbb{R}^\ell$ be nonempty open convex sets and let $f: X' \times Y \times E \to \mathbb{R}^m$ and $g: X' \times Y \times E \to \mathbb{R}^n$ be of class C_b^k, $k \geq 1$. Let $X \subset X'$ be such that the closure of X satisfies $\bar{X} \subset X'$ and assume that there is $T > 0$ such that for all $(x, y, \vartheta) \in X \times Y \times E$ the solution $(\varphi(t; x, y, \vartheta), \psi(t; x, y, \vartheta))$ of the differential equation (7.1) *remains in $X' \times Y$ for $t \in [0, T]$.*

a) *Let the flow of the differential equation* (7.1) *be inflowing with respect to Y, i.e., if Y has a boundary ∂Y, then it is piecewise of class C^1 and $n_Y(y) \cdot g(x, y, \vartheta) < 0$ for all $(x, y, \vartheta) \in X \times \partial Y \times E$, n_Y being an outer normal with respect to Y.*

b) *The flow of the differential equation* (7.1) *is outflowing with respect to X, i.e., if X has a boundary ∂X, then it is piecewise of class C^1 and $n_X(x) \cdot f(x, y, \vartheta) > 0$ for all $(x, y, \vartheta) \in \partial X \times Y \times E$, n_X being an outer normal with respect to X.*

c) i) *There are nonnegative constants ℓ_{12}, ℓ_{13}, ℓ_{21}, ℓ_{23} such that on the set $X \times Y \times E$,*

$$\left|\frac{\partial f}{\partial y}\right| \leq \ell_{12}, \quad \left|\frac{\partial f}{\partial \vartheta}\right| \leq \ell_{13},$$

$$\left|\frac{\partial g}{\partial x}\right| \leq \ell_{21}, \quad \left|\frac{\partial g}{\partial \vartheta}\right| \leq \ell_{23}.$$

ii) *There are constants $\ell_{22} < 0$ and γ_{11} such that the logarithmic norms of the Jacobians $-\partial f/\partial x$ and $\partial g/\partial y$ satisfy*

$$\mu\left(-\frac{\partial f}{\partial x}\right) \leq -\gamma_{11} \quad \text{and} \quad \mu\left(\frac{\partial g}{\partial y}\right) \leq \ell_{22}$$

on the set $X \times Y \times E$.

Hypothesis HDA

There is y^* such that $g(\cdot, y^*, \cdot)$ is bounded.

Hypothesis HDR

There is x^* such that $f(x^*, \cdot, \cdot)$ is bounded.

Hypothesis HDRF

For $\vartheta \in E$ there is a set $\Omega_\vartheta \subset X \times Y$ which is negatively invariant under the flow of the differential equation (7.1)*, i.e., for $x, y \in \Omega_\vartheta$ the solution satisfies $(\varphi(t; x, y, \vartheta), \psi(t; x, y, \vartheta)) \in \Omega_\vartheta$ for all $t \leq 0$.*

Hypothesis HDB

There is a constant γ_{22} such that the logarithmic norm of the Jacobian $-\partial g/\partial y$ satisfies

$$\mu\Big(-\frac{\partial g}{\partial y}\Big) \leq -\gamma_{22}.$$

Hypothesis HDAB

There is a constant ℓ_{11} such that the logarithmic norm of the Jacobian $\partial f/\partial x$ satisfies

$$\mu\Big(\frac{\partial f}{\partial x}\Big) \leq \ell_{11}.$$

B.2 Conditions

Condition CD

$$2\sqrt{\ell_{12}\ell_{21}} < \gamma_{11} - \ell_{22}.$$

Condition CDA

$$\ell_{22} + \delta < 0,$$

where

$$\delta = \frac{2\ell_{12}\ell_{21}}{\gamma_{11} - \ell_{22} + \sqrt{(\gamma_{11} - \ell_{22})^2 - 4\ell_{12}\ell_{21}}}.$$

Condition CDA(k)

$$\ell_{22} + \delta < k(\gamma_{11} - \delta).$$

Condition CDR

$$0 < \gamma_{11} - \delta.$$

Condition CDR(k)

$$k(\ell_{22} + \delta) < \gamma_{11} - \delta.$$

Condition CDAB($k - 1$)

$$\gamma_{11} - \ell_{22} - 2\delta > (k - 1)(\ell_{11} + \delta).$$

Bibliography

[1] O. D. Anosova, On invariant manifolds in singularly perturbed systems. *J. Dynam. Control Systems* **5** (1999), no. 4, 501–507.

[2] O. D. Anosova, Invariant manifolds in singularly perturbed systems. *Tr. Mat. Inst. Steklova* **236** (2002), 27–32; English transl. *Proc. Steklov Inst. Math.* **236** (2002), 19–24.

[3] U. M. Ascher, H. Chin, and S. Reich, Stabilization of DAEs and invariant manifolds. *Numer. Math.* **67** (1994), 131–149.

[4] B. Aulbach and D. Flockerzi, An existence theorem for invariant manifolds. *Z. Angew. Math. Phys.* **38** (1987), 151–171.

[5] B. Aulbach and C. Poetzsche, Invariant manifolds with asymptotic phase for non-autonomous difference equations. *Comput. Math. Appl.* **45** (2003), no. 6–9, 1385–1398.

[6] L. Barreira, M. Fan, C. Valls, and J. Zhang, Invariant manifolds for impulsive equations and non-uniform polynomial dichotomies. *J. Stat. Phys.* **141** (2010), no. 1, 179–200.

[7] L. Barreira, C. Silva, and C. Valls, Regularity of invariant manifolds for nonuniformly hyperbolic dynamics. *J. Dynam. Differential Equations* **20** (2008), no. 2, 281–299.

[8] L. Barreira and C. Valls, Smoothness of invariant manifolds for non-autonomous equations. *Commun. Math. Phys.* **259** (2005), no. 3, 639–677.

[9] L. Barreira and C. Valls, Smooth invariant manifolds in Banach spaces with non-uniform exponential dichotomy. *J. Funct. Anal.* **238** (2006), no. 1, 118–148.

[10] L. Barreira and C. Valls, Stable invariant manifolds for parabolic dynamics. *J. Funct. Anal.* **257** (2009), no. 4, 1018–1029.

[11] L. Barreira and C. Valls, Smooth stable invariant manifolds and arbitrary growth rates. *Nonlinear Anal. Ser. A* **72** (2010), no. 5, 2444–2456.

[12] F. Battelli and R. Johnson, On transversal smoothness of invariant manifolds. *Commun. Appl. Anal.* **5** (2001), no. 3, 383–401.

[13] F. Battelli and K. J. Palmer, Transverse intersection of invariant manifolds in singular systems. *J. Differential Equations* **177** (2001), no. 1, 77–120.

[14] F. Battelli and K. J. Palmer, Transversal periodic-to-periodic homoclinic orbits in singularly perturbed systems. *Discrete Contin. Dyn. Syst. Ser. B* **14** (2010), no. 2, 367–387.

[15] M. Beck and C. E. Wayne, Using global invariant manifolds to understand meta-stability in the Burgers equation with small viscosity. *SIAM Rev.* **53** (2011), no. 1, 129–153.

[16] R. Benitez and V. J. Bolos, Invariant manifolds of the Bonhoeffer-van der Pol oscillator. *Chaos Solitons Fractals* **40** (2009), no. 5, 2170–2180.

[17] W.-J. Beyn and W. Kless, Numerical Taylor expansions of invariant manifolds in large dynamical systems. *Numer. Math.* **80** (1998), no. 1, 1–38.

[18] S. Bianchini and L. V. Spinolo, Invariant manifolds for a singular ordinary differential equation. *J. Differential Equations* **250** (2011), no. 4, 1788–1827.

[19] M. Branicki and S. Wiggins, An adaptive method for computing invariant manifolds in non-autonomous, three-dimensional dynamical systems. *Physica D* **238** (2009), no. 16, 1625–1657.

[20] H. W. Broer, A. Hagen, and G. Vegter, A versatile algorithm for computing invariant manifolds. In *Model reduction and coarse-graining approaches for multiscale phenomena*, A. N. Gorban et al., eds., selected papers from the workshop held at the University of Leicester, Leicester, UK, August 24–26, 2005, Springer-Verlag, Berlin 2006, 17–37.

[21] H. W. Broer, H. M. Osinga, and G. Vegter, On the computation of normally hyperbolic invariant manifolds. In *Nonlinear dynamical systems and chaos*, Proceedings of the dynamical systems conference, Groningen 1995, Prog. Nonlinear Differ. Equ. Appl. 19, Birkhäuser, Basel 1996, 423–447.

[22] P. Brunovskyand I. Terescak, Regularity of invariant manifolds. *J. Dynam. Differential Equations* **3** (1991), no. 3, 313–337.

[23] J. Carr, *Applications of centre manifold theory*. Appl. Math. Sci. 35, Springer-Verlag, New York 1981.

[24] M. Chaperon, Invariant manifolds revisited. *Tr. Mat. Inst. Steklova* **236** (2002), 428–446; English transl. *Proc. Steklov Inst. Math.* **236** (2002), 415–433.

[25] M. Chaperon, Invariant manifold theory via generating maps. *C. R. Math. Acad. Sci. Paris* **346** (2008), no. 21–22, 1175–1180.

[26] D. Cheban and B. Schmalfuss, Invariant manifolds, global attractors, almost automorphic and almost periodic solutions of non-autonomous differential equations. *J. Math. Anal. Appl.* **340** (2008), no. 1, 374–393.

[27] S. N. Chow and K. Lu, C^k centre unstable manifolds. *Proc. Royal Soc. Edinburgh* **108**A (1988), 303–320.

[28] S. N. Chow and K. Lu, Invariant manifolds and foliations for quasiperiodic systems. *J. Differential Equations* **117** (1995), no. 1, 1–27.

[29] S. N. Chow, W. Liu, and Y. Yi, Center manifolds for invariant sets. *J. Differential Equations* **168** (2000), no. 2, 355–385.

[30] S. N. Chow, W. Liu, and Y. Yi, Center manifolds for smooth invariant manifolds. *Trans. Amer. Math. Soc.* **352** (2000), no. 11, 5179–5211.

[31] J. D. Cole, *Perturbation methods in applied mathematics*. Blaisdell Publishing Co., Waltham, Mass.-Toronto, Ont.-London 1968.

[32] W. A. Coppel and K. J. Palmer, Averaging and integral manifolds. *Bull. Austral. Math. Soc.* **2** (1970), 197–222.

[33] R. De la Llave, Invariant manifolds associated to invariant subspaces without invariant complements: a graph transform approach. *Math. Phys. Electron. J.* **9** (2003), Paper No. 3, 35 p.

[34] L. H. Duc and S. Siegmund, Hyperbolicity and invariant manifolds for planar nonautonomous systems on finite time intervals. *Int. J. Bifurcation Chaos Appl. Sci. Eng.* **18** (2008), no. 3, 641–674.

[35] F. Dumortier, Geometric singular perturbation theory: centre manifolds and blow up. In *International conference on differential equations*, Vol. 1, 2 (Berlin, 1999), World Sci. Publ., River Edge, NJ, 2000, 88–93.

[36] T. Eirola and J. von Pfaler, Numerical Taylor expansions for invariant manifolds. *Numer. Math.* **99** (2004), no. 1, 25–46.

[37] J. P. England, B. Krauskopf, and H. M. Osinga, Computing two-dimensional global invariant manifolds in slow-fast systems. *Int. J. Bifurcation Chaos Appl. Sci. Eng.* **17** (2007), no. 3, 805–822.

[38] N. Fenichel, Persistence and smoothness of manifolds for flows. *Indiana Univ. Math. J.* **21** (1971), 193–226.

[39] N. Fenichel, Asymptotic stability with rate conditions. *Indiana Univ. Math. J.* **23** (1974), 1109–1137.

[40] N. Fenichel, Asymptotic stability with rate conditions, II. *Indiana Univ. Math. J.* **26** (1977), 81–93.

[41] N. Fenichel, Geometric singular perturbation theory for ordinary differential equations. *J. Differential Equations* **31** (1979), 53–98.

[42] N. Fenichel, Hyperbolicity and exponential dichotomy for dynamical systems. In *Dynamics reported*, C. K. R. T. Jones et al., eds., Dynam. Report. Expositions Dynam. Systems (N.S.) 5, Springer-Verlag, Berlin 1996, 1–25.

[43] L. Z. Fishman, Invariant manifolds of multistep finite-difference methods for ordinary differential equations. *Dokl. Akad. Nauk* **356** (1997), no. 2, 173–175; English transl. *Dokl. Math.* **56** (1997), no. 2, 686–688.

[44] S. J. Fraser, Symbolic methods for invariant manifolds in chemical kinetics. *Int. J. Quantum Chem.* **106** (2006), no. 1, 228–243.

[45] B. M. Garay, Estimates in discretizing normally hyperbolic compact invariant manifolds of ordinary differential equations. *Comput. Math. Appl.* **42** (2001), no. 8–9, 1103–1122.

[46] A. N. Gorban and I. V. Karlin, *Invariant manifolds for physical and chemical kinetics*. Lecture Notes in Phys. 660, Springer-Verlag, Berlin 2005.

[47] J. Hadamard, Sur l'itération et les solutions asymptotiques des équations différentielles. *Bull. Soc. Math. France* **29** (1901), 224–228.

[48] E. Hairer and C. Lubich, Invariant tori of dissipatively perturbed Hamiltonian systems under symplectic discretization. *Appl. Numer. Math.* **29** (1999), no. 1, 57–71.

[49] E. Hairer, C. Lubich, and M. Roche, Error of Runge–Kutta methods for stiff problems studied via differential algebraic equations. *BIT* **28** (1988), 678–700.

[50] E. Hairer, C. Lubich, and M. Roche, *The numerical solution of differential-algebraic systems by Runge–Kutta methods*. Lecture Notes in Math. 1409, Springer-Verlag, Berlin 1989.

[51] E. Hairer, C. Lubich, and G. Wanner, *Geometric numerical integration. Structure-preserving algorithms for ordinary differential equations*. Springer Ser. Comput. Math. 31, 2nd edition, Springer-Verlag, Berlin 2006.

[52] E. Hairer, S. P. Nørsett, and G. Wanner, *Solving ordinary differential equations I: Nonstiff problems*. Springer Ser. Comput. Math. 8, Springer-Verlag, 2nd edition, Berlin 1993.

[53] E. Hairer and G. Wanner, *Solving ordinary differential equations II: Stiff and differential-algebraic problems*. 2nd edition, Springer Ser. Comput. Math. 14, Springer-Verlag, Berlin 1996.

[54] S. Hilger, Smoothness of invariant manifolds. *J. Funct. Anal.* **106** (1992), no. 1, 95–129.

[55] M. Hirsch, C. Pugh, and M. Shub, *Invariant manifolds*. Lecture Notes in Math. 583, Springer-Verlag, Berlin 1977.

[56] A. J. Homburg, Invariant manifolds near hyperbolic fixed points. *J. Difference Equ. Appl.* **12** (2006), no. 10, 1057–1068.

[57] A. J. Homburg, H. M. Osinga, and G. Vegter, On the computation of invariant manifolds of fixed points. *Z. Angew. Math. Phys.* **46** (1995), no. 2, 171–187.

[58] K. in 't Hout and C. Lubich, Periodic orbits of delay differential equations under discretisation. *BIT* **38** (1998), 72–91.

[59] G. Iooss and E. Lombardi, Approximate invariant manifolds up to exponentially small terms. *J. Differential Equations* **248** (2010), no. 6, 1410–1431.

[60] J. Jarnik and J. Kurzweil, On invariant sets and invariant manifolds of differential systems. *J. Differential Equations* **6** (1969), 247–263.

[61] T. Johnson and W. Tucker, A note on the convergence of parametrised non-resonant invariant manifolds. *Qual. Theory Dyn. Syst.* **10** (2011), no. 1, 107–121.

[62] C. K. R. T. Jones, Geometric singular perturbation theory. In *Dynamical systems* (Montecatini Terme, 1994), Lecture Notes in Math. 1609, Springer-Verlag, Berlin 1995, 44–118.

[63] A. Kelley, The stable, center-stable, center, center-unstable, unstable manifolds. *J. Differential Equations* **3** (1967), 546–570.

[64] U. Kirchgraber, Multistep methods are essentially one-step methods. *Numer. Math.* **48** (1986), 85–90.

[65] U. Kirchgraber, On the existence and geometry of invariant tori in the method of averaging. *J. Appl. Math. Phys.* **38** (1987), 213–225.

[66] U. Kirchgraber, F. Lasagni, K. Nipp, and D. Stoffer, On the application of invariant manifold theory, in particular to numerical analysis. In *Bifurcation and chaos: analysis, algorithms, applications*, Internat. Ser. Numer. Math. 97, Birkhäuser, Basel 1991, 189–197.

[67] U. Kirchgraber and K. Nipp, Geometric properties of RKMs applied to stiff ODE's of singular perturbation type. Research Report No. 89-02, Seminar für Angewandte Mathematik, ETH Zürich 1989.

[68] U. Kirchgraber and K. J. Palmer, *Geometry in the neighborhood of invariant manifolds of maps and flows and linearization*. Pitman Res. Notes Math. Ser. 233, Longman Scientific & Technical, Harlow 1990.

[69] H. W. Knobloch, Invariant manifolds for ordinary differential equations. In *Differential equations and mathematical physics*, Proc. Int. Conf., Birmingham/AL (USA) 1990, Math. Sci. Eng. 186, Academic Press, Inc., Boston, MA, 1992, 121–149.

[70] H. W. Knobloch and B. Aulbach, Singular perturbations and integral manifolds. *J. Math Phys. Sci.* **18** (1984), 415–424.

[71] H. W. Knobloch and D. Flockerzi, Invariant manifolds, zero dynamics and stability. In *Nonlinear synthesis*, Proc. IIASA Workshop, Sopron/Hung. 1989, Prog. Syst. Control Theory 9, Birkhäuser, Boston 1991, 132–140.

[72] H. W. Knobloch and F. Kappel, *Gewöhnliche Differentialgleichungen*. Mathematische Leitfäden, B. G. Teubner, Stuttgart 1974.

[73] M. Krupa and P. Szmolyan, Extending geometric singular perturbation theory to non-hyperbolic points – fold and canard points in two dimensions. *SIAM J. Math. Anal.* **33** (2001), 286–314.

[74] J. Kurzweil, Invariant manifolds of differential systems. *Z. Angew. Math. Mech.* **49** (1969), 11–14.

[75] P. A. Lagerstrom and R. G. Casten, Basic concepts underlying singular perturbation techniques. *SIAM Rev.* **14** (1972), 63–120.

[76] M. Lakner and M. Skapin-Rugelj, Global invariant manifolds. *Chaos Solitons Fractals* **26** (2005), no. 5, 1533–1540.

[77] O. Lanford III, *Lectures on dynamical systems*. Lecture Notes, ETH Zürich 1991.

[78] M. Lazzo and P. G. Schmidt, Periodic solutions and invariant manifolds for an even-order differential equation with power nonlinearity. *J. Dynam. Differential. Equations* **23** (2011), no. 1, 141–166.

[79] M. Y. Li and J. S. Muldowney, Dynamics of differential equations on invariant manifolds. *J. Differential Equations* **168** (2000), no. 2, 295–320.

[80] W. Liu, Geometric singular perturbations for multiple turning points: invariant manifolds and exchange lemmas. *J. Dynam. Differential. Equations* **18** (2006), no. 3, 667–691.

[81] J. Lorenz, Computation of invariant manifolds. In *Numerical analysis* 1991 (Dundee, 1991), ed. by D. F. Griffiths et al., Pitman Res. Notes Math. Ser. 260, Longman Sci. Tech., Harlow 1992, 118–127.

[82] J. Lorenz, Numerics of invariant manifolds and attractors. In *Chaotic numerics* (Deakin University, Geelong, Australia, 1993), Contemp. Math. 172, Amer. Math. Soc., Providence, RI, 1994, 185–202.

[83] C. Lubich, Integration of stiff mechanical systems by Runge–Kutta methods. *Z. Angew. Math. Phys.* **44** (1993), 1022–1053.

[84] C. Lubich, K. Nipp, and D. Stoffer, Runge–Kutta solutions of stiff differential equations near stationary points. *SIAM J. Numer. Anal.* **32** (1995), 1296–1307.

[85] C. Lubich and A. Ostermann, Hopf bifurcation of reaction-diffusion and Navier-Stokes equations under discretisation. *Numer. Math.* **81** (1998), 53–84.

[86] F. Ma and T. Kuepper, Numerical calculation of invariant manifolds for maps. *Numer. Linear Algebra Appl.* **1** (1994), no. 2, 141–150.

[87] J. Marsden und J. Scheuerle, The construction and smoothness of invariant manifolds by the deformation method. *SIAM J. Math. Anal.* **18** (1987), 1261-1274.

[88] R. März and A. R. Rodríguez-Santiesteban, Analyzing the stability behaviour of solutions and their approximations in case of index-2 differential-algebraic systems. *Math. Comput.* **71** (2002), 605–632.

[89] E. F. Mishchenko and N. Kh. Rozov, *Differential equations with small parameters and relaxation oscillations*. Math. Concepts Methods Sci. Engrg. 13, Plenum Press, New York 1980.

[90] A. Murua, Private Communication, 2000.

[91] K. Nipp, An extension of Tikhonov's theorem in singular perturbations for the planar case. *Z. Angew. Math. Phys.* **34** (1983), no. 3, 277–290.

[92] K. Nipp, Invariant manifolds of singularly perturbed ordinary differential equations. *Z. Angew. Math. Phys.* **36** (1985), no. 2, 309–320.

[93] K. Nipp, Breakdown of stability in singularly perturbed autonomous systems. I. Orbit equations. *SIAM J. Math. Anal.* **17** (1986), no. 3, 512–532.

[94] K. Nipp, Breakdown of stability in singularly perturbed autonomous systems. II. Estimates for the solutions and application. *SIAM J. Math. Anal.* **17** (1986), no. 5, 1068–1085.

[95] K. Nipp, An algorithmic approach for solving singularly perturbed initial value problems. In *Dynamics reported*, Vol. 1, Dynam. Report. Ser. Dynam. Systems Appl. 1, Wiley, Chichester 1988, 173–263.

[96] K. Nipp, Smooth attractive invariant manifolds of singularly perturbed ODEs. Report No. 92–13, Seminar für Angewandte Mathematik, ETH–Zürich, Zürich 1992.

[97] K. Nipp, Numerical integration of differential algebraic systems and invariant manifolds. *BIT* **42** (2002), 408–439.

[98] K. Nipp, Invariant manifold results for some particular situations. Preprint, 2006.

[99] K. Nipp and D. Stoffer, Attractive invariant manifolds for maps: Existence, smoothness and continuous dependence on the map. Research Report SAM-92-11, ETH-Zürich, Zürich 1992.

[100] K. Nipp and D. Stoffer, Invariant manifolds and global error estimates of numerical integration schemes applied to stiff systems of singular perturbation type – Part I: RK-methods. *Numer. Math.* **70** (1995), 245–257.

[101] K. Nipp and D. Stoffer, Invariant manifolds and global error estimates of numerical integration schemes applied to stiff systems of singular perturbation type – Part II: Linear multistep methods. *Numer. Math.* **74** (1996), 305–323.

[102] R. E. O'Malley, Jr., *Introduction to singular perturbations*. Applied Mathematics and Mechanics 14, Academic Press, New York, London 1974.

[103] H. M. Osinga, Computing invariant manifolds: variations on the graph transform. PhD thesis, University of Groningen, Groningen 1996.

[104] G. Osipenko, E. Ershov, and J. H. Kim, *Lectures on invariant manifolds of perturbed differential equations and linearization*. State Technical University, St. Petersburg 1996.

[105] O. Perron, Über Stabilität und asymptotisches Verhalten der Lösungen eines Systems endlicher Differenzengleichungen. *J. Reine Angew. Math.* **161** (1929), 41–64.

[106] C. Poetzsche and M. Rasmussen, Computation of nonautonomous invariant and inertial manifolds. *Numer. Math.* **112** (2009), no. 3, 449–483.

[107] M. Prizzi, Invariant manifolds for singularly perturbed parabolic equations. *Rend. Ist. Mat. Univ. Trieste* **26** (1994), no. 1–2, 151–210.

[108] C. C. Pugh, Invariant manifolds. In *Actes du Congrès International des Mathématiciens* (Nice, 1970), Tome 2, Gauthier-Villars, Paris 1971, 925–928.

[109] R. Purfürst, *Invariante Mannigfaltigkeiten und Hopf-Bifurkation bei singulär gestörten Systemen*, Report der AdW der DDR (1982).

[110] A. G. Ramm, Invariant manifolds for dissipative systems. *J. Math. Phys.* **50** (2009), no. 4, 042701, 11 p.

[111] K. P. Rybakowski, An abstract approach to smoothness of invariant manifolds. *Appl. Anal.* **49** (1993), no. 1–2, 119–150.

[112] T. Sahai and A. Vladimirsky, Numerical methods for approximating invariant manifolds of delayed systems. SIAM J. Appl. Dyn. Syst. **8** (2009), no. 3, 1116–1135 (electronic only).

[113] K. Sakamoto, Invariant manifolds in singular perturbation problems for ordinary differential equations. *Proc. Roy. Soc. Edinburgh Sect.* A **116** (1990), no. 1–2, 45–78.

[114] K. Sakamoto, Smooth linearization of vector fields near invariant manifolds. *Hiroshima Math. J.* **24** (1994), no. 2, 331–355.

[115] A. Sawant and A. Acharya, Model reduction via parametrized locally invariant manifolds: Some examples. *Comput. Methods Appl. Mech. Eng.* **195** (2006), no. 44–47, 6287–6311.

[116] B. Schmalfuss and K. R. Schneider, Invariant manifolds for random dynamical systems with slow and fast variables. *J. Dynam. Differential Equations* **20** (2008), no. 1, 133–164.

[117] M. Shub, *Global stability of dynamical systems*. Springer-Verlag, New York 1987.

[118] D. Stoffer, General linear methods: connection to one step methods and invariant curves. *Numer. Math.* **64** (1993), 395–407.

[119] D. Stoffer, On the global error of linear multistep methods. In *Geometric behaviour of numerical integration methods for ODEs*, Habilitationsschrift ETH-Zürich, Zürich 1994.

[120] D. Stoffer, On the qualitative behaviour of symplectic integrators. Part I. Perturbed linear systems. *Numer. Math.* **77** (1997), no. 4, 535–547.

[121] D. Stoffer, On the qualitative behaviour of symplectic integrators. II. Integrable systems. *J. Math. Anal. Appl.* **217** (1998), no. 2, 501–520.

[122] D. Stoffer, On the qualitative behaviour of symplectic integrators. Part III. Perturbed integrable systems. *J. Math. Anal. Appl.* **217** (1998), no. 2, 521–545

[123] D. Stoffer and K. Nipp, Invariant curves for variable step size integrators. *BIT* **31** (1991), no. 1, 169–180; Erratum: "Invariant curves for variable step size integrators", *ibid.* **32** (1992), no. 2, 367–368.

[124] T. Ström, On logarithmic norms. *SIAM J. Num. Anal.* **12** (1975), no. 5, 741–753.

[125] A. M. Stuart and A. R. Humphries, *Dynamical systems and numerical analysis*. Cambridge Monogr. Appl. Comput. Math. 2, Cambridge University Press, Cambridge 1996.

[126] T. Stumpp, Asymptotic expansions and attractive invariant manifolds of strongly damped mechanical systems. *Z. Angew. Math. Mech.* **88** (2008), no. 8, 630–643.

[127] S.-K. Tin, N. Kopell, and C. K. R. T. Jones, Invariant manifolds and singularly perturbed boundary value problems. *SIAM J. Numer. Anal.* **31** (1994), no. 6, 1558–1576.

[128] D. Tognola, Invariant manifolds, passage through resonance, stability and a computer assisted application to a synchronous motor. Diss. ETH No. 12744, Zürich 1998.

[129] S. Wiggins, *Normally hyperbolic invariant manifolds in dynamical systems*. Appl. Math. Sci. 105, Springer-Verlag New York, NY, 1994.

[130] Y. Yi, A generalized integral manifold theorem. *J. Differential Equations* **102** (1993), no. 1, 153–187.

[131] J. Zhang and X. Yang, Exponential dichotomy and invariant manifolds on different time scales. *J. Jilin Univ. Sci.* **49** (2011), no. 2, 237–239 (in Chinese).

[132] W. N. Zhang, Generalized exponential dichotomies and invariant manifolds for differential equations. *Adv. in Math. (China)* **22** (1993), no. 1, 1–45.

[133] Y.-K. Zou and W.-J. Beyn, Invariant manifolds for nonautonomous systems with application to one-step methods. *J. Dynam. Differential Equations* **10** (1998), no. 3, 379–407.

Index